Dynamic Testing and Seismic Qualification Practice

Dynamic Testing and Seismic Qualification Practice

Clarence W. de Silva
Carnegie-Mellon University

LexingtonBooks
D.C. Heath and Company
Lexington, Massachusetts
Toronto

Library of Congress Cataloging in Publication Data

De Silva, Clarence W.
Dynamic testing and seismic qualification practice.

Includes index.
1. Vibration testing. 2. Earthquake engineering.
I. Title.
TA355.D38 624.1'762'0287 80-8879
ISBN 0-669-04393-1 AACR2

Published simultaneously in Canada.

Printed in the United States of America.

International Standard Book Number: 0-669-04393-1

Library of Congress Catalog Card Number: 80-8879

To my wife, Clarice—for her understanding
and the wonderful gift of love

Contents

Figures

Tables

Preface

This book is directed primarily at technicians, engineers, and other practicing professionals in the field of seismic qualification and dynamic testing. The main objective has been to introduce the basic theory and practical aspects of dynamic testing and analysis as applied particularly to seismic qualification of equipment and components. At present, technical information on this subject is widely scattered in the literature and is sometimes not accessible to the general public. Furthermore, the nomenclature used in seismic qualification applications often is inconsistent. While covering the fundamental aspects of the subject at an introductory level, for the benefit of the beginner in the field, the book attempts to present the necessary theory and practical information concisely and systematically, employing a rational nomenclature. Thus, the book also will serve as a convenient reference tool for the practicing professional. It is assumed that the reader is familiar with the field of mechanical engineering, as most analytical developments in the book assume a junior-level knowledge of engineering and applied mathematics.

The dynamic-system modeling and analysis concepts presented in the book are important for two primary reasons. Dynamic testing can be better understood and conducted if the underlying theory is properly understood and utilized in test program development. In addition, because system analysis frequently is employed in seismic qualification of devices, it is important to understand system modeling and analysis techniques and their equivalence (or relationship) to various dynamic testing methods.

The book emphasizes that at least a partial knowledge of the dynamic characteristics of an object can significantly facilitate its dynamic testing and analysis, and so gray-box testing and analysis are encouraged. In this respect, pretest information aquisition and exploratory testing are very useful.

Digital Fourier analysis, spectral analysis, and various techniques of signal processing, widely used in processing test data and analytical responses for presentation and evaluation are discussed in sufficient detail.

The response spectrum method, used extensively in practice to represent excitation inputs in dynamic testing and analysis related to seismic qualification, is covered, as are power spectral density, Fourier spectrum, and time-history representations of excitation inputs. Particular attention is given to stochastic concepts, because dynamic environments normally encountered in practice are random to some degree.

The last five chapters of the book are devoted primarily to practical aspects of dynamic testing and seismic qualification. Chapter 8 is devoted to instrumentation in dynamic testing, covering the general categories of signal generation, sensing and monitoring, conditioning and recording equipment. The discussion of each category gives the functions, the principle of operation, and the basic theory of representative instruments. The method developed for selecting shaker specifications in seismic qualification tests (Chapter 9) and the development of an optimal rectilinear test (Chapter 10) are based on previous work of the author.

Rather than distract the reader by including a large number of literature references within the text, a categorized bibliography is provided at the end of the book. These sources should be reviewed to gain a more thorough knowledge of the various topics discussed in the book.

Acknowledgments

The author gratefully acknowledges the assistance of many individuals in the preparation of this book. F. Loceff, K.M. Vashi, and A.J. Ayoob, of Westinghouse Electric Corporation, Pittsburgh, introduced the author to their extensive seismic qualification activities. Nancy Witt typed portions of the preliminary manuscript and assisted in preparing the proposal for this project. The final manuscript was typed by Vivian Achwal and Asoka Liyanaarachchi. The assistance of R.A. Thiga, dean of the Graduate School, University of Petroleum and Minerals, was most encouraging. The figures in the book were drawn by A. Shehab-Eldeen.

On principle, it is wrong to try founding a theory on observable magnitudes alone. It is the theory which decides what we can observe.

—Einstein

Dynamic Testing and Seismic Qualification Practice

1 Introduction

Dynamic testing, which is also known as vibration testing, is usually performed by applying a vibration excitation to the test object and monitoring its dynamic response and the performance of its intended functions. The technology of dynamic testing has gradually evolved since World War II, at which time the method was used primarily in aircraft testing. Subsequently, dynamic testing has been used successfully in testing a wide spectrum of products, ranging from small printed circuit boards and microprocessor chips to large missiles, space vehicles, ground vehicles, agricultural machinery, structural systems, and the equipment of nuclear power plants.

Dynamic testing may be useful at several stages in the development, production, and utilization of a product. In the initial design stage, a design weakness and possible improvements could be determined through dynamic testing of a scale model, a preliminary design, or a partial product. In the production stage, the quality of workmanship of the final product may be evaluated using nondestructive dynamic testing. Another important application of dynamic testing is detemining the adequacy of a product (with a good reputation) for a specific application or range of applications. Seismic qualification of a product for use in a specific nuclear power plant, for example, would require estabilishing the capability of that product to withstand a prespecified seismic disturbance that is characteristic of the geographic location of the nuclear power plant. This may be achieved by analytical means or by dynamic tests.

In this book, we shall present the fundamental concepts of dynamic testing and describe the application of dynamic testing in seismic qualification. We shall also investigate some analytical methods employed in seismic qualification. An attempt will be made to present the underlying theory as completely as is feasible.

Dynamic Testing

Dynamic testing typically consists of applying an excitation to the test object and monitoring its response and operation. Although free vibration testing (in which only an initial excitation is applied and removed so that the test object is allowed to vibrate freely) also comes under the general topic of dynamic testing, what concerns us most in this book is the forced vibration type of dynamic test. Tests consisting of object excitations caused by explosions, natural causes, operating vibrations, and the like, are not considered in the book.

Dynamic tests can be divided into two categories, depending on their utilization: determination tests, and evaluation tests. The purpose of determination tests is to obtain specific dynamic information regarding the test object. Resonance frequencies (or natural frequencies), damping parameters, mode shapes, and frequency-response functions of the test object are some of the information that can be obtained from a determination test. Determination tests are usually conducted at low excitation levels.

Evaluation tests are used to study whether the test object can withstand a specified dynamic environment (proof testing) or to determine the worst dynamic environment that the test object could withstand without any malfunction or structural failure (fragility testing). In evaluation tests, it is usually necessary to monitor functional performance (functional operability) as well as structural integrity of the test object. Evaluation tests are conducted at relatively high excitation levels.

The fundamental test setup for dynamic testing of an object is indicated in figure 1-1. The test-input signal is prespecified, based on the test requirements. Input signal is generated according to the specifications, using a signal generator. This signal is further modified and amplified, using a signal conditioner, and is used subsequently to excite the shaker. The shaker generates a dynamic excitation in accordance with the input signal. Shaker motion is applied at a suitable mounting location of the test object, and the resulting dynamic response is monitored. In evaluation tests, any possible malfunctions or structural failures also should be monitored. Note that, in this basic arrangement, there is no feedback from the response monitor to the signal generator; that is, there is no guarantee that the signal from the signal generator is always applied to the test object in its original form. Because of various dynamic interactions present in the test setup (for example, interactions between the test object and the shaker table), the generated signal will be distorted before it reaches the test object. A controlled test should include some form of feedback loop, which would incorporate modifications into the input signal depending on monitored response of the test object.

Before main dynamic tests are performed, it may be necessary to conduct a series of exploratory tests in order to obtain a sufficient knowledge regarding the dynamic characteristics of the test object. Such information is useful in planning the main tests. Exploratory tests essentially fall into the category of determination tests. Main dynamic tests are conducted primarily to determine the capability of the test object to withstand a prespecified vibration (dynamic) environment. In this sense, main dynamic tests are evaluation tests.

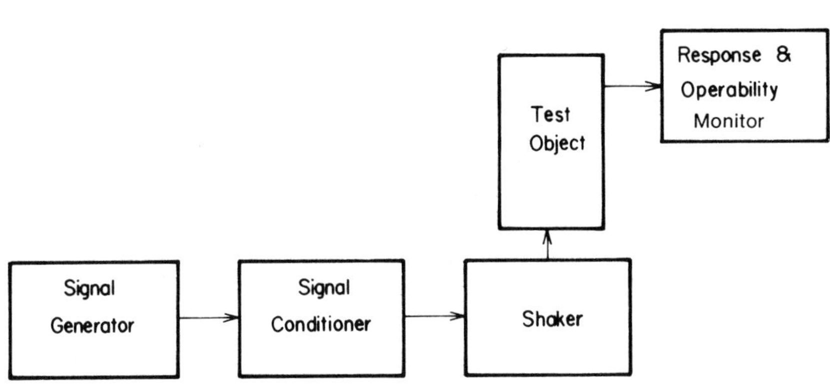

Figure 1-1. Basic Test Setup in Dynamic Testing

Seismic Qualification

Seismic qualification of an object is the procedure for establishing that the structural integrity of the object will be maintained and that no operational malfunctions will result during and after the application of a specified earthquake excitation. Equipment to be used in essential services or in safety-related functions, which is to be installed in areas of high seismic activity, is usually subjected to seismic qualification, as required by some regulatory agency.

The three basic methods of seismic qualification are (1) analysis, (2) dynamic testing, and (3) combined analysis and testing. Seismic qualification by analysis is done by using an appropriate analytical model of the object to be qualified. Once the model is developed, a digital-computer simulation (or a purely analytical solution in rare circumstances) is carried out in order to estimate the object's response to a prespecified excitation input applied at the mounting locations. Structural integrity could be investigated in this manner by using deformation and stress considerations at various critical locations in the object. In addition, possible operational malfunctions and failure in components could be predicted by this method, provided a satisfactory and direct analytical correlation between the component response and its operational performance is available. In general, qualification by analysis should be supplemented by judicious use of past experience concerning the operation of the object or equipment.

Frequently, dynamic testing is required in addition to analysis. Results from a dynamic analysis can be extremely useful in planning a dynamic test. The development of the test environment (excitation input signal), for example, could be done using analytical means. In this respect, combined analysis and testing is generally the most desirable method of seismic qualification.

Seismic ground motions are random, or stochastic. Consequently, probabilistic representations of input excitation sometimes are used in analytical methods, and random-signal generators are employed in dynamic tests. Alternatively, main characteristics of the specified seismic environment can be represented by a deterministic signal possessing equivalent amplitude (signal level), frequency content, and decay rate (damping) characteristics. Such deterministic signals are frequently used in dynamic tests as well as in analytical simulations for seismic qualification. Specification of the seismic environment is undoubtedly one of the major considerations in a seismic-qualification program.

Regulatory Considerations

Early seismic qualification programs (in the 1960s) were conducted on the basis of very crude test requirements. Capability to withstand a 0.2g (g = one acceleration due to gravity) earthquake acceleration was a standard requirement, for example, particularly for electric utility equipment. It was necessary to establish that all devices in the equipment would remain functionally operable and that their structural integrity would be maintained when subjected to that level of acceleration. Static analysis was the method most frequently used in seismic qualification by analysis.

Regulatory requirements for seismic qualification evolved primarily with respect to their application in the nuclear industry. Public consciousness of the safety and reliability of nuclear power plants has focused on the hazards of possible radioactive contamination and exposure following an accident. Furthermore, the reliability of operation of modern power-generation and transmission systems is vital from the viewpoint of the socioeconomic impact of power shortages and failures. Before 1971, there were no specific licensing requirements for nuclear power plant seismic qualification in the form of U.S. standards from the Atomic Energy Commission (now the U.S. Nuclear Regulatory Commission, or NRC). From 1971 to 1974, most seismic qualification programs were guided by IEEE Standard 344-1971, "Recommended Practices for Seismic Qualification of Class IE Equipment for Nuclear Power Generating Stations," with NRC approval. This standard was revised in 1975. The primary regulatory guides issued by the NRC for seismic qualification of nuclear power plant equipment are Regulator Guide 1.89, "Qualification of Class IE Equipment for Nuclear Power Plants," issued in November 1974; Regulatory Guide 1.92, "Combining Modal

Responses and Spatial Components in Response Analysis," issued in December 1974 and revised in February 1976; and Regulatory Guide 1.100, "Seismic Qualification of Electric Power Plants," issued in March 1974 and revised in August 1977. For seismic qualification by analysis of safety-related equipment, the nuclear industry has since adopted the ASME Boiler and Pressure Vessel Code, Section III, Nuclear Power Plant Component Code. Other codes, guides, and standards that are useful in seismic qualification include the following:

NRC Regulatory Guide 1.61, "Damping Values for Seismic Design of Nuclear Power Plants"

IEEE Standard 317-1976, "Standard for Electric Penetration Assemblies in Containment Structures for Nuclear Power Generating Stations"

IEEE Standard 323-1974, "Standard for Qualifying Class IE Equipment for Nuclear Power Generating Stations"

IEEE Standard 380-1975, "Definitions of Terms Used in IEEE Standards for Nuclear Power Stations"

IEEE Standard 501-1978, "Standard Seismic Testing of Relays"

IEEE Standard 533-1979, "Standard for Qualification of Class IE Lead Storage Batteries for Nuclear Power Generating Stations,"

IEEE Standard 382-1972, "IEEE Trial-Use Guide for Type Test of Class I Electric Valve Operators for Nuclear Power Generating Stations" (Rev. 6, 1978)

IEEE Standard 383-1974, "IEEE Standard for Type Test of Class IE Electric Cables, Field Splices, and Connections for Nuclear Power Generating Stations"

IRIG Standard 106-66, "Communication and Telemetry Standard of the Intermediate Range Instrumentation Group"

In these standards, Class IE refers to the safety classification of the electrical equipment that is essential both in emergency operation and shutdown of a nuclear power plant and in preventing any significant radioactive contamination.

Scope of This Book

The main objective of this book is to provide an introductory-level reference tool for both technical and managerial personnel working in the area of dynamic testing as applied to seismic qualification of equipment. Basic analytical principles presented in the book will be applicable to seismic qualification both by analysis and by dynamic testing.

Four aspects of dynamic testing need careful consideration: (1) available testing methods, (2) specification of the test environment, (3) test instrumentation, and (4) Data aquisition, processing, and interpretation. These four topics will be addressed in the book with sufficient detail and rigor but keeping the beginner in the field in mind.

2 Analytical Models of Dynamic Systems

A system can be defined as a selected set of interacting components or elements, each possessing an input-output (or cause-effect, or causal) relationship. A dynamic system is one whose response variables are functions of time. A model is some form of representation of a practical system. An analytical model (or mathematical model) comprises a set of equations or the equivalent that approximately represents the system. Sometimes, a set of curves, digital data stored in a computer memory, and other numerical data—rather than a set of equations—might be termed an analytical model if such data represent the system of interest.

Analytical models are very useful in predicting the dynamic behavior (response) of a system when it is subjected to a certain excitation (input). In the context of dynamic testing, analytical models are commonly used to develop the input signal applied to the shaker and to study dynamic effects and interactions in the test object, the shaker table, and their interfaces. In seismic qualification by analysis (see chapter 7), a suitable analytical model replaces the test specimen.

Model Development

There are two broad categories of system models: lumped-parameter models and continuous-parameter models. Lumped-parameter models are more common than continuous-parameter models, but continuous-parameter elements sometimes are included in otherwise lumped-parameter models in order to improve the model accuracy.

In lumped-parameter models, various characteristics in the system are lumped into representative elements. A coil spring, for example, has a mass, an elastic (spring) effect, and an energy-dissipation characteristic, each of which is distributed over the entire coil. In an analytical model, these individual characteristics can be approximated by a separate mass element, a spring element, and a damper element, which are interconnected in some parallel-series configuration.

We shall restrict our discussion to mechanical and electrical systems, although the analogies also could be extended to hydraulic and thermal systems. The basic system elements can be divided into two groups: energy-storage elements and energy-dissipation elements. Table 2-1 shows the linear

Table 2-1
Some Linear Constitutive Relations

Category	Linear Constitutive Relations	
	Mechanical	Electrical
Energy Storage Elements	Mass 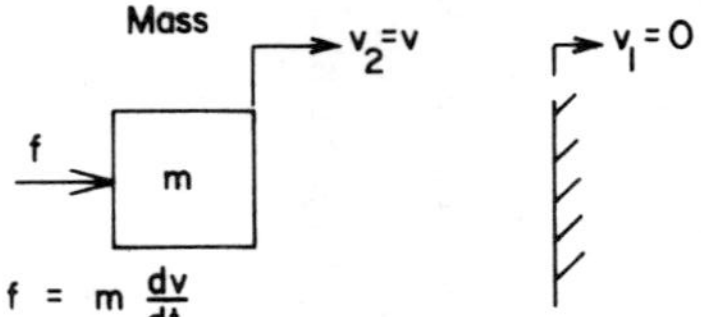 f, m, $v_2 = v$, $v_1 = 0$ $f = m \frac{dv}{dt}$ Newton's Law	Capacitor 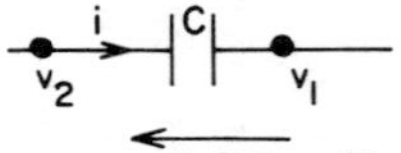 i, C, v_2, v_1 $v = v_2 - v_1$ $i = C \frac{dv}{dt}$
	Spring 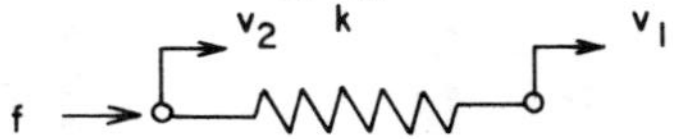 f, v_2, k, v_1 $v = v_2 - v_1$ $v = \frac{1}{k} \frac{df}{dt}$ Hooke's Law	Inductor 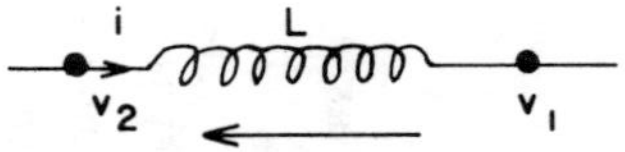 i, L, v_2, v_1 $v = v_2 - v_1$ $v = L \frac{di}{dt}$
Energy Dissipation Elements	Damper 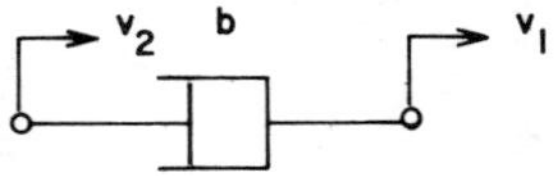 v_2, b, v_1 $v = v_2 - v_1$ $f = b v$	Resistance 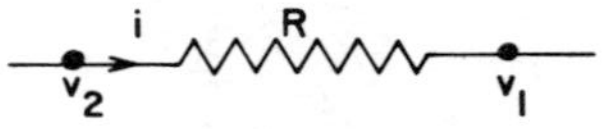 i, R, v_2, v_1 $v = v_2 - v_1$ $i = \frac{1}{R} v$ Ohm's Law

relationships that describe the behavior of translatory mechanical elements and electrical elements. These relationships are known as constitutive relations. In particular, Newton's second law is considered the constitutive relation for the mass element, although some analysts prefer to treat it separately. The analogy used in table 2-1 between mechanical and electrical elements is known as the force-current analogy. This analogy appears more logical than a force-voltage analogy, as is clear from table 2-2. This follows from the fact that both force and current are through-variables, which are analogous to fluid flow through a pipe, and both velocity and voltage are across-variables, which vary across the flow direction, as in the case of pressure. The correspondence between the parameter pairs given in table 2-2 follows from the relations in table 2-1. Note that the rotational mechanical elements possess constitutive relations between torque and angular velocity, which can be treated as a generalized force and a generalized velocity, respectively. In fluid systems, basic elements corresponding to capacitance (capacity), inductance (fluid inertia), and resistance (fluid friction) exist. Constitutive relations between pressure difference and mass flow rate can be written for these elements. In thermal systems, generally only two elements—capacitance and resistance—can be identified. Constitutive relations exist between temperature difference and heat transfer rate in this case.

The main steps in developing a lumped-parameter dynamic model are as follows:

1. Identify all major energy-storage and energy-dissipation characteristics of the dynamic system and represent them by suitable elements that are appropriately interconnected.
2. Write down the constitutive relations for each element.
3. Identify a set of independent variables to represent the system response.
4. Develop differential equations in terms of the system variables.

Step 3 is achieved by using the conditions of constraints (for geometric fit) and force balance. In this context, it should be guaranteed that, at a common junction of two or more elements, the motion (velocity) is common and all internal forces and external forces add up to zero. Step 4 is performed by eliminating any auxiliary or dependent variables.

As an example, suppose that a dynamic system is represented by the lumped-parameter model shown in figure 2-1(a). By the application of step 2, the following constitutive equations are obtained for the linear damper (of damping constant b) and the mass m:

$$f_1 = bv_1 \tag{2.1}$$

Table 2-2
Force-Current Analogs

	Mechanical	*Electrical*
System-response variables		
Through-variables	Force f	Current i
Across-variables	Velocity v	Voltage v
System parameters	M	C
	k	$1/L$
	b	$1/R$

$$f - f_2 = m\frac{dv}{dt} \tag{2.2}$$

Excitation force $f(t)$ is the system input and is assumed to be a known function of time. Velocity v of mass is taken as the system response (or output). This is the only independent variable of system response. This is observed by writing the condition of geometric fit at A.

$$v_1 = v \tag{2.3}$$

(a)

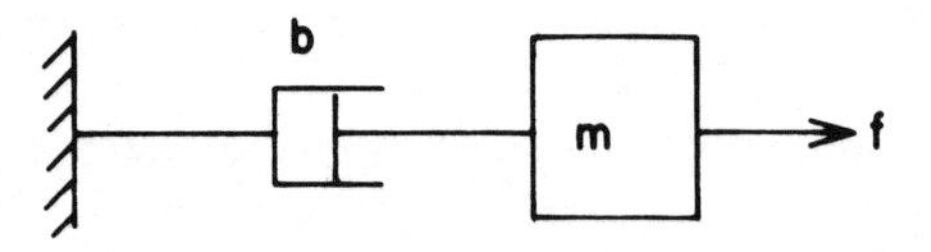

(b)

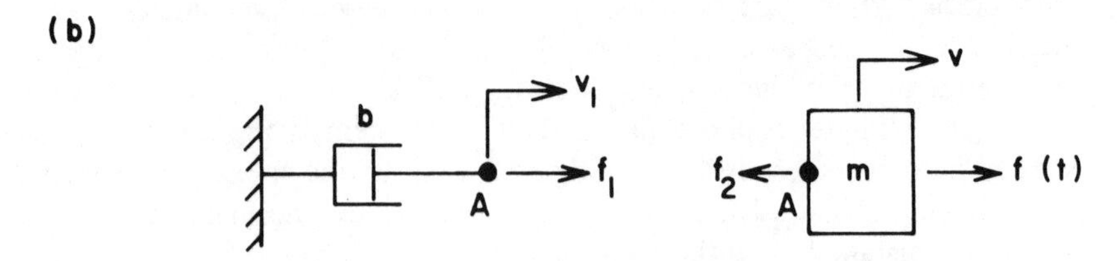

Figure 2-1. Lumped-Parameter Analytical Model Example: (a) Model, (b) Analysis

and the condition of force balance at A:

$$f_1 - f_2 = 0 \tag{2.4}$$

It follows that v_1, f_1, and f_2 are all auxiliary or dependent variables. Thus step 3 is completed. By straightforward algebraic manipulation of equations (2.1) through (2.4), the differential equation of motion is obtained (step 4):

$$m \frac{dv}{dt} + bv = f(t) \tag{2.5}$$

If the constitutive relations are linear, the analytical model is linear. If the parameters of a system (for example, m, k, b) do not change with time, it is a constant-parameter system (or a time-invariant system). Real systems are generally both nonlinear and time-variant. If only small changes in the system-response variables are of interest—about the steady state, for example— then a linear model is usually adequate. Constant parameters may be assumed, however, if the time duration of interest is small in comparison to the time required for a significant change to take place in the parameters (for instance, because of component wearout).

A general characteristic of lumped-parameter systems is that they can be represented by ordinary differential equations (ODE). If a system is represented by a single ODE, the order of the ODE (corresponding to the highest derivative) is also the system order. This is equal to the number of independent energy-storage elements present in the system model. A system modeled as in figure 2-1(a), for example, has a first-order system (differential) equation. Also, the model has just one energy-storage element. It follows that the system is first-order. The (number of) degrees of freedom of a system is equal, in most cases, to the number of independent generalized coordinates that are required to describe the motion of the inertia elements in the system. An exception to this occurs when the system has holonomic constraints.[1]

Continuous-parameter systems are represented by partial differential equations (PDE). In continuous mechanical systems, the inertia, spring, and damping effects are distributed continuously (or piecewise continuously) in space. Examples are bars in longitudinal vibration and beams and plates in transverse vibration. A continuous system can be interpreted as a limiting case of a set of interconnected lumped elements. In addition to the time variable, continuous systems require other independent variables to represent geometric location of the system continuum in space (spatial variables). As an example, consider a slender beam in transverse vibration. In figure 2-2, $v(x, t)$ represents the transverse displacement in location x at time t. The excitation load consists of a set of point forces $F_k(t)$ located at x_k, a

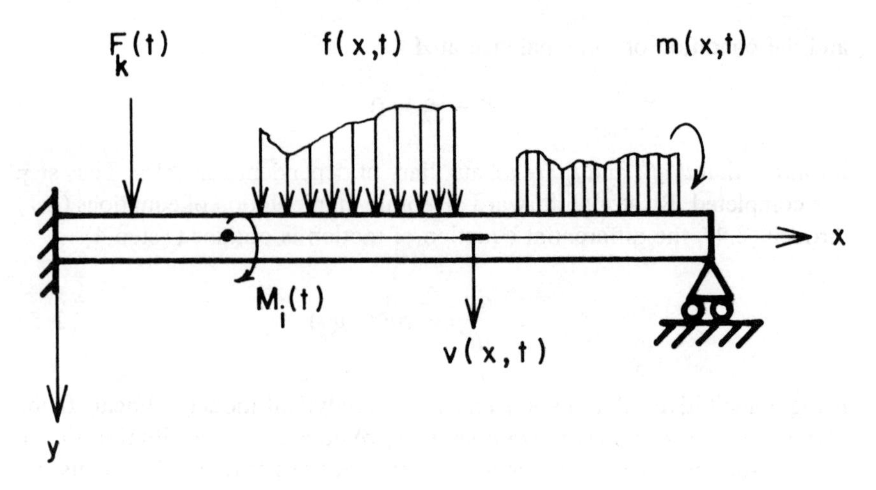

Figure 2-2. Bernoulli-Euler Beam Model

distributed forcing function $f(x, t)$ per unit length, a set of point moments $M_i(t)$ located at x_i and a distributed moment $m(x, t)$ per unit length. The response of this system is given by the Bernoulli-Euler PDE:

$$\frac{\partial^2}{\partial x^2} EI(x) \frac{\partial^2 v}{\partial x^2} + \rho A \frac{\partial^2 v}{\partial t^2} = \sum_k \delta(x - x_k) F_k(t) + f(x, t) + \sum_i \frac{d\delta}{dx}(x - x_i) M_i(t) - \frac{\partial m}{\partial x}(x, t) \tag{2.6}$$

in which $\delta(x)$ is the well-known Dirac delta function (see the section "Impulse-Response Function" later in this chapter) and where

$I(x)$ = second moment of area of the beam cross-section about its neutral axis.
$A(x)$ = area of beam cross section
ρ = mass density of the beam
E = Young's modulus

To solve this PDE, which is fourth-order in x and second-order in t, we will require four independent boundary conditions and two initial conditions (at $t = 0$). A special case of this problem is analyzed in the section "Bernoulli-Euler Beam Dynamics" later in this chapter.

System-Response Analysis

The behavior of a dynamic system when subjected to a certain excitation may be studied by analyzing a model of the system. This is commonly known as system-response analysis. System response may be studied either in the time domain, where the independent variable of system response is time, or in the frequency domain, where the independent variable of system response is frequency. Time-domain analysis and frequency-domain analysis are equivalent. Variables in the two domains are connected through Fourier (integral) transform. The preference of one domain over the other depends on such factors as the nature of the excitation input, the type of analytical model available, the time duration of interest, and the quantities that need to be determined.

Impulse-Response Function

The basic block-diagram representation of a single-input–single-output system is shown in figure 2-3. If the system is linear, then the principle of superposition holds. More specifically, if y_1 is the system response to excitation $u_1(t)$ and y_2 is the response to excitation $u_2(t)$, then $\alpha y_1 + \beta y_2$ is the system response to input $\alpha u_1(t) + \beta u_2(t)$ for any constants α and β and any time functions $u_1(t)$ and $u_2(t)$. This is true for both time-variant-parameter linear systems and constant-parameter linear systems.

A unit pulse of width $\Delta\tau$ starting at time $t = \tau$ is shown in figure 2-4(a). Its area is unity. A unit impulse is the limiting case of a unit pulse when $\Delta\tau \to 0$. Unit impulse acting at time $t = \tau$ is denoted by $\delta(t - \tau)$ and is graphically represented as in figure 2-4(b). In mathematical analysis, this is known as the Dirac delta function. It is mathematically defined by two conditions:

$$\begin{aligned} \delta(t-\tau) &= 0 \quad \text{for} \quad t \neq \tau \\ &\to \infty \quad \text{at} \quad t = \tau \end{aligned} \tag{2.7}$$

and

$$\int_{-\infty}^{\infty} \delta(t-\tau)\,dt = 1 \tag{2.8}$$

The Dirac delta function has the well-known and useful properties:

$$\int_{-\infty}^{\infty} f(t)\ \delta(t-\tau)\,dt = f(\tau) \tag{2.9}$$

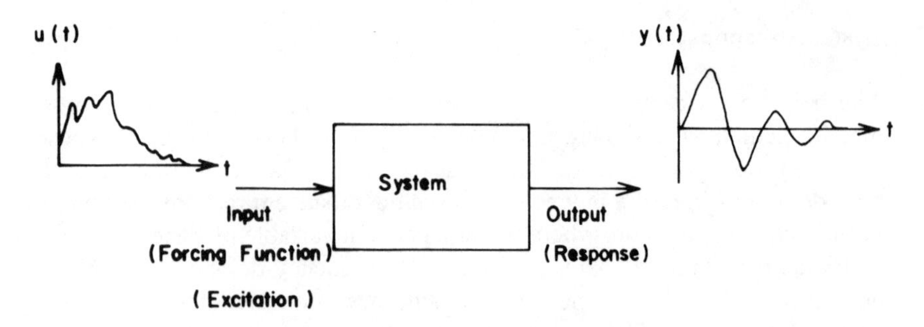

Figure 2-3. Block Diagram Representation of Single-Input–Single Output Systems

and

$$\int_{-\infty}^{\infty} \frac{d^n f(t)}{dt^n} \delta(t-\tau)\, dt = \left. \frac{d^n f(t)}{dt^n} \right|_{t=\tau} \tag{2.10}$$

for any well-behaved time function $f(t)$.

The system output to a unit-impulse input acted at time $t = 0$ is known as the impulse-response function and is denoted by $h(t)$. The system output to an arbitrary input may be expressed in terms of its impulse-response function.

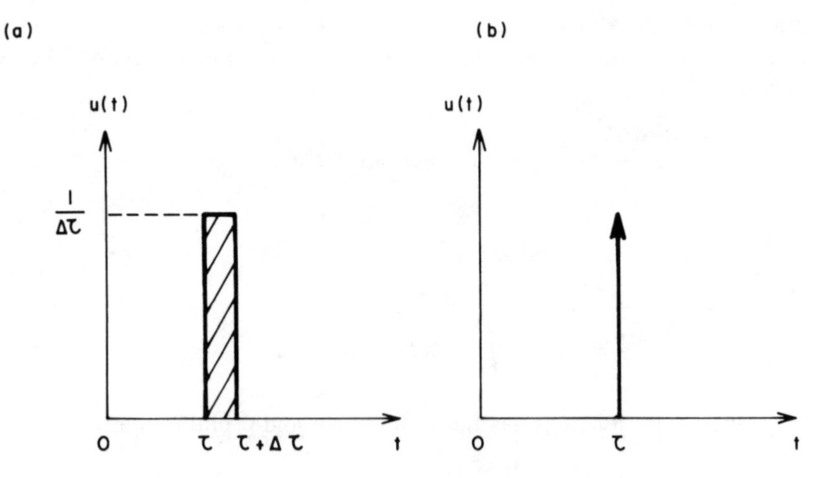

Figure 2-4. Illustrations of (a) Unit Pulse, (b) Unit Impulse

Without loss of generality we shall assume that the system input $u(t)$ starts at $t = 0$; that is,

$$u(t) = 0 \quad \text{for} \quad t < 0 \tag{2.11}$$

For physically realizable systems, the response does not depend on the future value of the input. Consequently,

$$y(t) = 0 \quad \text{for} \quad t < 0 \tag{2.12}$$

and

$$h(t) = 0 \quad \text{for} \quad t < 0 \tag{2.13}$$

Furthermore, if the system is a constant-parameter system, then the response does not depend on the time origin used for the input. Mathematically, this is stated as follows: if the response to input $u(t)$ satisfying equation (2.11) is $y(t)$, which satisfies equation (2.12), then the response to input $u(t-\tau)$, which satisfies

$$u(t-\tau) = 0 \quad \text{for} \quad t < \tau \tag{2.14}$$

is $y(t-\tau)$, and it satisfies

$$y(t-\tau) = 0 \quad \text{for} \quad t < \tau \tag{2.15}$$

This situation is illustrated in figure 2-5. It follows that the delayed-impulse input $\delta(t-\tau)$, having time delay τ, produces the delayed response $h(t-\tau)$.

A given input $u(t)$ can be divided approximately into a series of pulses of width $\Delta\tau$ and magnitude $u(\tau).\Delta\tau$. In figure 2-6, for $\Delta\tau \to 0$, the pulse shown by the shaded area becomes an impulse acting at $t = \tau$, having the magnitude $u(\tau).d\tau$. This impulse is given by $\delta(t-\tau)u(\tau)d\tau$. In a linear, constant-parameter system, it produces the response $h(t-\tau)u(\tau)d\tau$. By integrating over the entire time duration of the input $u(t)$, the overall response $y(t)$ is obtained as

$$y(t) = \int_0^\infty h(t-\tau)u(\tau)d\tau \tag{2.16}$$

Equation (2.16) is known as the convolution integral. In view of equation (2.13), it follows that $h(t-\tau) = 0$ for $\tau > t$. Consequently, the upper limit of integration in equation (2.16) could be made t without affecting the result. Similarly, in view of equation (2.11), the lower limit of integration in equation (2.16) could be made $-\infty$. Furthermore, by introducing the change

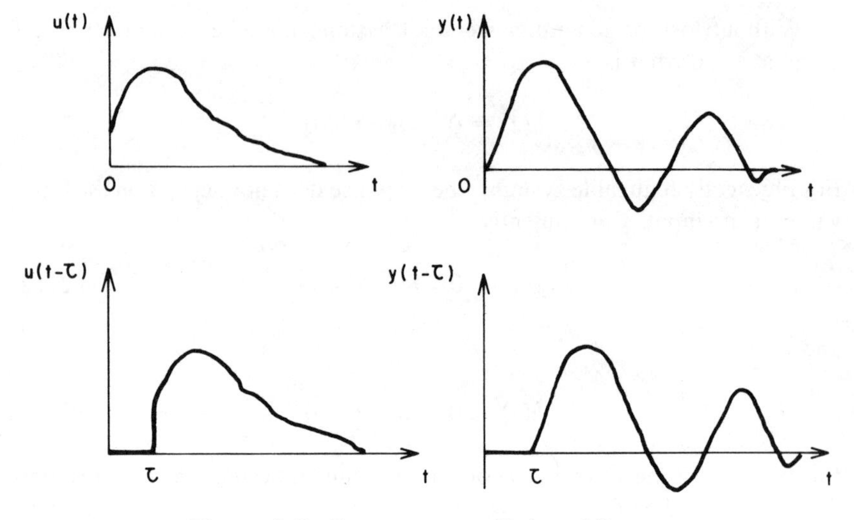

Figure 2-5. Response to a Delayed Input

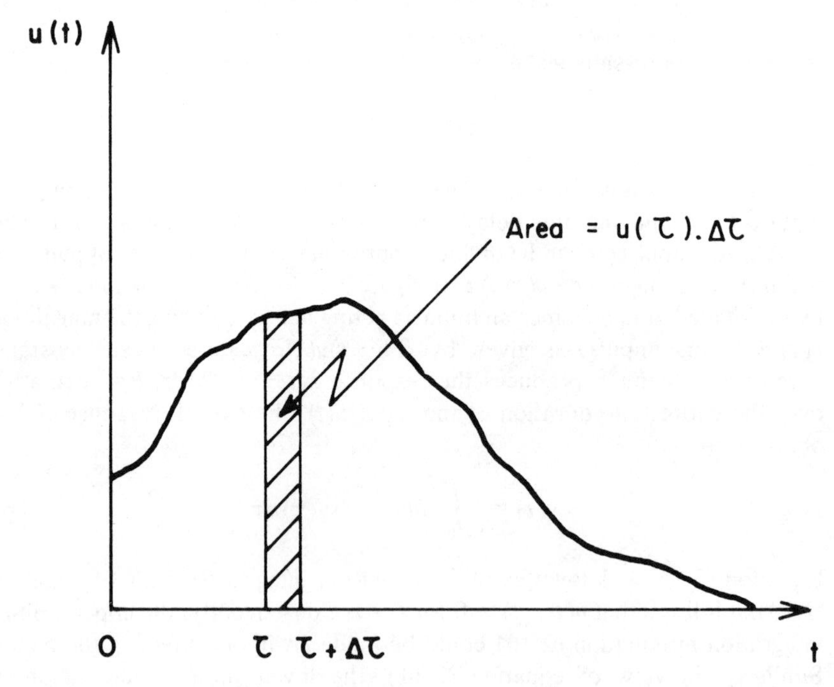

Figure 2-6. General Input Treated as a Continuous Series of Impulses

of variable $\tau \to t - \tau$, an alternative version of the convolution integral is obtained. Several valid versions of the convolution integral (or response equation) for linear, constant-parameter systems are as follows:

$$y(t) = \int_0^\infty h(\tau)u(t-\tau)d\tau \tag{2.17}$$

$$y(t) = \int_{-\infty}^\infty h(t-\tau)u(\tau)d\tau \tag{2.18}$$

$$y(t) = \int_{-\infty}^\infty h(\tau)u(t-\tau)d\tau \tag{2.19}$$

$$y(t) = \int_{-\infty}^t h(t-\tau)u(\tau)d\tau \tag{2.20}$$

$$y(t) = \int_{-\infty}^t h(\tau)u(t-\tau)d\tau \tag{2.21}$$

$$y(t) = \int_0^t h(t-\tau)u(\tau)d\tau \tag{2.22}$$

$$y(t) = \int_0^t h(\tau)u(t-\tau)d\tau \tag{2.23}$$

In fact, the lower limit of integration in the convolution integral could be any value satisfying $\tau \leq 0$, and the upper limit could be any value satisfying $\tau \geq t$. The use of a particular pair of integration limits depends on whether the functions $h(t)$ and $u(t)$ implicitly satisfy the conditions given by equations (2.12) and (2.13) or these conditions have to be imposed on them by means of the proper integration limits. It should be noted that the two versions given by equations (2.22) and (2.23) take these conditions into account explicitly and therefore are valid for all inputs and impulse-response functions.

It should be cautioned that the response given by the convolution integral assumes a zero initial state and is known as the zero-state response because the impulse response itself assumes a zero initial state. Also, as t increases ($t \to \infty$), this solution approaches the steady-state response denoted by y_{ss}.

System-Transfer Function

The Laplace transform of a piecewise-continuous function $f(t)$ is denoted by $\hat{F}(s)$ and is given by,

$$\hat{F}(s) = \int_0^\infty f(t)\exp(st)dt \tag{2.24}$$

in which s is a complex independent variable known as the Laplace variable, given by

$$s = \sigma + j\omega \tag{2.25}$$

and $j = \sqrt{-1}$. Laplace transform operation is represented as $\mathscr{L}f(t) = \hat{F}(s)$. The "hats," or circumflexes, over the symbols are used to distinguish them from the corresponding Fourier transforms (see next section). The inverse Laplace transform operation is represented by $f(t) = \mathscr{L}^{-1}\hat{F}(s)$ and is expressed as

$$f(t) = \frac{1}{2\pi j} \int_{\sigma - j\infty}^{\sigma + j\infty} \hat{F}(s) \exp(st)\,ds \tag{2.26}$$

The integration is performed along a vertical line parallel to the complex axis, located at σ on the real axis in the complex Laplace s-plane. For a given piecewise-continuous function $f(t)$, the Laplace transform exists if the integral in equation (2.24) converges. A sufficient condition for this is

$$\int_0^\infty |f(t)| \exp(-\sigma t)\,dt < \infty$$

Convergence is guaranteed by choosing a sufficiently large positive σ.

By use of Laplace transformation, the convolution integral equation can be converted into an algebraic relationship. To illustrate this, consider equation (2.17). By definition, its Laplace transform is written as

$$\hat{Y}(s) = \int_0^\infty \int_0^\infty h(\tau) u(t - \tau)\,d\tau \exp(-st)\,dt$$

Since the integration with respect to t is performed while keeping τ constant, we have $dt = d(t - \tau)$. Consequently,

$$\hat{Y}(s) = \int_{-\tau}^\infty u(t - \tau) \exp[-s(t - \tau)]\,d(t - \tau) \int_0^\infty h(\tau) \exp(-s\tau)\,d\tau$$

The lower limit of the first integration can be made equal to zero, in view of equation (2.14). Again using the definition of Laplace transformation, the foregoing relation can be expressed as

$$\hat{Y}(s) = \hat{H}(s)\,\hat{U}(s) \tag{2.27}$$

in which

$$\hat{H}(s) = \mathscr{L}h(t) = \int_0^\infty h(t) \exp(-st)\,dt \tag{2.28}$$

Some useful Laplace transform relations are given in table 2-3. Transfer function of a system is denoted by $\hat{H}(s)$ and is defined by equation (2.27). More specifically, system transfer function is given by the ratio of the Laplace transformed output and the Laplace transformed input, with zero initial conditions. In view of equation (2.28), it is clear that system transfer function can be expressed as the Laplace transform of impulse-response function. Transfer function of a linear and constant-parameter system is a unique function that completely represents the system. A physically realizable linear, constant-parameter system possesses a unique transfer function, even if the Laplace transforms of a given input and the correspond-

Table 2-3
Important Laplace Transform Relations

$\mathscr{L}^{-1}\hat{F}(s) = f(t)$	$\mathscr{L}f(t) = \hat{F}(s)$
$\dfrac{1}{2\pi j}\displaystyle\int_{\sigma+j\infty}^{\sigma+j\infty} F(s)\exp(st)ds$	$\displaystyle\int_{0}^{\infty} f(t)\exp(-st)dt$
$k_1 f_1(t) + k_2 f_2(t)$	$k_1\hat{F}_1(s) + k_2\hat{F}_2(s)$
$\exp(-at)f(t)$	$F(s+a)$
$f^{(n)}(t) = \dfrac{d^n f(t)}{dt^n}$	$s^n\hat{F}(s) - s^{n-1}f(0^+) - s^{n-2}f^1(0^+) - \cdots - f^{n-1}(0^+)$
$\displaystyle\int_{-\infty}^{t} f(t)dt$	$\dfrac{\hat{F}(s)}{s} + \dfrac{\int_{-\infty}^{0} f(t)dt}{s}$
Impulse function $\delta(t)$	1
Step function $\mathscr{U}(t)$	$\dfrac{1}{s}$
t^n	$\dfrac{n!}{s^{n+1}}$
$\exp(-at)$	$\dfrac{1}{s+a}$
$\sin\omega t$	$\dfrac{\omega}{s^2+\omega^2}$
$\cos\omega t$	$\dfrac{s}{s^2+\omega^2}$

ing output do not exist. This is clear from the fact that the transfer function is a system model and does not depend on the system input.

Consider the nth-order linear, constant-parameter system given by

$$a_n \frac{d^n y}{dt^n} + a_{n-1} \frac{d^{n-1} y}{dt^{n-1}} + \cdots + a_0 y = b_0 u + b_1 \frac{du(t)}{dt} + \cdots + b_m \frac{d^m u(t)}{dt^m} \tag{2.29}$$

For physically realizable systems, $m < n$. By applying Laplace transformation and integrating by parts, it may be verified that

$$\mathscr{L} \frac{d^k f(t)}{dt^k} = s^k \hat{F}(s) - s^{k-1} f(0) - s^{k-2} \frac{df(0)}{dt} - \cdots + \frac{d^{k-1} f(0)}{dt^{k-1}} \tag{2.30}$$

The initial conditions are set to zero in obtaining the transfer function. This results in

$$\hat{H}(s) = \frac{b_0 + b_1 s + \cdots + b_m s^m}{a_0 + a_1 s + \cdots a_n s^n} \tag{2.31}$$

for $m < n$. Note that equation (2.31) contains all the information contained in equation (2.29). Consequently, transfer function may be considered an analytical model of the system. Transfer function may be employed to determine the total response of a system for a given input, even though it is defined in terms of the response under zero initial conditions. This is quite logical because the analytical model of a system is independent of the system's initial conditions.

The denominator polynomial of a transfer function is the system's characteristic polynomial. Its roots are known as poles or eigenvalues of the system. If all eigenvalues have negative real parts, the system is said to be stable. Response of a stable system is bounded (that is, remains finite) when the input is bounded. The zero-input response of a stable system approaches zero with time.

Frequency-Response Function

The Fourier integral transform of the impulse-response function is given by

$$H(f) = \int_{-\infty}^{\infty} h(t) \exp(-j2\pi f t)\, dt \tag{2.32}$$

This is known as the frequency-response function of a system.

Fourier transform operation is represented as $\mathscr{F}\, h(t) = H(f)$. In view of equation (2.13), the lower limit of integration in equation (2.32) could be made zero. Then, from equation (2.24), with $s = j2\pi f$, we obtain

$$H(f) = \hat{H}(j2\pi f) \tag{2.33}$$

It should be noted that frequency-response function, like transfer function, is a complete representation of a linear, constant-parameter system. In view of equations (2.11) and (2.12), we obtain

$$U(f) = \hat{U}(j2\pi f) \tag{2.34}$$

$$Y(f) = \hat{Y}(j2\pi f) \tag{2.35}$$

Then, from equation (2.27),

$$Y(f) = H(f)U(f) \tag{2.36}$$

If the Fourier integral transform of a function exists, then its Laplace transform also exists. The converse is not generally true, however, because of poor convergence of the Fourier integral in comparison to the Laplace integral. This arises from the fact that the factor $\exp(-\sigma t)$ is not present in the Fourier integral. For physically realizable linear, constant-parameter systems, $H(f)$ exists even if $U(f)$ and $Y(f)$ do not exist for a particular input. The experimental determination of $H(f)$, however, requires system stability. For the nth-order system given by equation (2.29), the frequency-response function is determined by applying equation (2.33) to equation (2.31) as

$$H(f) = \frac{b_0 + b_1 j2\pi f + \cdots + b_m(j2\pi f)^m}{a_0 + a_1 j2\pi f + \cdots + a_n(j2\pi f)^n} \tag{2.37}$$

This generally is a complex function of f that has a magnitude denoted by $|H(f)|$ and a phase angle denoted by $\angle H(f)$.

A further interpretation of frequency-response function can be given. Consider a harmonic input having cyclic frequency f, expressed by

$$u(t) = u_0 \cos 2\pi f t \tag{2.38}$$

In analysis, it is convenient to use the complex input

$$u(t) = u_0(\cos 2\pi f t + j \sin 2\pi f t) = u_0 \exp(j2\pi f t) \tag{2.39}$$

and take only the real part of the final result. Note that equation (2.39) does not implicitly satisfy the requirement of equation (2.11). Therefore, the appropriate versions of the convolution integral for this case are equations (2.16) and (2.21) through (2.23). By substituting equation (2.39) into (2.21), for instance, we obtain

$$y(t) = \mathrm{Re}\left[\int_{-\infty}^{t} h(\tau) u_0 \exp[j2\pi f(t-\tau)]\, d\tau\right]$$

or

$$y(t) = \mathrm{Re}\left[u_0 \exp(j2\pi ft) \int_{-\infty}^{t} h(\tau) \exp(-j2\pi f\tau)\, d\tau\right] \tag{2.40}$$

in which Re[ ] denotes the real part. As $t \to \infty$, the integral in equation (2.40) becomes the frequency-response function $H(f)$, and the response $y(t)$ becomes the steady-state response y_{ss}. Accordingly,

$$y_{ss} = \mathrm{Re}[H(f) u_0 \exp(j2\pi ft)]$$

or

$$y_{ss} = u_0 |H(f)| \cos(2\pi ft + \phi) \tag{2.41}$$

for a harmonic excitation, in which the phase-lead angle $\phi = \angle H(f)$. It follows from equation (2.41) that, when a harmonic excitation is applied to a stable linear, constant-parameter system having frequency-response function $H(f)$, its steady-state response will also be harmonic at the same frequency, but with an amplification factor of $|H(f)|$ in its amplitude and a phase lead of $\angle H(f)$. Consequently, the frequency-response function of a stable system can be experimentally determined using a sine-sweep test or a sine-dwell test. With these methods, a harmonic excitation is applied at the system input, and the amplification factor and the phase-lead angle in the corresponding response are determined at steady state. The frequency of excitation is varied continuously for a sine sweep and in steps for a sine dwell. Sweep rate should be slow enough and dwell times should be long enough to guarantee steady-state conditions at the output. The pair of plots of $|H(f)|$ and $\angle H(f)$ against f completely represents complex frequency-response function. This pair is commonly known as the Bode plot or diagram. In these plots, logarithmic scales are normally used for frequency f and magnitude $|H(f)|$.

System representation using the frequency-response function or Bode diagram is known as the frequency-domain representation. In the time-domain representation, in contrast, the system is represented by differential equations.

Response to a Support Motion

An important consideration in seismic-qualification analysis and testing of equipment is the response to a support motion. To illustrate the method of analysis, consider the linear, single-degree-of-freedom system consisting of mass m, spring constant k, and damping constant b subjected to support motion (displacement) $u(t)$. Vertical and horizontal configurations of this system are shown in figure 2-7. Both configurations possess the same equation of motion, provided the support motion $u(t)$ and the mass response (displacement) y are measured from the fixed points that correspond to the initial, static-equilibrium position of the system. In the vertical configuration, the compressive force in the spring exactly balances the weight of the mass when it is in static equilibrium. In the horizontal configuration, the spring is unstretched when in static equilibrium. It may be easily verified that the equation of motion is given by

$$m\ddot{y} + b\dot{y} + ky = ku(t) + b\dot{u}(t) \tag{2.42}$$

in which $(\dot{\ }) = d/dt$ and $(\ddot{\ }) = d^2/dt^2$. It is convenient to define two parameters ω_n and ζ, which are undamped natural frequency and damping ratio, respectively, through $\omega_n = \sqrt{k/m}$ and $2\zeta\omega_n = b/m$. This results in the equivalent equation of motion,

$$\ddot{y} + 2\zeta\omega_n\dot{y} + \omega_n^2 = \omega_n^2 u(t) + 2\zeta\omega_n\dot{u}(t) \tag{2.43}$$

There are several ways to determine the response y from equation (2.43) once the excitation function $u(t)$ is specified. The procedure adopted here is to solve the modified equation,

$$\ddot{y} + 2\zeta\omega_n\dot{y} + \omega_n^2 y = \omega_n^2 u(t) \tag{2.44}$$

This can be identified as the equation of motion of the single-degree-of-freedom system shown in figure 2-8, response y being measured from the static-equilibrium position of the mass, as before.

Impulse Response. Many important characteristics of a system can be studied by analyzing the system response to an impulse or a step-input excitation. Such characteristics include system stability, damping properties, and natural frequencies. In this way, an idea of the system response to an

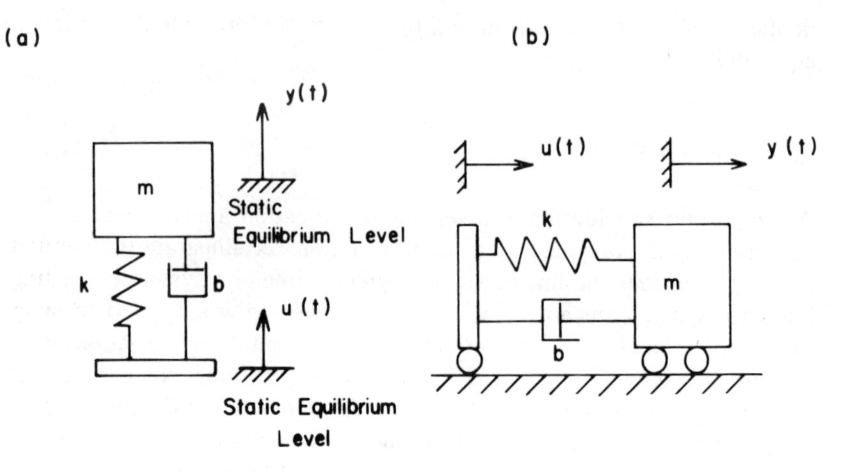

Figure 2-7. Analysis of System Response to Support Motions

arbitrary input is gained. The responses to standard inputs or test inputs also serve as the basis for system comparison. It is usually possible to determine the degree of nonlinearity in a system by exciting it with two input intensity levels separately and checking whether the proportionality is retained at the output or whether limit cycles are encountered by the response (when the excitation is harmonic).

The transfer function for equation (2.44) is

$$\hat{H}(s) = \frac{\omega_n^2}{s^2 + 2\zeta\omega_n s + \omega_n^2} \tag{2.45}$$

The characteristic equation of this system is given by

$$s^2 + 2\zeta\omega_n s + \omega_n^2 = 0 \tag{2.46}$$

The eigenvalues (poles) are given by its roots. Three possible cases exist.

Case 1 ($\zeta < 1$). This is the case of complex eigenvalues λ_1 and λ_2. Since the coefficients of the characteristic equation are real, the complex roots should occur in conjugate pairs. Hence,

$$\lambda_1, \lambda_2 = -\zeta\omega_n \pm j\omega_d \tag{2.47}$$

in which

$$\omega_d = \sqrt{1 - \zeta^2}\,\omega_n \tag{2.48}$$

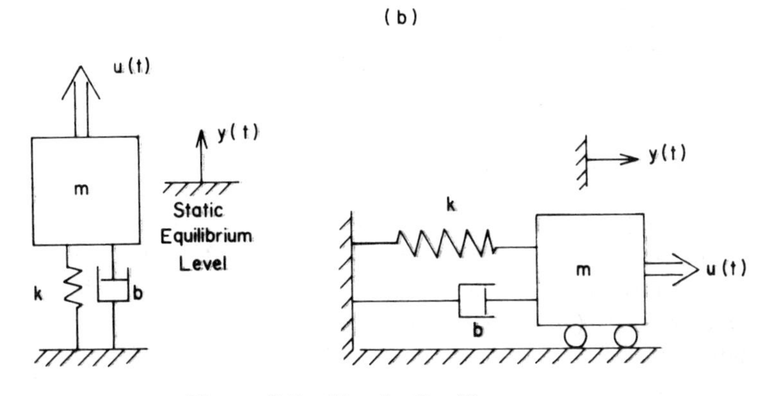

Figure 2-8. Simple Oscillator

Case 2 ($\zeta > 1$). This case corresponds to the possibility of real and unequal eigenvalues,

$$\lambda_1, \lambda_2 = -\zeta\omega_n \pm \sqrt{\zeta^2 - 1}\,\omega_n$$
$$= -a, -b \quad (2.49)$$

for $a \neq b$, in which

$$ab = \omega_n^2 \quad (2.50)$$

and

$$a + b = 2\zeta\omega_n \quad (2.51)$$

Case 3 ($\zeta = 1$). In this case, the eigenvalues are real and equal:

$$\lambda_1 = \lambda_2 = -\omega_n \quad (2.52)$$

In all three cases, the real parts of the eigenvalues are negative. Consequently, the second-order systems shown in figures 2-7 and 2-8 are always stable.

The impulse-response functions $h(t)$ corresponding to the three cases are determined by taking the inverse Laplace transform (table 2-3) of equation (2.45) for $\zeta < 1$, $\zeta > 1$, and $\zeta = 1$, respectively. The following results are obtained:

$$y_{\text{impulse}}(t) = h(t) = \frac{\omega_n}{\sqrt{1 - \zeta^2}} \exp(-\zeta\omega_n t) \sin \omega_d t \qquad \text{for } \zeta < 1 \quad (2.53)$$

$$y_{\text{impulse}}(t) = h(t) = \frac{ab}{(b-a)}[\exp(-at) - \exp(-bt)] \qquad \text{for } \zeta > 1 \tag{2.54}$$

$$y_{\text{impulse}}(t) = h(t) = \omega_n^2 \exp(-\omega_n t) \qquad \text{for } \zeta = 1 \tag{2.55}$$

An explanation concerning the dimensions of $h(t)$ is appropriate at this juncture. Since the transfer function of equation (2.44) is dimensionless, it follows that $y(t)$ has the same dimensions as $u(t)$. Since $h(t)$ is the response to a unit impulse $\delta(t)$, it follows that they have the same dimensions. The magnitude of $\delta(t)$ is represented by a unit area in the $u(t)$ versus t plane. Consequently, $\delta(t)$ has the dimensions of (1/time) or (frequency). It follows that $h(t)$ also has the dimensions of (1/time) or (frequency).

The Riddle of Zero Initial Conditions. For a second-order system, zero initial conditions correspond to $y(0) = 0$ and $\dot{y}(0) = 0$. It is clear from equations (2.53) through (2.55) that $h(0) = 0$, but $\dot{h}(0) \neq 0$, which appears to violate the zero-initial-conditions assumption. This situation is characteristic in system response to impulses and their derivatives. This may be explained as follows. When an impulse is applied to a system at rest (zero initial state), the highest derivative of the system differential equation becomes infinity momentarily. As a result, the next lower derivative becomes finite (nonzero) at $t = 0^+$. The remaining lower derivatives maintain their zero values at that instant. When an impulse is applied to the system shown in figure 2-8, for example, the acceleration $\ddot{y}(t)$ becomes infinity, and the velocity $\dot{y}(t)$ takes a nonzero (finite) value shortly after its application ($t = 0^+$). The displacement $y(t)$, however, would not have sufficient time to change at $t = 0^+$. The impulse input is therefore equivalent to a velocity initial condition in this case. This initial condition is determined by integrating equation (2.44):

$$\dot{y}(t) + 2\zeta\omega_n y(t) + \omega_n^2 \int_{0^-}^{t} y(t)dt = \omega_n^2 \int_{0^-}^{t} \delta(t)dt \tag{2.56}$$

In view of equation (2.8), the right-hand side of equation (2.56) becomes ω_n^2 for $t > 0$. Also, as $t \to 0^+$, the second and third terms on the left-hand side of equation (2.56) vanish. This gives

$$\dot{y}(0^+) = \omega_n^2 \tag{2.57}$$

The use of the initial conditions $\dot{y}(0^+) = \omega_n^2$ and $y(0^+) = 0$ in the homogeneous (unforced or free) solution of equation (2.44)—that is, the solution with $u(t) = 0$—gives the same results obtained using the inverse Laplace method—that is, equations (2.53) through (2.55).

The impulse-response functions given by equations (2.53) through (2.55) are plotted in figure 2-9 for some representative damping ratios. It should be noted that, for $0 < \zeta < 1$, the angular frequency of damped vibrations is ω_d, which is less than the undamped natural frequency ω_n [also see equation (2.48)].

Step Response. Unit step function is defined by

$$\begin{aligned} \mathscr{U}(t) &= 1 \quad \text{for} \quad t > 0 \\ &= 0 \quad \text{for} \quad t \leq 0 \end{aligned} \tag{2.58}$$

Unit impulse function $\delta(t)$ may be interpreted as the time derivative of $\mathscr{U}(t)$,

$$\delta(t) = \frac{d\mathscr{U}(t)}{dt} \tag{2.59}$$

Note that equation (2.59) reestablishes the fact that, for nondimensional $\mathscr{U}(t)$, the dimension of $\delta(t)$ is $(\text{time})^{-1}$. Since $\mathscr{L}\mathscr{U}(t) = 1/s$ (see table 2-3), the unit step response of system equation (2.44) can be obtained by taking the inverse Laplace transform of

$$\hat{Y}_{\text{step}}(s) = \frac{1}{s} \frac{\omega_n^2}{(s^2 + 2\zeta\omega_n s + \omega_n^2)} \tag{2.60}$$

which follows from equation (2.27).

To facilitate using table 2-3, partial fractions of equation (2.60) are determined in the form

$$\frac{a_1}{s} + \frac{a_2 + a_3 s}{(s^2 + 2\zeta\omega_n s + \omega_n^2)}$$

in which the constants a_1, a_2, and a_3 are determined by comparing the numerator polynomial

$$\omega_n^2 = a_1(s^2 + 2\zeta\omega_n s + \omega_n^2) + s(a_2 + a_3 s)$$

Then $a_1 = 1$, $a_2 = -2\zeta\omega_n$, and $a_3 = -1$.

Alternatively, the convolution integral may be used. In particular, from equation (2.23), we obtain

$$y_{\text{step}}(t) = \int_0^t h(\tau)\,d\tau \tag{2.61}$$

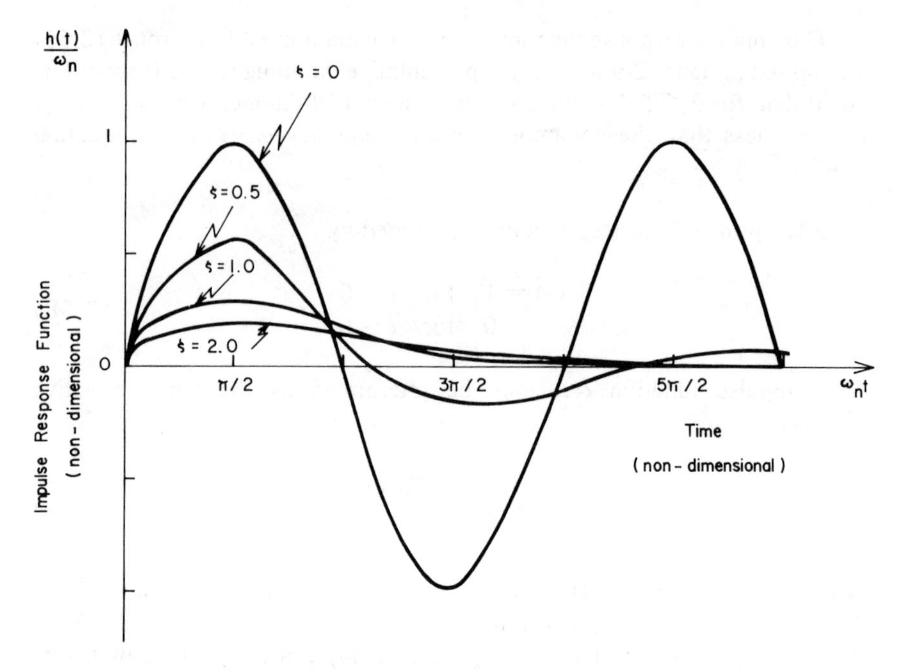

Figure 2-9. Impulse-Response Functions for a Simple Oscillator

because, for a delayed unit step,

$$\begin{aligned} \mathscr{U}(t-\tau) &= 1 \quad \text{for} \quad \tau < t \\ &= 0 \quad \text{for} \quad \tau \geq t \end{aligned} \tag{2.62}$$

Thus, unit step response is the time integral of impulse-response function in general. From either method, the following results are obtained:

$$y_{\text{step}}(t) = 1 - \frac{1}{\sqrt{1-\zeta^2}} \exp(-\zeta\omega_n t) \sin(\omega_d t + \phi) \quad \text{for} \quad \zeta < 1 \tag{2.63}$$

$$y_{\text{step}} = 1 - \frac{1}{(b-a)} [b\ exp(-at) - a \exp(-bt)] \quad \text{for} \quad \zeta > 1 \tag{2.64}$$

$$y_{\text{step}} = 1 - (\omega_n t + 1) \exp(-\omega_n t) \quad \text{for} \quad \zeta = 1 \tag{2.65}$$

In equation (2.63),

$$\cos\phi = \zeta \tag{2.66}$$

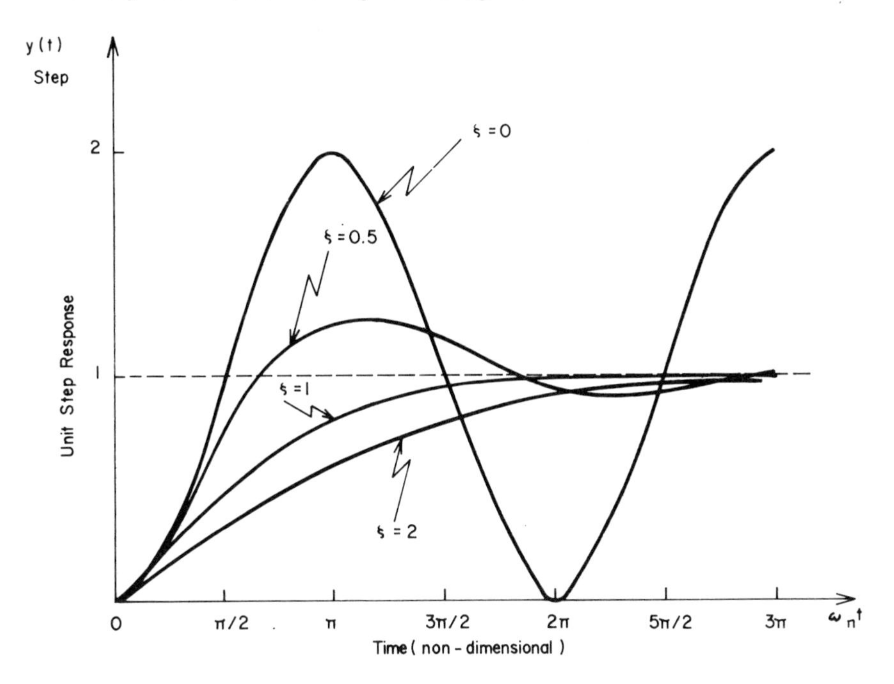

Figure 2-10. Unit Step Response for a Simple Oscillator

Note that, since a step input does not cause the highest derivative of the system equation to approach infinity at $t = 0^+$, the initial conditions that are required to solve the system equation remain unchanged at $t = 0^+$, provided that there are no derivative terms on the input side of the system equation. If there are derivative terms of the input, then, for example, a step becomes an impulse and the situation changes. This is the case when there are s terms in the numerator of the system transfer function [see equation (2.43), for example]. Roots of the numerator polynomial of a transfer function are called system zeros, as opposed to system poles. System zeros can have an effect on the response. The present system—given by equation (2.45)—has no zeros. Its unit step response—, that is plots of equations (2.63) through (2.65)—, is shown in figure 2-10 for some representative damping ratios.

In view of equation (2.59), the step response of equation (2.43) is obtained by using the principle of superposition, as the sum of the unit step response and $(2\zeta/\omega_n)$ times the unit impulse response of equation (2.44). Thus, from equations (2.53) through (2.55) and (2.63) through (2.65), we obtain

$$y(t) = 1 - \frac{\exp(-\zeta\omega_n t)}{\sqrt{1-\zeta^2}}[\sin(\omega_d t + \phi) - 2\zeta \sin \omega_d t] \qquad \text{for } \zeta < 1 \tag{2.67}$$

$$y(t) = 1 + \frac{1}{(b-a)}[a \exp(-at) - b \exp(-bt)] \qquad \text{for } \zeta > 1 \qquad (2.68)$$

$$y(t) = 1 + (\omega_n t - 1) \exp(-\omega_n t) \qquad \text{for } \zeta = 1 \qquad (2.69)$$

Note that the system given by equation (2.43) has a system zero at $s = -\omega_n/(2\zeta)$. This has modified the system response because of its input-derivative action.

Liebnitz's Rule. The time derivative of an integral whose limits of integration are also functions of time may be obtained using Liebnitz's rule. It is expressed as

$$\frac{d}{dt}\int_{a(t)}^{b(t)} f(\tau, t)d\tau = f[b(t), t]\frac{db(t)}{dt} - f[a(t), t]\frac{da(t)}{dt} + \int_{a(t)}^{b(t)} \frac{\partial f}{\partial t}(\tau, t)d\tau \qquad (2.70)$$

By repeated application of Liebnitz's rule to equation (2.23), we can determine the ith derivative of the output variable,

$$\frac{d^i y(t)}{dt^i} = \left[h(t) + \frac{dh(t)}{dt} + \cdots + \frac{d^{i-1}h(t)}{dt^{i-1}}\right]u(0) + \left[h(t) + \frac{dh(t)}{dt} + \cdots + \frac{d^{i-2}h(t)}{dt^{i-2}}\right]\frac{du(0)}{dt} + \cdots + h(t)\frac{d^{i-1}u(0)}{dt^{i-1}} + \int_0^t h(\tau)\frac{d^i u(t-\tau)}{dt^i}d\tau \qquad (2.71)$$

From this, it follows that the zero-state response to input $[d^i u(t)]/dt^i$ is $[d^i y(t)]/dt^i$, provided that all lower-order derivatives of $u(t)$ vanish at $t = 0$. This result verifies the fact, for instance, that the first derivative of the unit step response gives the impulse-response function.

Frequency-Response Functions of the Second Order Systems

In view of equation (2.33), the frequency-response function corresponding to the system equation (2.44) is given by

$$H(f) = \frac{f_n^2}{f_n^2 - f^2 + j2\zeta f_n f} \tag{2.72}$$

Its magnitude and phase angle are plotted against frequency in figure 2-11. From equation (2.41), it follows that, if the sinusoidal input given by

$$u(t) = u_0 \cos 2\pi f t \tag{2.73}$$

is applied to the system given by equation (2.44), its steady-state response (output) may be given by

$$y_{ss}(t) = u_0 \frac{f_n^2}{\sqrt{(f_n^2 - f^2)^2 + 4\zeta^2 f_n^2 f^2}} \cos(2\pi f t - \phi) \tag{2.74}$$

in which the phase-lag angle is

$$\phi = \tan^{-1} \frac{2\zeta f_n f}{(f_n^2 - f^2)} \tag{2.75}$$

Resonant frequency is the excitation frequency at which the amplication magnitude $|H(f)|$ is maximum. Using straightforward calculus on equation (2.72), it can be shown that the resonant frequence f_r for this system may be expressed by

$$f_r = \sqrt{1 - 2\zeta^2}\, f_n \tag{2.76}$$

It follows that the resonant frequency is less than the damped natural frequency given by equation (2.48) unless the damping is zero when f_r, f_d, and f_n coincide. An important fact to remember is that sine input tests determine frequency-response functions and, from them, the resonant frequencies of a test object. By applying an impulse or a step input and observing the frequency of oscillation, however, damped natural frequency is obtained.

For the system given by equation (2.43), the frequency-response function is

$$H(f) = \frac{f_n^2 + j2\zeta f_n f}{f_n^2 - f^2 + j2\zeta f_n f} \tag{2.77}$$

Its magnitude and phase angle are plotted against frequency in figure 2-12.

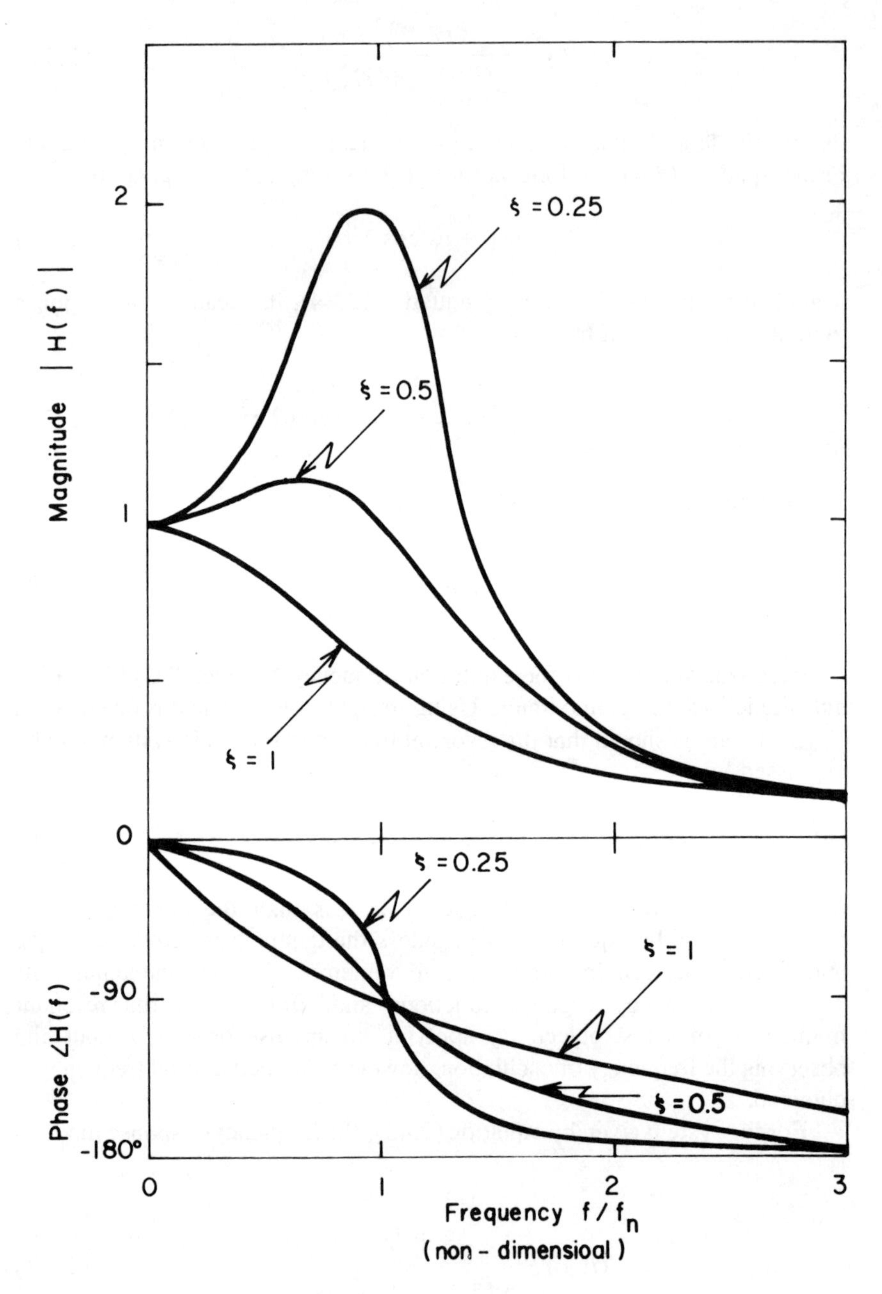

Figure 2-11. Frequency Response of a Simple Oscillator

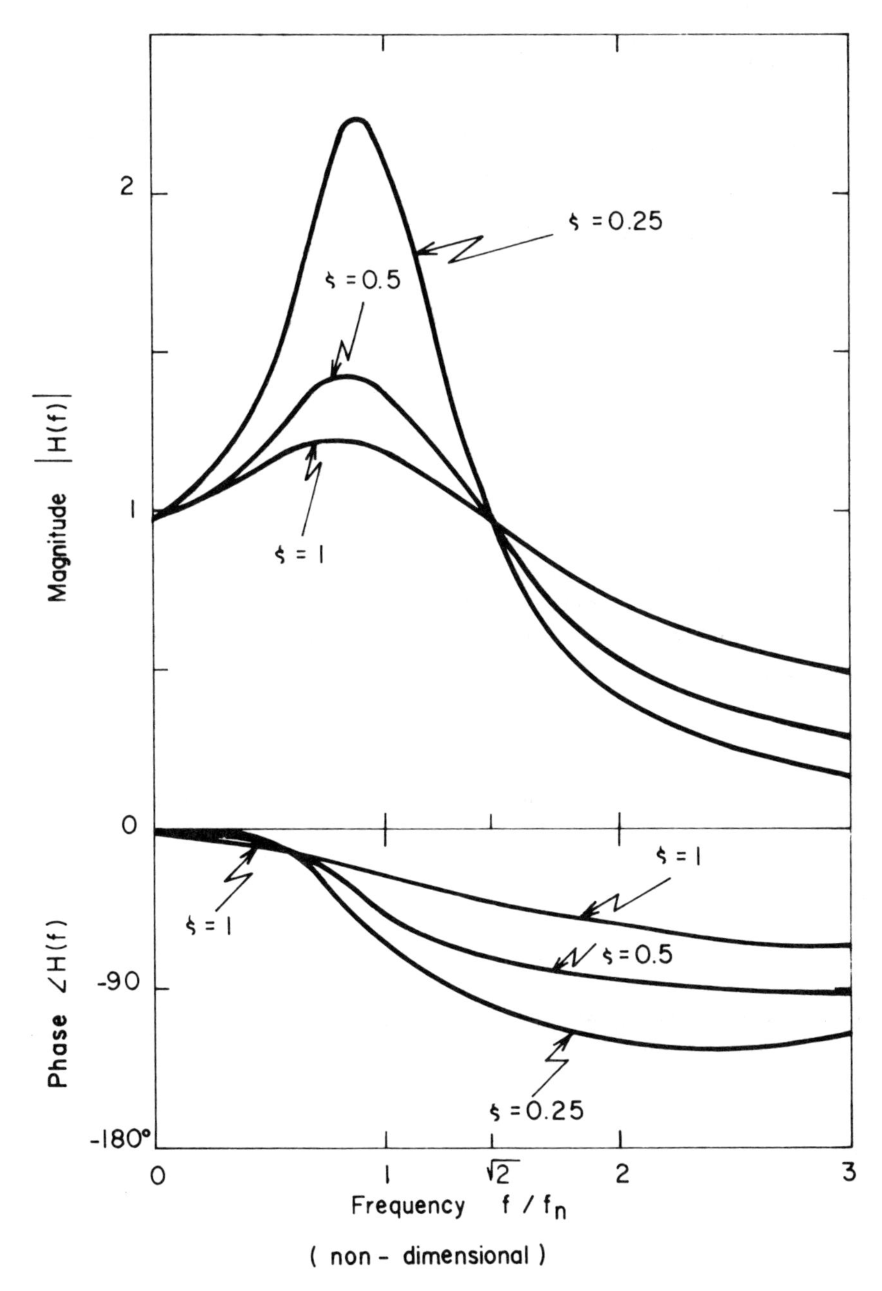

Figure 2-12. Frequency Response of a Simple Oscillator That Has a System Zero

For the sinusoidal input given by equation (2.73), the steady-state response of this system may be given by

$$y_{ss}(t) = u_0 \sqrt{\frac{f_n^2(f_n^2 + 4\zeta^2 f^2)}{(f_n^2 - f^2)^2 + 4\zeta^2 f_n^2 f^2}} \cos(2\pi f t - \phi) \tag{2.78}$$

in which the phase lag is given by

$$\phi = \tan^{-1}\left[\frac{2\zeta f_n f}{(f_n^2 - f^2)}\right] - \tan^{-1}\left[\frac{2\zeta f}{f_n}\right] \tag{2.79}$$

Transmissibility

Transmissibility, or force transmissibility, is defined as the ratio of the amplitude of force transmitted to the supporting structure of a system to the amplitude of the excitation force applied to the system. This is a measure of the amount of vibration transmitted from a system to its supporting structure. For good vibration isolation, we desire low transmissibility. If transmissibility is denoted by T, then the percentage isolation I is expressed as

$$I = (1 - T)100\% \tag{2.80}$$

An expression for the transmissibility function may be determined by considering a single-degree-of-freedom system (simple oscillator) supported on a rigid structure and excited by a harmonic force given by the real part of

$$F(t) = F_0 \exp(j\omega t) \tag{2.81}$$

The system may be represented as in figure 2-8, with $u(t)$ replaced by $F(t)$. Its equation of motion is

$$m\ddot{y} + b\dot{y} + ky = F_0 \exp(j\omega t) \tag{2.82}$$

The steady-state response is given by the real part of

$$y_{ss} = \frac{F_0}{[m(j\omega)^2 + bj\omega + k]} \exp(j\omega t) \tag{2.83}$$

The force transmitted to the supporting structure at steady state is the real part of $ky_{ss} + b\dot{y}_{ss}$. In view of equation (2.83), this force is given by the real part of

$$F_T = \frac{[k + bj\omega]}{[m(j\omega)^2 + bj\omega + k]} F_0 \exp(j\omega t) \qquad (2.84)$$

The transmissibility function is given by

$$T(\omega) = \frac{|F_T|}{|F(t)|} \qquad (2.85)$$

or

$$T(\omega) = \frac{\sqrt{k^2 + b^2\omega^2}}{\sqrt{(k - m\omega^2)^2 + b^2\omega^2}} \qquad (2.86)$$

Using the definitions of natural frequency ω_n and damping ratio ζ, transmissibility may be expressed as

$$T(\omega) = \frac{\sqrt{1 + [2\zeta\omega/\omega_n]^2}}{\sqrt{[1 - (\omega/\omega_n)^2]^2 + [2\zeta\omega/\omega_n]^2}} \qquad (2.87)$$

It is interesting to note that $T(\omega)$ is also equal to the magnitude of the frequency-response function of the system shown in figure 2-7 [see equation (2.77)]. This means that, if the supporting structure were free to move, the ratio of the amplitude of the support motion to the amplitude of the motion of the mass is given by equation (2.87). This ratio is known as motion transmissibility. It follows that the force transmissibility and the motion transmissibility are equal. Transmissibility plots for a simple oscillator are given by the magnitude curves in Figure 2-12. The following observations can be made by referring to these plots. If $\omega/\omega_n < \sqrt{2}$, then $T > 1$, and hence the input disturbances are amplified in this frequency range. It follows that $\omega/\omega_n > \sqrt{2}$ is the only frequency range in which positive vibration isolation is possible. In this region, T incrases with damping (ζ). This means that satisfactory vibration isolation is achieved with very little or no damping. Close to $\omega/\omega_n = 1$—actually, at $\omega = \omega_r$—a resonance condition exists. In this neighborhood, T can be very large. This high transmissibility near resonance can be reduced by increasing damping. Thus, there is a trade-off in selecting damping for the resonance region and the isolation region. Since T decreases with increasing ω/ω_n in the isolation region for a given excitation frequency ω, vibration isolation can be improved (that is, T decreases) by decreasing ω_n. This may be achieved by either decreasing k or increasing m. The latter method, however, increases the dead weight of the vibrating system (and hence the weight supported by the structure), which is not generally advisable. The former approach—of decreasing the support-spring stiffness—is the more favorable method of vibration isolation. In practice, this is achieved by using soft springs with light damping. A disadvantage of soft springs is that they could introduce rocking and associated instability to the vibrating system, which could be counteracted by adding mass (inertia blocks) to the base of the system.

For low damping, equation (2.87) may be approximated by

$$T(\omega) = \frac{1}{|(\omega/\omega_n)^2 - 1|} \tag{2.88}$$

This expression may be used to select the support-spring stiffness for a given machinery. First, T is determined from equation (2.80) for the required degree of vibration isolation. Then, ω_n is computed from equation (2.88) for a given frequency of excitation. Since the system mass m is known, k can be determined from ω_n.

Mechanical Impedance and Mobility

We have noted in the previous section that force-transmissibility function is a special frequency-transfer function in which both input and output are force variables. Similarly, velocity-transmissibility function is a frequency-transfer function in which both input and output are motion variables. Mechanical impedance corresponds to a complex frequency-transfer function in which the input is a velocity variable and the output is a force variable. Mobility corresponds to a complex frequency-transfer function in which the input is a force variable and the output is a velocity variable. Mobility is sometimes called mechanical admittance and is the inverse of mechanical impedance. When the same location is considered for both input and output, we get point impedance and point mobility. When the output location is different from the input location, we get transfer impedance and transfer mobility.

Point-impedance and point-mobility functions for the three basic mechanical elements—mass, linear spring, and linear damper—are determined from their constitutive relations (see table 2-1). In each case, the corresponding transfer function is obtained by replacing d/dt by s, and the frequency-transfer function is obtained by replacing s by $j\omega$. The expressions are listed in table 2-4.

A useful procedure in obtaining impedance or mobility at a point in a

Table 2-4
Impedance and Mobility Functions of Some Basic Mechanical Elements

Element	*Mechanical Impedance*	*Mobility*
Mass	$j\omega m$	$\frac{1}{j\omega m}$
Linear spring	$\frac{k}{j\omega}$	$\frac{j\omega}{k}$
Linear damper	b	$\frac{1}{b}$

complex mechanical system with many interconnected elements is (1) adding impedances of parallel elements and (2) adding mobilities of series elements. Parallel elements are those that have the same velocity (across-variable). Series elements are those that have the same force (through-variable). The rationale for the foregoing procedure is that impedance has velocity in its denominator, whereas mobility has force in its denominator. The following two examples illustrate this procedure.

For the system shown in figure 2-8, the point impedance at the mass can be obtained directly from the equation of motion [equation (2.82)]. Since the equation motion relates the output displacement y to the input force, the corresponding relationship between force and velocity ($\dot{y}$) is obtained by differentiating in the time domain or multiplying by $j\omega$ in the frequency domain. It follows that the point impedance at the mass is

$$Z(\omega) = \frac{m(j\omega)^2 + bj\omega + k}{j\omega}$$

which can be written as the sum

$$Z(\omega) = mj\omega + b + \frac{k}{j\omega} \tag{2.89}$$

Equation (2.89) can be obtained directly using rule 1, stated earlier. This should be clear from figure 2-8, because the mass, the damper, and the spring have a common velocity (that is, they are connected in parallel) in this system.

As a second example, consider the support-excited system shown in figure 2-7. It should be noted that the support velocity $u(t)$ is caused by a force $F(t)$. From Newton's second law, we have $m\ddot{y} = F(t)$. The transfer function between y and u is obtained from equation (2.42) as

$$\frac{Y(\omega)}{U(\omega)} = \frac{k + bj\omega}{m(j\omega)^2 + bj\omega + k}$$

From this, the point impedance at the support is obtained as

$$Z(\omega) = \frac{m(j\omega)^2 Y(\omega)}{(j\omega) U(\omega)} = \frac{m(j\omega)^2 (k + bj\omega)}{j\omega[m(j\omega)^2 + bj\omega + k]}$$

or

$$Z(\omega) = \frac{mj\omega(k + bj\omega)}{[m(j\omega)^2 + bj\omega + k]} \tag{2.90}$$

Alternatively, since the spring and the damper in the system (figure 2-7) are connected in parallel (that is, they have a common velocity), their combined impedance is $k/j\omega + b$. Since this combination is connected in series with the mass (that is, they have a common force), their combined mobility can be obtained by adding the individual mobilities,

$$M(\omega) = \frac{1}{Z(\omega)} = \frac{1}{mj\omega} + \frac{1}{k/j\omega + b}$$

which simplifies to

$$M(\omega) = \frac{1}{Z(\omega)} = \frac{m(j\omega)^2 + bj\omega + k}{mj\omega(k + bj\omega)}$$

which is identical to the result given by equation (2.90).

State-Space Representation

In this chapter, we have introduced two fundamental ways of representing constant-parameter, linear systems: (1) in the time domain, by a set of differential equations; and (2) in the frequency domain, by a complex frequency-response function (or, in the Laplace domain, by a transfer function). The first representation is general and is also used in time-varying-parameter, nonlinear systems.

More than one variable might be needed to represent the response of a system. There also could be more than one input variable. System equations are the set of differential equations relating the response variables to the input variables. This set of system equations generally is coupled, so that more than one response variable appears in each differential equation. A particularly useful time-domain representation for a system is a state model. In this representation (state-space representation), a nth-order system is represented by n first-order differential equations, which generally are coupled. This is of the general form:

$$\frac{dq_1}{dt} = f_1(q_1, q_2, \ldots, q_n, r_1, r_2, \ldots, r_m, t)$$

$$\frac{dq_2}{dt} = f_2(q_1, q_2, \ldots, q_n, r_1, r_2, \ldots, r_m, t)$$

$$\vdots$$

$$\frac{dq_n}{dt} = f_n(q_1, q_2, \ldots, q_n, r_1, r_2, \ldots, r_m, t) \qquad (2.91)$$

The n state variables can be expressed as the state vector,

$$\boldsymbol{q} = [q_1, q_2, \ldots, q_n]^T \qquad (2.92)$$

which is a column vector. Note that $[\ ]^T$ denotes matrix or vector transpose. The space formed by all possible state vectors of a system is the state space.

State vector of a system is a least set of variables required to completely determine the state of the system at all instants of time. They may or may not have physical interpretation. State vector is not unique; many choices are possible for a given system. Output (response) variables of a system can be completely determined from any such choice of state variables. Since state vector is a least set, a given state variable cannot be expressed as a linear combination of the remaining state variables in that state vector.

The m variables $r_1, r_2, \ldots, r_m$ in equations (2.91) are the input variables, and they can be expressed as the input vector:

$$\boldsymbol{r} = [r_1, r_2, \ldots, r_m]^T \qquad (2.93)$$

Now equation (2.91) can be written in the vector notation,

$$\dot{\boldsymbol{q}} = \boldsymbol{f}(\boldsymbol{q}, \boldsymbol{r}, t) \qquad (2.94)$$

When t is not present explicitly in the function $\boldsymbol{f}$, the system is said to be autonomous.

Linearization

Equilibrium states of the system given by equation (2.94) correspond to

$$\dot{\boldsymbol{q}} = 0 \tag{2.95}$$

Consequently, the equilibrium states $\overline{\boldsymbol{q}}$ are obtained by solving the set of n algebraic equations

$$\boldsymbol{f}(\boldsymbol{q}, \boldsymbol{r}, t) = 0 \tag{2.96}$$

for a special steady input $\overline{\boldsymbol{r}}$. Usually a system operates in the neighborhood of one of its equilibrium states. This state is known as its operating point. The steady state of a system is also an equilibrium state.

Suppose that a slight excitation is given to a system operating at an equilibrium state. If the system response builds up and deviates further from the equilibrium state, the equilibrium state is said to be unstable. If the system returns to the original operating point, the equilibrium state is stable. If it remains at the new state without either returning to the equilibrium state or building up the response, the equilibrium state is said to be neutral.

To study the stability of various equilibrium states of a nonlinear system, it is first necessary to linearize the system model about these equilibrium states. Linear models are also useful in analyzing nonlinear systems when it is known that the variations of the system response about the system operating point are small in comparison to the maximum allowable variation (dynamic range). Equation (2.94) can be linearized for small variations $\delta\boldsymbol{q}$ and $\delta\boldsymbol{r}$ about an equilibrium point $(\overline{\boldsymbol{q}\boldsymbol{r}})$ by employing up to only the first derivative term in the Taylor series expansion of the nonlinear function $\boldsymbol{f}$. The higher-order terms are negligible for small $\delta\boldsymbol{q}$ and $\delta\boldsymbol{r}$. This method yields

$$\delta\dot{\boldsymbol{q}} = \frac{\partial \boldsymbol{f}}{\partial \boldsymbol{q}}(\overline{\boldsymbol{q}}, \overline{\boldsymbol{r}}, t)\,\delta\boldsymbol{q} + \frac{\partial \boldsymbol{f}}{\partial \boldsymbol{r}}(\overline{\boldsymbol{q}}, \overline{\boldsymbol{r}}, t)\,\delta\boldsymbol{r} \tag{2.97}$$

State vector and input vector for the linearized system are denoted by

$$\partial\boldsymbol{q} = \boldsymbol{x} = [x_1, x_2, \ldots, x_n]^T \tag{2.98}$$

$$\delta \boldsymbol{r} = \boldsymbol{u} = [u_1, u_2, \ldots, u_m]^T \tag{2.99}$$

Linear system matrix $\boldsymbol{A}(t)$ and input gain matrix $\boldsymbol{B}(t)$ are defined as

$$\boldsymbol{A}(t) = \frac{\partial \boldsymbol{f}}{\partial \boldsymbol{q}}(\bar{\boldsymbol{q}}, \bar{\boldsymbol{r}}, t) \tag{2.100}$$

$$\boldsymbol{B}(t) = \frac{\partial \boldsymbol{f}}{\partial \boldsymbol{r}}(\bar{\boldsymbol{q}}, \bar{\boldsymbol{r}}, t) \tag{2.101}$$

Then the linear state model can be expressed by

$$\dot{\boldsymbol{x}} = \boldsymbol{A}\boldsymbol{x} + \boldsymbol{B}\boldsymbol{u} \tag{2.102}$$

If the system is a constant-parameter system, or if it can be assumed as such for the time period of interest, then $\boldsymbol{A}$ and $\boldsymbol{B}$ become constant matrices.

Time variation of the state vector of a linear, constant-parameter system can be obtained using the Laplace transform method. The Laplace transform of equation (2.102) is given by

$$s\hat{\boldsymbol{X}}(s) - \boldsymbol{x}(0) = \boldsymbol{A}\hat{\boldsymbol{X}}(s) + \boldsymbol{B}\hat{\boldsymbol{U}}(s) \tag{2.103}$$

Consequently,

$$\boldsymbol{x}(t) = \mathscr{L}^{-1}(s\boldsymbol{I} - \boldsymbol{A})^{-1}\boldsymbol{x}(0) + \mathscr{L}^{-1}(s\boldsymbol{I} - \boldsymbol{A})^{-1}\boldsymbol{B}\hat{\boldsymbol{U}}(s) \tag{2.104}$$

in which $\boldsymbol{I}$ denotes the identity (unit) matrix. The square matrix $(s\boldsymbol{I} - \boldsymbol{A})^{-1}$ is known as the resolvent matrix. Its inverse Laplace transform is the state-transition matrix:

$$\boldsymbol{\Phi}(t) = \mathscr{L}^{-1}(s\boldsymbol{I} - \boldsymbol{A})^{-1} \tag{2.105}$$

It can be shown that $\boldsymbol{\Phi}(t)$ is equal to the matrix exponential

$$\boldsymbol{\Phi}(t) = \exp(\boldsymbol{A}t) = \boldsymbol{I} + \boldsymbol{A}t + \frac{1}{2!}\boldsymbol{A}^2t^2 + \cdots \tag{2.106}$$

The state-transition matrix maybe analytically determined as a closed matrix function by the direct use of inverse transformation on each term of the resolvent matrix, using eqution (2.105), or as a series solution, using equation (2.106). We can reduce the infinite series given in equation (2.106) into a finite matrix polynomial of order $n - 1$ by using the Cayley-Hamilton theorem. This theorem states that a matrix satisfies its own characteristic equation. The characteristic polynomial of $\boldsymbol{A}$ can be expressed as

$$\begin{aligned}\Delta(\lambda) &= \det(\boldsymbol{A} - \lambda\boldsymbol{I}) \\ &= a_n\lambda^n + a_{n-1}\lambda^{n-1} + \cdots + a_0\end{aligned} \tag{2.107}$$

in which det( ) denotes determinant. The notation

$$\Delta(\boldsymbol{A}) = a_n\boldsymbol{A}^n + a_{n-1}\boldsymbol{A}^{n-1} + \cdots + a_0\boldsymbol{I} \tag{2.108}$$

is used. Then, by the Cayley-Hamilton equation,

$$0 = a_n\boldsymbol{A}^n + a_{n-1}\boldsymbol{A}^{n-1} + \cdots + a_0\boldsymbol{I} \tag{2.109}$$

To get a polynomial expansion for $\exp(\boldsymbol{A}t)$, we write

$$\exp(\boldsymbol{A}t) = S(\boldsymbol{A}) \cdot \Delta(\boldsymbol{A}) + \alpha_{n-1}\boldsymbol{A}^{n-1} + \alpha_{n-2}\boldsymbol{A}^{n-2} + \cdots + \alpha_0\boldsymbol{I} \tag{2.110}$$

in which $S(\boldsymbol{A})$ is an appropriate infinite series. Since $\Delta(\boldsymbol{A}) = 0$ by the Cayley-Hamiltion theorem, however, we have

$$\exp(\boldsymbol{A}t) = \alpha_{n-1}\boldsymbol{A}^{n-1} + \alpha_{n-2}\boldsymbol{A}^{n-2} + \cdots + \alpha_0\boldsymbol{I} \tag{2.111}$$

Now it is just a matter of determining the coefficients $\alpha_0, \alpha_1, \ldots, \alpha_{n-1}$, which are functions of time. This is done as follows. From equation (2.110),

$$\exp(\lambda t) = S(\lambda) \cdot \Delta(\lambda) + \alpha_{n-1}\lambda^{n-1} + \alpha_{n-2}\lambda^{n-2} + \cdots + \alpha_0 \tag{2.112}$$

If $\lambda_1, \lambda_2, \ldots, \lambda_n$ are the eigenvalues of $\boldsymbol{A}$, however, then, by definition,

$$\Delta(\lambda_i) = \det(\boldsymbol{A} - \lambda_i\boldsymbol{I}) = 0 \quad \text{for} \quad i = 1, 2, \ldots, n \tag{2.113}$$

Thus, from equation (2.112), we obtain

$$\exp(\lambda_i t) = \alpha_{n-1}\lambda_i^{n-1} + \alpha_{n-2}\lambda_i^{n-2} + \cdots + \alpha_0 \quad \text{for} \quad i = 1, 2, \ldots, n \tag{2.114}$$

If the eigenvalues are all distinct, equation (2.114) represents a set of n independent algebraic equations from which the n unknowns $\alpha_0, \alpha_1, \ldots,$

α_{n-1} could be determined. Since the product in the Laplace domain is a convolution integral in the time domain, and vice versa [see, for example, equation (2.27)], the second term on the right-hand side of equation (2.104) can be expressed as a matrix convolution integral. This gives

$$\boldsymbol{x}(t) = \boldsymbol{\Phi}(t)\boldsymbol{x}(0) + \int_0^t \boldsymbol{\Phi}(t-\tau)\boldsymbol{B}\,\boldsymbol{u}(\tau)\,d\tau \tag{2.115}$$

The first part of this solution is the zero-input response; the second part is the zero-state response.

State variables are not necessarily measurable and generally are not system outputs. Linearized relationship between state variables and system output (response) variables $\boldsymbol{y}(t)$ may be expressed as

$$\boldsymbol{y}(t) = \boldsymbol{C}\boldsymbol{x}(t) \tag{2.116}$$

in which the output vector is

$$\boldsymbol{y} = [y_1, y_2, \ldots, y_p]^T \tag{2.117}$$

and $\boldsymbol{C}$ denotes the output (measurement) gain matrix. When $m > 1$ and $p > 1$, the system is said to be a multi-input–multi-output (MIMO) system. This is represented by the block diagram in figure 2-13. Note that, in this case, we have a transfer matrix $\hat{\boldsymbol{H}}(s)$ given by

$$\hat{\boldsymbol{H}}(s) = \boldsymbol{C}(s\boldsymbol{I} - \boldsymbol{A})^{-1}\boldsymbol{B} \tag{2.118}$$

which satisfies

$$\hat{\boldsymbol{Y}}(s) = \hat{\boldsymbol{H}}(s)\hat{\boldsymbol{U}}(s) \tag{2.119}$$

Since

$$(s\boldsymbol{I} - \boldsymbol{A})^{-1} = \frac{\operatorname{adj}(s\boldsymbol{I} - \boldsymbol{A})}{\det(s\boldsymbol{I} - \boldsymbol{A})} \tag{2.120}$$

in which adj( ) denotes the adjoint, it is seen that, as stated before, the poles, or eigenvalues, of the system (matrix $\boldsymbol{A}$) are given by the solution of its characteristic equation:

$$\det(s\boldsymbol{I} - \boldsymbol{A}) = 0 \tag{2.121}$$

which should be compared with equation (2.111). If all eigenvalues of $\boldsymbol{A}$ have negative real parts, then the state-transition matrix $\boldsymbol{\Phi}(t)$ in equation

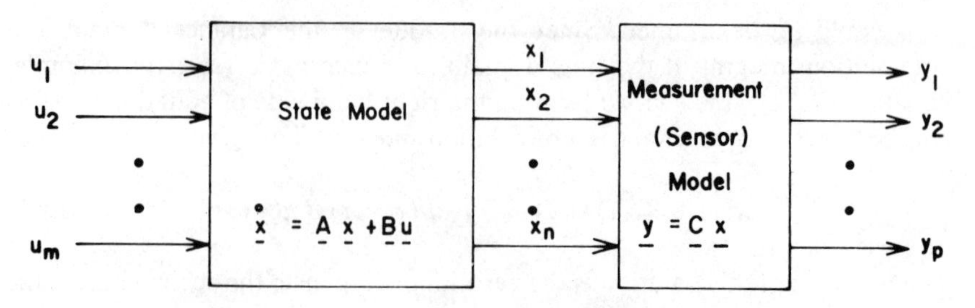

Figure 2-13. Block Diagram Representation of Multi-Input–Multi-Output Systems

(2.115) will be bounded as $t \to \infty$, which means that the linear system is stable.

Illustrative Examples

As a first example, consider the second-order nonlinear autonomous system

$$\dot{q}_1 = q_1 q_2 - r(t)$$

$$\dot{q}_2 = q_1 - q_2$$

If the input at equilibrium state is $\bar{r}$, from equation (2.96), we have

$$\bar{q}_1 \bar{q}_2 - \bar{r} = 0$$

$$\bar{q}_1 - \bar{q}_2 = 0$$

The corresponding equilibrium states are

$$\bar{\mathbf{q}} = [\sqrt{\bar{r}}, \sqrt{\bar{r}}\,]^T \quad \text{and} \quad [-\sqrt{\bar{r}}, -\sqrt{\bar{r}}]^T$$

Note that $\bar{r} > 0$ for a physically realizable system. Using equations (2.97) through (2.99), the state model is linearized about an equilibrium state. Thus

$$\dot{x}_1 = \bar{q}_2 x_1 + \bar{q}_1 x_2 - u(t)$$

$$\dot{x}_2 = x_1 - x_2$$

which corresponds to

$$A = \begin{bmatrix} \bar{q}_2 & \bar{q}_1 \\ 1 & -1 \end{bmatrix} \quad \text{and} \quad B = \begin{bmatrix} -1 \\ 0 \end{bmatrix}$$

The eigenvalues of A are given by the characteristic equation:

$$\det \begin{bmatrix} -\bar{q}_2 & -\bar{q}_1 \\ -1 & +1 \end{bmatrix} = 0$$

or

$$\lambda^2 + (1 - \bar{q}_2)\lambda - (\bar{q}_1 + \bar{q}_2) = 0$$

For the roots of this second-order equation to have negative real parts, we must have $1 - \bar{q}_2 > 0$ and $\bar{q}_1 + \bar{q}_2 < 0$. Accordingly, the equilibrium state

$$[-\sqrt{\bar{r}}, -\sqrt{\bar{r}}]^T$$

is stable, and

$$[\sqrt{\bar{r}}, \sqrt{\bar{r}}]^T$$

is unstable.

As a second example, consider a test setup for a delicate instrument enclosed in a massless casing, which is modeled as shown in figure 2-14. A velocity input $[u(t)]$ is applied to the shaker table using a linear actuator. We shall determine a state model to analyze the velocity (v) of the instrument. From the state model, we shall determine the frequency-response function relating u and v.

Since the springs A and B can be combined into one, there are only three independent energy-storage elements (m, $A + B$, k_2) in this model. Therefore, we expect a third-order state model. Also, it is a single-input $[u(t)]$, single-output (v) system. An equivalent model is shown in figure 2-15. In general, it is convenient to use the velocities of independent masses and the forces in independent springs as the state variables. We define forces as follows:

$$f_1 = \text{compressive force in } k_1$$

$$f_2 = \text{compressive force in } k_2$$

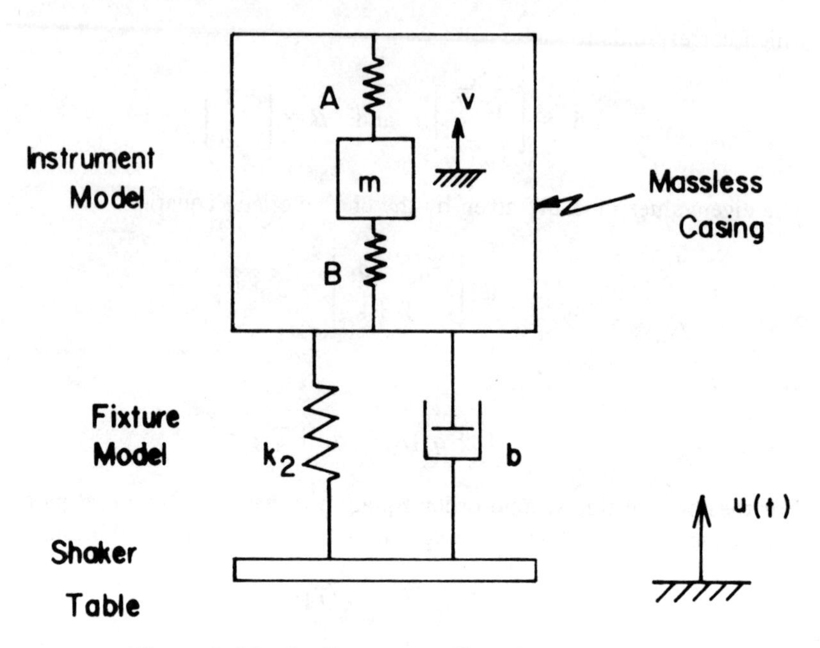

Figure 2-14. An Instrument Test-Setup Model

The state vector is

$$\boldsymbol{x} = [v, f_1, f_2]^T \quad \text{in which}$$

$$v_1 = \text{velocity at the junction } P.$$

Newton's law for m is

$$m\dot{v} = f_1$$

Hooke's law for k_1 and k_2 is

$$\dot{f}_1 = k_1(v_1 - v)$$

$$\dot{f}_2 = k_2(u - v_1)$$

The damper equation is

$$f_1 - f_2 = b(u - v_1)$$

Finally, we eliminate the auxiliary variable v_1 to obtain the state model:

$$\dot{v} = \frac{1}{m} f_1$$

$$\dot{f}_1 = -k_1 \left[v + \frac{1}{b}(f_1 - f_2) - u \right]$$

$$\dot{f}_2 = \frac{k_2}{b}(f_1 - f_2)$$

which corresponds to

$$\boldsymbol{A} = \begin{bmatrix} 0 & \frac{1}{m} & 0 \\ -k_1 & -\frac{k_1}{b} & \frac{k_1}{b} \\ 0 & \frac{k_2}{b} & -\frac{k_2}{b} \end{bmatrix} \quad \text{and} \quad \boldsymbol{B} = \begin{bmatrix} 0 \\ k_1 \\ 0 \end{bmatrix}$$

The out equation is simply

$$y = v$$

which corresponds to

$$\boldsymbol{C} = [1 \;\; 0 \;\; 0]$$

From equation (2.118), the system transfer function is

$$H(s) = [1 \; 0 \; 0] \begin{bmatrix} s & -\frac{1}{m} & 0 \\ k_1 & s + \frac{k_1}{b} & -\frac{k_1}{b} \\ 0 & -\frac{k_2}{b} & s + \frac{k_2}{b} \end{bmatrix}^{-1} \begin{bmatrix} 0 \\ k_1 \\ 0 \end{bmatrix}$$

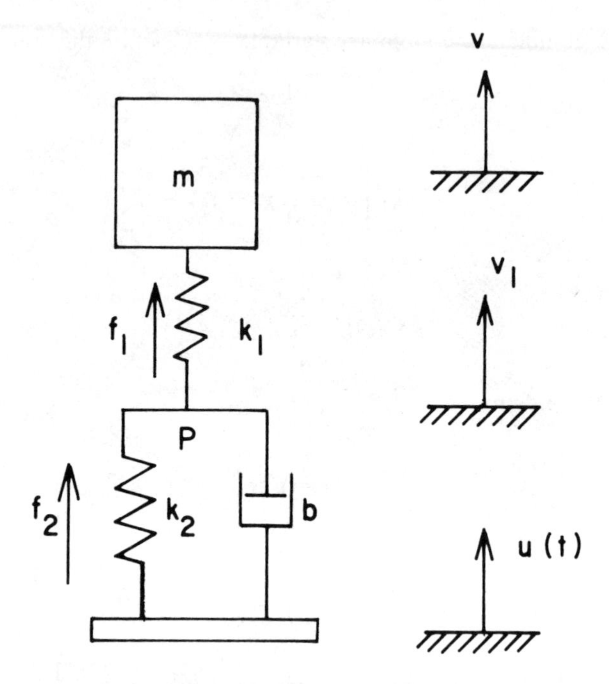

Figure 2-15. Equivalent Model for Figure 2-14

Alternatively, $\hat{H}(s)$ may be obtained by solving the three simultaneous equations:

$$s\hat{V}(s) = \frac{1}{m}\hat{F}_1(s)$$

$$s\hat{F}_1(s) = -k_1\left[\hat{V}(s) + \frac{1}{b}\{\hat{F}_1(s) - \hat{F}_2(s)\} - \hat{U}(s)\right]$$

$$s\hat{F}_2(s) = \frac{k_2}{b}[\hat{F}_1(s) - \hat{F}_2(s)]$$

to obtain the ratio $\hat{V}(s)/\hat{U}(s)$. From either method,

$$\hat{H}(s) = \frac{k_1 b(bs + k_2)}{ms^2[b^2 s + b(k_1 + k_2)] + kb(bs + k_2)}$$

The frequency-response function is obtained by simply substituting $j\omega$ for s [equation (2.33)].

Modal Analysis of Lumped-Parameter Systems

Consider the free motion ($\boldsymbol{u} = 0$) of the nth order, constant-parameter, linear system given by equation (2.102). From equation (2.115), we notice that the solution is given by

$$\boldsymbol{x} = \boldsymbol{\Phi}(t)\boldsymbol{x}(0) \tag{2.122}$$

We stated that $\boldsymbol{\Phi}(t)$ is given by the matrix-expotential expansion equation (2.106). To discuss the rationale for this exponential response further, we begin by assuming a homogeneous solution of the form

$$\boldsymbol{x} = \boldsymbol{X} \exp(\lambda t) \tag{2.123}$$

By substituting equation (2.123) in the homogeneous equation of motion [that is, equation (2.102) with $\boldsymbol{u} = 0$], the following matrix-eigenvalue problem results:

$$(\boldsymbol{A} - s\boldsymbol{I})\boldsymbol{X} = 0 \tag{2.124}$$

We shall assume that the n eigenvalues ($\lambda_1, \lambda_2, \ldots, \lambda_n$) of $\boldsymbol{A}$ are distinct. Then the corresponding eigenvectors $\boldsymbol{X}_1, \boldsymbol{X}_2, \ldots, \boldsymbol{X}_n$ are linearly independent vectors; that is, any one eigenvector cannot be expressed as a linear combination of the rest of the eigenvectors in the set. Thus, the general solution for free dynamics is

$$\boldsymbol{x}(t) = \boldsymbol{X}_1 \exp(\lambda_1 t) + \boldsymbol{X}_2 \exp(\lambda_2 t) + \cdots + \boldsymbol{X}_n \exp(\lambda_n t) \tag{2.125}$$

Each of the n eigenvectors has an unknown parameter. The total of n unknowns is determined using n initial conditions:

$$\boldsymbol{x}(0) = \boldsymbol{x}_0 \tag{2.126}$$

Mode Shapes of Nonoscillatory Systems

Since the eigenvectors are independent, if the initial state is set at $\boldsymbol{x}_0 = X_i$, then the subsequent motion should not have any $\boldsymbol{X}_j$ terms with $j \neq i$ in equation (2.125). Otherwise, when we set $t = 0$, $\boldsymbol{X}_i$ becomes a linear combination of the remaining eigenvectors, which contradicts the linear independence. Hence, the motion due to this eigenvector initial condition is given by $\boldsymbol{x}(t) = \boldsymbol{X}_i \exp(\lambda_i t)$, which is parallel to $\boldsymbol{X}_i$ throughout the motion. Thus, $\boldsymbol{X}_i$ gives the mode shape of the system corresponding to the eigenvalue λ_i.

Mode Shapes of Oscillatory Systems

The analysis in the preceding section is valid for real eigenvalues and eigenvectors. In vibratory systems, λ_i and $\boldsymbol{X}_i$ generally are complex. Let

$$\lambda_i = \sigma_i + j\omega_i \tag{2.127}$$

$$\boldsymbol{X}_i = \boldsymbol{R}_i + j\boldsymbol{I}_i \tag{2.128}$$

For real systems, there exist corresponding complex conjugates:

$$\overline{\lambda}_i = \sigma_i - j\omega_i \tag{2.129}$$

$$\overline{\boldsymbol{X}}_j = \boldsymbol{R}_i - j\boldsymbol{I}_i \tag{2.130}$$

Equations (2.127) through (2.130) represent the *i*th mode of the system. The corresponding damped natural frequency is ω_i, and the damping parameter is σ_i. The net contribution of the *i*th mode to the solution—equation (2.125)—is

$$(\boldsymbol{R}_i \cos \omega_i t - \boldsymbol{I}_i \sin \omega_i t) 2 \exp(\sigma_i t)$$

In the second example discussed in the earlier section on state-space representation, we saw that only some of the state variables in $\boldsymbol{x}(t)$ correspond to displacements of the masses (or spring forces). These can be extracted through an output relationship of equation (2.116). The contribution of the *i*th mode to displacement variables is

$$\boldsymbol{Y}_i = \boldsymbol{C}[\boldsymbol{R}_i \cos \omega_i t - \boldsymbol{I}_i \sin \omega_i t] 2 \exp(\sigma_i t) \tag{2.131}$$

If equation (2.131) can be expressed in the form

$$\boldsymbol{Y}_i = \boldsymbol{S}_i \sin(\omega_i t + \phi_i) \exp(\sigma_i t) \tag{2.132}$$

in which $\boldsymbol{S}_i$ is a constant vector that is defined up to one unknown, then it is possible to excite the system so that every independent mass element undergoes oscillations in phase (hence, passing through the equilibrium state simultaneously) at a specific frequency ω_i. This type of motion is known as normal mode motion. The vector $\boldsymbol{S}_i$ gives the mode shape corresponding to the natural frequency ω_i. Normal mode motion is possible for undamped systems and for certain classes of damped systems. The initial state that is required to excite the *i*th mode is $\boldsymbol{x}_0 = \boldsymbol{R}_i$. The corresponding displacement and velocity initial conditions are obtained from equation (2.131):

$$\boldsymbol{Y}_i(0) = \boldsymbol{C}\boldsymbol{R}_i \tag{2.133}$$

$$\dot{\boldsymbol{Y}}_i(0) = \boldsymbol{C}(\boldsymbol{R}_i\sigma_i - \boldsymbol{I}_i\omega_i) \tag{2.134}$$

Note that the constant factor 2 has been ignored because $\boldsymbol{X}_i$ is known up to one unknown complex parameter.

Illustrative Example

A torsional dynamic model of a pipeline segment is shown in figure 2-16(a). Free-body diagrams in figure 2-16(b) show internal torques acting at sectioned inertia junctions for free motion. A state model is obtained using the generalized velocities (angular velocities Ω_i) of the inertia elements and the generalized forces (torques T_i) as state variables. A minimum set required for complete representation determines the system order. There are two inertia elements and three spring elements—a total of five energy-storage elements. The three springs are not independent, however. The motion of two springs completely determines the motion of the third. This indicates that the system is a fourth-order system. We obtain the model as follows: Newton's law gives

$$I_1\dot{\Omega}_1 = -T_1 + T_2$$

$$I_2\dot{\Omega}_2 = -T_2 - T_3$$

Hooke's law gives

$$\dot{T}_1 = k_1\Omega_1$$

$$\dot{T}_2 = k_2(\Omega_2 - \Omega_1)$$

Torque T_3 is determined in terms of T_1 and T_2, using the displacement relation for the inertia I_2:

$$\frac{T_1}{k_1} + \frac{T_2}{k_2} = \frac{T_3}{k_3}$$

The state vector is chosen as

$$\boldsymbol{x} = [\Omega_1, \Omega_2, T_1, T_2]^T$$

(a)

I_1 I_2

k_1 k_2 k_3

T_1 T_2 T_3

Ω_1 Ω_2

(b)

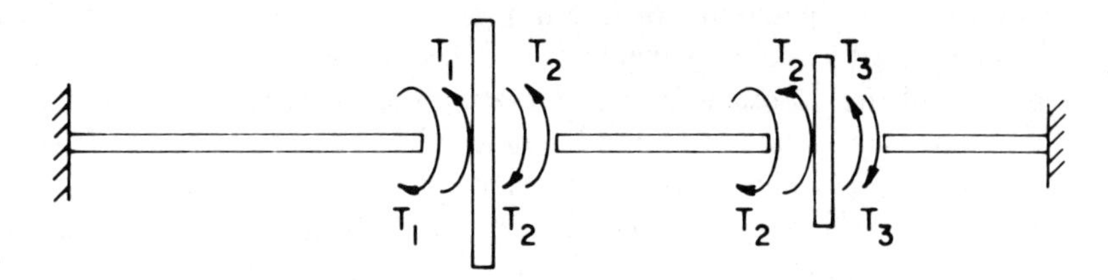

Figure 2-16. Examples of (a) Dynamic Model of a Pipeline Segment, (b) Free Body Diagrams

The corresponding system matrix is

$$A = \begin{bmatrix} 0 & 0 & -\dfrac{1}{I_1} & \dfrac{1}{I_1} \\ 0 & 0 & -\dfrac{1}{I_2}\left(\dfrac{k_3}{k_1}\right) & -\dfrac{1}{I_2}\left(1+\dfrac{k_3}{k_2}\right) \\ k_1 & 0 & 0 & 0 \\ -k_2 & k_1 & 0 & 0 \end{bmatrix}$$

The output-displacement vector is

$$y = \left[\frac{T_1}{k_1}, \frac{T_1}{k_1} + \frac{T_2}{k_2}\right]$$

which corresponds to the output-gain matrix,

$$C = \begin{bmatrix} 0 & 0 & \dfrac{1}{k_1} & 0 \\ \\ 0 & 0 & \dfrac{1}{k_1} & \dfrac{1}{k_2} \end{bmatrix}$$

For the special case given by $I_1 = I_2 = I$ and $k_1 = k_3 = k$, the system eigenvalues are

$$\lambda_1, \overline{\lambda}_1 = \pm j\omega_1 = \pm j\sqrt{\frac{k}{I}}$$

$$\lambda_2, \overline{\lambda}_2 = \pm j\omega_2 = \pm j\sqrt{\frac{k + 2k_2}{I}}$$

and the corresponding eigenvectors are

$$\boldsymbol{X}_1, \overline{\boldsymbol{X}}_1 = \boldsymbol{R}_1 \pm j\boldsymbol{I}_1 = \frac{\alpha_1}{2}[\omega_1, \omega_1, \mp jk_1, 0]^T$$

$$\boldsymbol{X}_2, \overline{\boldsymbol{X}}_2 = \boldsymbol{R}_2 \pm j\boldsymbol{I}_2 = \frac{\alpha_2}{2}[\omega_2, -\omega_2, \mp jk_1, \pm 2jk_2]^T$$

In view of equation (2.131), the modal contributions to the displacement vector are

$$\boldsymbol{Y}_1 = \begin{bmatrix} 1 \\ \\ 1 \end{bmatrix} \alpha_1 \sin \omega_1 t$$

and

$$\boldsymbol{Y}_2 = \begin{bmatrix} 1 \\ \\ -1 \end{bmatrix} \alpha_2 \sin \omega_2 t$$

These equations are of the form given by equation (2.132). The mode shapes are given by the vectors $\boldsymbol{S}_1 = [1, 1]^T$ and $\boldsymbol{S}_2 = [1, -1]^T$, which are illustrated in figure 2-17. In general, each modal contribution introduces two unknown

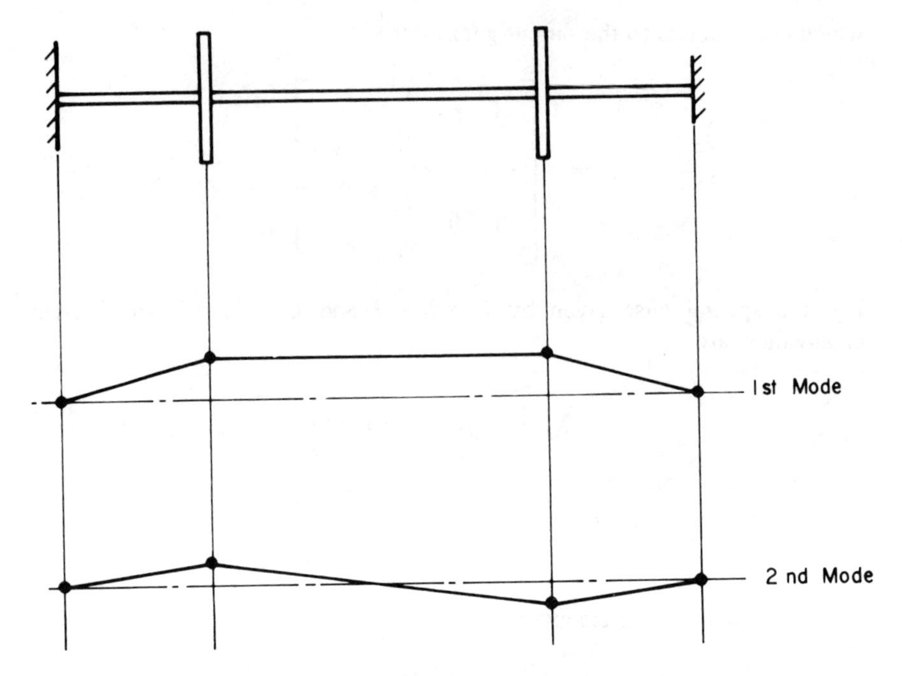

Figure 2-17. Mode Shapes of the System in Figure 2-16

parameters a_i and ϕ_i, into the free response (homogeneous solution), where ϕ_i are the phase angles associated with the sinusoidal terms. For an n-degree-of-freedom (order-$2n$) system, this results in $2n$ unknowns, which require the $2n$ initial conditions $x(0)$.

Modal Analysis of Continuous Systems

In real systems, inertial and elastic effects usually are found continuously distributed in one, two, or three dimensions. Lumped-parameter (discrete) models are used to simplify the analysis. Sometimes, however, a lumped-parameter model might not be adequate to represent the continuous system with sufficient accuracy. Continuous-system models are useful in such cases. Some models can be hybrids, with continuous as well as discrete elements.

The decision to use a continuous-system model should depend on (1) the nature and spatial size of the component and (2) the desired degree of accuracy. Dynamic response of continuous systems is represented by partial differential equations (PDE). In addition to time t, spatial coordinates appear as independent variables in these PDE.

Bernoulli-Euler Beam Dynamics

Transverse vibration of beam elements in physical systems is often represented by the Bernoulli-Euler beam equation. This is satisfactory for slender beams at relatively low frequencies of interest. Transverse motion $v(x, t)$ of a Bernoulli-Euler beam subjected to a distributed load $f(x, t)$ per unit length (figure 2.18) is given by [see equation (2.6)]

$$\frac{\partial^2}{\partial x^2} EI \frac{\partial^2 v}{\partial x^2} + \rho A \frac{\partial^2 v}{\partial t^2} = f(x, t) \tag{2.135}$$

Internal damping in the beam material has been neglected. External damping can be included through the forcing function.

First, consider the case of free motion [$f(x, t) = 0$]. In this case, modal vibration corresponds to in-phase motion of each particle of the beam at the same natural frequency. Then the spatial dependence and the time dependence of $v(x, t)$ are separable for each mode:

$$v(x, t) = V(x)\hat{\Theta}(t) \tag{2.136}$$

By substituting equation (2.136) in (2.135), we obtain

$$\frac{1}{\rho A V(x)} \frac{d^2}{dx^2} EI \frac{d^2 V(x)}{dx^2} = -\frac{1}{\hat{\Theta}(t)} \frac{d^2 \hat{\Theta}(t)}{dt^2} \tag{2.137}$$

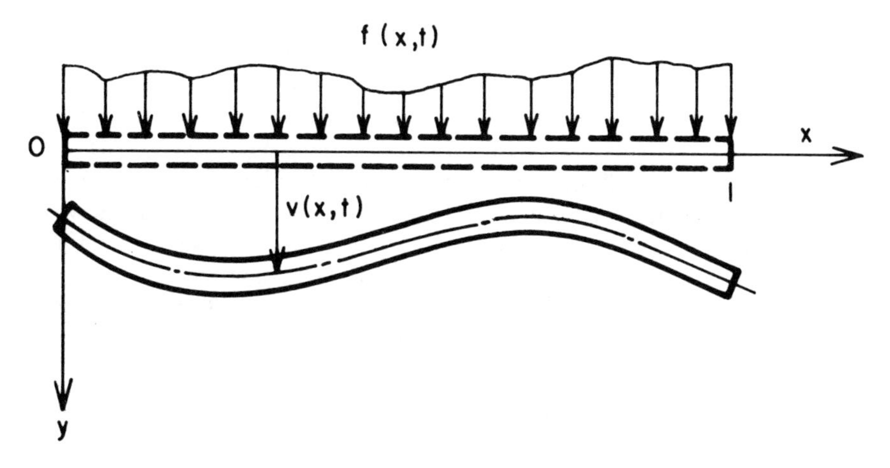

Figure 2-18. Bernoulli-Euler Beam Subjected to a Distributed Load

Since the left-hand side of equation (2.137) is a function of x only and the right-hand side is a function of t only, each expression should be equal to the same constant (denoted by ω^2), which is verified to be positive. Hence,

$$\frac{d^2}{dx^2} EI(x) \frac{d^2 V(x)}{dx^2} - \omega^2 \rho A(x) V(x) = 0 \tag{2.138}$$

and

$$\frac{d^2 \hat{\Theta}(t)}{dt^2} + \omega^2 \hat{\Theta}(t) = 0 \tag{2.139}$$

To verify for most boundary conditions that the constant is positive, we multiply equation (2.138) by $V(x)$ and integrate twice, by parts, to obtain

$$\int_0^l EI\left(\frac{d^2 V}{dx^2}\right)^2 dx + \left[V(x) \frac{d}{dx} EI \frac{d^2 V(x)}{dx^2} - EI \frac{dV(x)}{dx} \frac{d^2 V(x)}{dx^2} \right]_0^l = \omega^2 \int_0^l \rho A V(x)^2 dx \tag{2.140}$$

The second term on the left-hand side of equation (2.140) vanishes for most boundary conditions. Since both integral terms in equation (2.140) are positive, it is established that ω^2 is a positive constant, at least for those boundary conditions.

The independent solutions $V_i(x)$ of equation (2.138) are the mode shapes. These are continuous functions because they define the modal motion of an infinite set of mass elements in space. This is analogous to eigenvectors in lumped-parameter systems, which consist of only a finite number of mass elements. Because of this analogy, mode shapes of continuous systems often are called eigenfunctions. Note that, since equation (2.138) is homogenous, $V_i(x)$ is determined only up to a single unknown (its amplitude). Hence, the solution requires only three boundary conditions. The fourth boundary condition determines, for a nontrivial solution, the natural frequencies ω_i that are assumed to be distinct. The corresponding solution of the harmonic equation (2.139) is

$$\hat{\Theta}_i(t) = a_i \sin(\omega_i t + \phi_i) \tag{2.141}$$

Since the assumed solution is equation (2.136), the net free motion (homogeneous solution) is given by the sum of all independent solutions:

$$v_h(x, t) = \sum_i a_i V_i(x) \sin(\omega_i t + \phi_i) \tag{2.142}$$

The unknown constant in $V_i(x)$ has been incorporated into a_i. The constants a_i and ϕ_i are determined by first expressing the initial conditions [for example, $v(x, 0)$ and $\dot{v}(x, 0)$] as series expansions in $V_i(x)$ and comparing them with the corresponding expressions obtained from equation (2.142). The similarity of equations (2.125) and (2.142) should be noted, along with the fact that a continuous system has an infinite set of mode shapes (eigenfunctions) and corresponding natural frequencies. The index i in the summation of equation (2.142) ranges from 1 to ∞. In numerical computations, this infinite series has to be truncated. This is acceptable because the contribution from the higher modes is insignificant in most practical situations, where excitation frequencies are not very high.

An orthogonality relationship for the mode shapes can be derived as follows. The eigenfunction equation (2.138), corresponding to the ith mode, is given by

$$\frac{d^2}{dx^2} EI(x) \frac{d^2}{dx^2} V_i(x) - \omega_i^2 \rho A(x) = 0 \tag{2.143}$$

Equation (2.143) is multiplied by $V_j(x)$ and integrated, by parts, in two steps over the beam length, using boundary conditions. This gives

$$\int_0^l EI(x) \frac{d^2}{dx^2} V_i(x) \frac{d^2 V_j(x)}{dx^2} dx = \omega_i^2 \int_0^l \rho A(x) V_i(x) V_j(x) dx \tag{2.144}$$

If $i = j$, the left-hand side of equation (2.144) evaluates to a constant value, which depends on the particular mode i. If $i \neq j$, the left-hand side remains unchanged if i and j are interchanged in equation (2.144). If the result is subtracted from equation (2.144) the left-hand side vanishes. Since $\omega_i \neq \omega_j$ for different modes, however, we obtain the orthogonality condition,

$$\begin{aligned} \int_0^l \rho A(x) V_i(x) V_j(x) dx &= 0 \quad \text{for} \quad i \neq j \\ &= m_i \quad \text{for} \quad i = j \end{aligned} \tag{2.145}$$

State Model for Forced Motion. To obtain a state model for forced motion $[f(x, t) \neq 0]$, the forced response is first expressed as a series expansion in eigenfunctions:

$$v(x, t) = \sum_{i=1}^{\infty} V_i(x)\theta_i(t) \tag{2.146}$$

Mathematically, this expansion is valid, because $V_i(x)$, $i = 1, 2, \ldots, \infty$ is a complete set of orthogonal eigenfunctions [equation (2.145)], and forms a basis for the Hilbert space.[2] Also, note that the generalized coordinates $\theta_i(t)$ depend on the forcing function $[f(x, t)]$ and are not equal to $\hat{\Theta}(t)$, given in equation (2.141) for free motion.

First, equation (2.146) is substituted in (2.135); then, equation (2.143) is applied; and, finally, the result is multiplied by $V_k(x)$ and integrated over $(0, l)$, using orthogonality [equation (2.145)]. This gives

$$\ddot{\theta}_k(t) + \omega_k^2\,\theta_k(t) = r_k(t) \qquad k = 1, 2, \ldots \tag{2.147}$$

in which

$$r_k(t) = \frac{1}{m_k}\int_0^l f(x, t)\,V_k(x)\,dx \tag{2.148}$$

The second-order equations (2.147) can be written in the state-space form by defining, for example, the state variables:

$$\begin{aligned} x_{2i-1} &= \theta_i \\ x_{2i} &= \dot{\theta}_i \qquad i = 1, 2, \ldots \end{aligned} \tag{2.149}$$

Note that the infinite series in equation (2.146) must be truncated at the nth mode for numerical computations. The corresponding system matrix is

$$A = \begin{bmatrix} 0 & 1 & & & \\ -\omega_1^2 & 0 & & \mathbf{0} & \\ & & \ddots & & \\ & & & 0 & 1 \\ & \mathbf{0} & & -\omega_n^2 & 0 \end{bmatrix}_{2n\times 2n} \tag{2.150}$$

Consider the case of the forcing function consisting of m point loads $F_j(t)$, located at x_j on the beam. Then,

$$f(x, t) = \sum_{j=1}^{m} F_j(t)\,\delta(x - x_j) \qquad (2.151)$$

From equation (2.148) in conjunction with (2.9), we have

$$r_k(t) = \frac{1}{m_k} \sum_{j=1}^{m} F_j(t)\, V_k(x_j) \qquad (2.152)$$

If the input variables are defined as

$$u_j(t) = F_j(t) \qquad j = 1, 2, \ldots, m$$

the corresponding input-gain matrix becomes

$$B = \begin{bmatrix} 0 & \cdots & 0 \\ V_1(x_1)/m_1 & \cdots & V_1(x_m)/m_1 \\ \vdots & \vdots & \vdots \\ 0 & \cdots & 0 \\ V_n(x_1)/m_n & \cdots & V_n(x_m)/m_n \end{bmatrix}_{2n \times m} \qquad (2.153)$$

This completes the state-model derivation for the forced motion of a Bernoulli–Euler beam.

Notes

1. S.H. Crandall, D.C. Karnopp, E.F. Kurtz, Jr., and D.C. Pridmore-Brown, *Dynamics of Mechanical and Electromechanical Systems* (New York: McGraw-Hill, 1968), pp. 115–128.
2. C.H. Page, *Physical Mathematics* (New York: Van Nostrand, 1955).

3 Signal Processing and Digital Fourier Analysis

The most uncertain part of a seismic-qualification program is the simulation of the seismic environment. The primary difficulty results from the fact that the probability of accurately predicting the recurrence of an earthquake at a given equipment site during the design life of the equipment is very small; and the probability of accurately predicting the nature of the ground motions if an earthquake were to occur is still smaller. The best we can do is make a conservative estimate for the nature of the ground motions resulting from the strongest earthquake that is reasonably expected. The resulting time history, known as a design earthquake, is usually based on the records of past seismic motions if they are available for that region, geographic information, subsoil conditions on site, and the characteristics of the equipment and its supporting structure. In particular, the estimated time-history components (in three orthogonal directions) should have the same amplitude, phasing, frequency content, and time-decay (transient) characteristics of typical seismic motions for that site, if known. This means that frequency-domain representation of seismic ground motions usually provides better insight about their characteristics than time-domain representation—that is, a time history, does. These two representations were discussed in chapter 2. Fortunately, frequency-domain information can be derived from time-domain data by using Fourier transform techniques.

Until recently, signal processing that was required in dynamic testing was performed almost exclusively through analog methods. With these methods, the measured signal usually is converted into an electical signal, which, in turn, is passed through a series of electrical circuits to achieve the required processing. Alternatively, motion or pressure signals could be used in conjunction with mechanical or hydraulic circuits to perform analog processing. Present complex test programs require the capability to process many measurements quickly and accurately. The performance of analog signal analyzers is limited by hardware costs, size, data-handling capacity, and computational accuracy. Digital processing for synthesis and analysis of vibration-test input signals and for interpretation and evaluation of test results began to replace the classical analog methods in the late 1960s. Currently, special-purpose digital analyzers, with real-time digital Fourier analysis capability, are commonly used in dynamic-testing applications. The advantages of incorporating digital processing in dynamic testing include flexibility and convenience in the type of signal that can be analyzed;

availability of complex processing; accuracy and reliability; reduction in operational costs; practically unlimited repeatability of processing; and reduction in overall size and weight of the analyzer.

Fourier analysis is the heart of digital signal processing in dynamic testing. Digital Fourier analysis, as the term implies, deals with Fourier analysis of digital data using a digital computer. Since the introduction of the concepts of Fourier series and the Fourier integral in 1807 by Joseph Fourier in his solution of the celebrated diffusion equation, Fourier analysis has found wide application in numerous fields, including mathematics, physics, and engineering. Until the 1950s, however, the majority of these applications were limited to determining analytical solutions to specific problems or to a class of problems. With the advent of the digital computer, numerical techniques related to Fourier transformation began to receive wide attention from researchers in many applied disciplines. Based on economic considerations, and because of the low-speed digital hardware and software available at the time, these early applications were limited to off-line digital computations associated with large-scale systems in which the initial investments could be easily recovered and to special situations in which the performance accuracy was the most important criterion (for example, aerospace applications). Real-time digital Fourier analysis is now a reality because of recent developments in high-speed microprocessor technology and, more important, as a result of an algorithm, rediscovered by Cooley and Tukey,[1] that efficiently computes the discrete Fourier transform (DFT).

This algorithm, widely known as the fast Fourier transform (FFT), has revolutionized digital signal processing as a result of the considerable reduction in computer time associated with these computations. It is appropriate to mention that similar algorithms had been developed by several workers in the early part of this century, but their significance was adequately recognized only after the publication of Cooley and Tukey's paper. Fast algorithms are particularly useful in real-time signal processing, in which the signals from the test object are analyzed on-line continuously at the same speed as, or faster than, they are generated and measured, so that test performance can be improved simultaneously, using the computed results.

Fourier Analysis in Dynamic Testing

In dynamic testing, Fourier analysis is used in three principal ways: (1) to determine the frequency response of the test object in prescreening tests; (2) to represent the dynamic environment by its Fourier spectrum or its power spectral density so that a test-input signal can be generated to represent it; and (3) to monitor the Fourier spectrum of the response at key locations in the test object and at control locations of the test table and use the information diagnostically or in controlling the excitation input.

Dynamic testing is usually accomplished using a shaker apparatus, as shown by the operational diagram in figure 3-1. The test object is secured to the shaker table in a manner representative of its installation during actual service. In-service operating conditions are simulated while the table is actuated by applying a suitable input signal to an electromagnetic, electromechanical, or hydraulic actuator. Often, more than one signal is required to simulate three-dimensional characteristics of the dynamic environment. The test-input signal is either stored on analog magnetic tape or generated in real time by a signal generator. The capability of the test object or a similar unit to withstand a predefined dynamic environment is evaluated by monitoring the dynamic response (accelerations, velocities, displacements, strains, and so forth) and functional operability variables (for example, temperatures, pressures, flow rates, voltages, currents).

Digital Fourier analysis of the response signals will aid in detecting existing defects or impending failures in various components of the test equipment. The digital Fourier results, particularly for the control-sensor output, also are useful in the feedback control of the shaker, in frequency-band equalization in real time of the input signal, and in synthesizing future test signals.

Fourier Transform Fundamentals

In mathematical analysis, it is common to define new variables in terms of the variables of the given problem in order to simplify the solution of a complex problem. This is the basis of transform analysis. The purpose is to transform the given problem from one domain to a different domain through a transform relationship. The solution of the problem is usually simpler in the new domain, and occasionally the transormation can reveal special characteristics of a physical problem. The Fourier integral transform (FIT), also known as the continuous Fourier transform, is the most useful transform technique in dynamic testing. This versatility results from the fact that, when applied to the response of linear, time-invariant-parameter dynamic systems, it represents a reversible transformation between time and frequency. In the time domain, the system response is described as a function of the independent variable *time*; in the frequency domain, the system response is described as a function of the independent variable *frequency*. If the time response is periodic, it can be represented by a sum of sinusoidal components at the fundamental frequency (reciprocal of the period) and its higher harmonics. This results in a complex Fourier series expansion (FSE). Consequently, the FSE is simply a special case of the FIT.

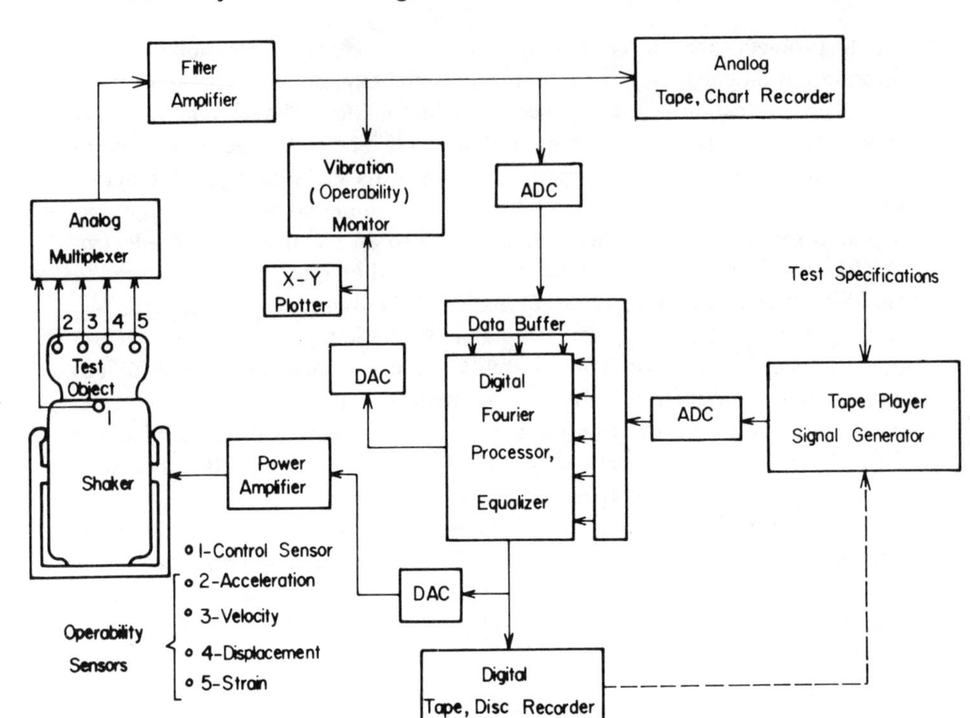

Figure 3-1. Computer-Aided Dynamic-Testing Schematic

The relationship between the frequency and time domains can be explained further. If a linear, time-invariant system is excited using a sine wave of a given frequency, the system output (response) at steady state will oscillate at the same frequency but with an amplification and a phase change. If this is repeated for a range of frequencies, curves of amplification versus frequency and phase shift versus frequency are obtained. This pair of curves is unique for a given system and, when the two are combined into a single complex function of frequency, is known as the frequency-response function (or frequency-transfer function) of the system. It possesses complete information regarding the system. The determination of the system response to an arbitrary input signal (function of time) is a tedious task. If its FIT is known, however, the steady-state response of the system is obtained by simply multiplying it by the frequency-response function. A periodic signal can be decomposed into sinusoidal components of frequencies that are simple multiples of the fundamental (lowest) frequency. Consequently, its spectrum is discrete, with uniform intervals. A nonperiodic signal can be considered a signal of infinite period. Its spectrum, therefore, is generally continuous. The input signals used in dynamic testing are usually nonperiodic.

When the digital computer is employed to determine the FIT of a piecewise-continuous signal, it is necessary that the signal be sampled at discrete values of time and that the discrete Fourier transform (DFT) be applied to the resulting data sequence. Unfortunately, many people who work in the application fields often fail to observe the analogy between the FIT and DFT; consequently, the vast body of knowledge available with regard to the FIT is not exploited in the digital Fourier analysis. This has led to duplication of the analytical effort, has resulted in inconsistent definitions, and has brought about difficulties for the inexperienced user in interpreting the digitally computed results. The complex Fourier series expansion of a periodic function may also be obtained using the DFT techniques. This procedure also suffers from a lack of consistency. Real-time application of the digital Fourier methods in dynamic testing necessitates economical, high-speed processing, with restrictive use of the computer memory. The fast Fourier transform (FFT) revolutionized the field of digital Fourier analysis in the mid 1960s by reducing the number of arithmetic operations required for the discrete Fourier transformation of an N-point data sequence by a factor of $2N/\ln_2 N$. Since then, a greater emphasis has been placed on the development of efficient numerical algorithms based on sound mathematical theory. Furthermore, a clear understanding of the underlying mathematical concepts is necessary for the proper interpretation of the results from a particular numerical procedure.

Fourier Integral Transform (FIT)

The Fourier integral transform $X(f)$ of a signal $x(t)$ is given by the forward transformation relationship:

$$X(f) = \int_{-\infty}^{\infty} x(t) \exp(-j2\pi ft)\,dt \tag{3.1}$$

in which $j = \sqrt{-1}$. Equation (3.1) is multiplied by $\exp(j2\pi f\tau)$ and integrated over $f(-\infty, \infty)$ using the well-known property of the Dirac delta function, or unit impulse function, $\delta(t)$,

$$\int_{-\infty}^{\infty} \exp(j2\pi f(\tau - t)\,df = \delta(\tau - t) \tag{3.2}$$

to obtain the inverse transformation relationship

$$x(t) = \int_{-\infty}^{\infty} X(f) \exp(j2\pi ft)\,df \tag{3.3}$$

It has been assumed that the Fourier integral on the right-hand side of equation (3.1) exists. A sufficient but not necessary condition for the existence of the FIT is

$$\int_{-\infty}^{\infty} |x(t)|\, dt < \infty \tag{3.4}$$

It should be noted that, in the generalized case, both functions $x(t)$ and $X(f)$ are complex. In real situations, such as a seismic-ground-motion time history, however, the signal $x(t)$ is a real function and the frequency spectrum $X(f)$ is complex. It is customary to represent the frequency spectrum $X(f)$ by its magnitude $|X(f)|$ and phase angle $\angle X(f)$ rather than in terms of the real part and the imaginary part.

Fourier Series Expansion (*FSE*)

If the signal $x(t)$ is periodic of period T, then it has the Fourier series expansion

$$x(t) = \frac{1}{T} \sum_{n=-\infty}^{\infty} A_n \exp(j2\pi nt/T) \tag{3.5}$$

This is analogous to equation (3.3). Now equation (3.5) is multiplied by $\exp(-j2\pi mt/T)$ and integrated over $t(0, T)$ using the orthogonality property

$$\int_{0}^{T} \exp[j2\pi(n-m)t/T]\, dt = T\delta_{mn} \tag{3.6}$$

in which the Kronecker delta $\delta_{mn} = 0$ for $m \neq n$ and is 1 for $m = n$. This results in

$$A_m = \int_{0}^{T} x(t) \exp(-j2\pi mt/T)\, dt \tag{3.7}$$

Equation (3.7) should be compared with its FIT counterpart equation (3.1).

At this juncture, it is shown that the FSE is a special case of the FIT. To accomplish this, the case in which $X(f)$ is a sum of equidistant impulses separated by the frequency interval $\Delta F = 1/T$ is considered:

$$X(f) = \frac{1}{T} \sum_{n=-\infty}^{\infty} A_n \delta(f - n/T) \tag{3.8}$$

Equation (3.8) is substituted into equation (3.3) to obtain the corresponding time function

$$x(t) = \frac{1}{T} \sum_{n=-\infty}^{\infty} A_n \exp(j2\pi nt/T) \tag{3.9}$$

Consequently, an FSE identical to equation (3.5) is obtained. The FSE [equation (3.9) or equation (3.5)] is the inverse FIT of the sum of equidistant impulses [equation (3.8)]. Furthermore, since $x(t)$ is a general periodic signal, equation (3.8) represents the FIT of a general periodic function of period T having harmonic complex amplitudes A_n/T.

Discrete (Finite) Fourier Transform (DFT)

The discrete (finite) Fourier transform is a transform in its own right, relating an N-element sequence of sampled digital data $\{x_n\} = [x_0, x_1, \ldots, x_{N-1}]$ in the discrete time domain to an N-element sequence $\{X_m\} = [X_0, X_1, \ldots, X_{N-1}]$ in the discrete frequency domain. The forward transform is

$$X_m = \Delta T \sum_{n=0}^{N-1} x_n \exp(-j2\pi mn/N) \qquad m = 0, 1, \ldots, N-1 \qquad (3.10)$$

in which ΔT is the sample time step. Equation (3.10) is multiplied by $\exp(j2\pi mr/N)$ and summed over $m(0, N-1)$ using the orthogonality property

$$\sum_{m=0}^{N-1} \exp([j2\pi m(r-n)/N] = N\delta_{rn}$$

to obtain the inverse transform given by

$$x_n = \frac{1}{N\Delta T} \sum_{m=0}^{N-1} x_m \exp(j2\pi mn/N) \qquad (3.11)$$

It should be noted that $N.\Delta T = T$ is the record length of the sampled signal.

Table 3-1 summarizes the transform relationships discussed thus far.

Unification of the Three Fourier Transform Types

By discrete Fourier transformation of a set of sampled data from a signal, we cannot expect to get a set of points in its exact Fourier spectrum. Because of sampling of the signal, some information will be lost. Clearly, we should be able to reduce the error by decreasing the sample step (ΔT). Similarly, we do not expect to get the exact Fourier series coefficients by discrete Fourier transformation of sampled data from a periodic function. It is very important to study the nature of these errors, which are commonly known as aliasing distortions.

Table 3-1
Unified Fourier Transform Relationships

Description	*Fourier Integral Transform (FIT)*	*Discrete Fourier Transform (DFT)*	*Fourier Seires Expansion (FSE)*
Forward transform	$X(f) = \int_{-\infty}^{\infty} x(t) \exp(-j2\pi ft)\,dt$	$X_n = \Delta T \sum_{m=0}^{N-1} x_m \exp(-j2\pi nm/N)$ $n = 0, 1, \ldots, N-1$	$A_n = \int_0^T x(t) \exp(-j2\pi nt/T)\,dt$ $n = 0, 1, \ldots$
Inverse transform	$x(t) = \int_{-\infty}^{\infty} X(f) \exp(j2\pi ft)\,df$	$x_m = \Delta F \sum_{n=0}^{N-1} X_n \exp(j2\pi nm/N)$ $m = 0, 1, \ldots, N-1$	$x(t) = \Delta F \sum_{n=-\infty}^{\infty} A_n \exp(j2\pi nt/T)$
Orthogonality	$\int_{-\infty}^{\infty} \exp[j2\pi f(\tau - t)]\,df = \delta(\tau - t)$	$\frac{1}{N} \sum_{n=0}^{N-1} \exp[j2\pi n(r-m)/N] = \delta_{rm}$	$\frac{1}{T} \int_0^T \exp[j2\pi(r-n)t/T]\,dt = \delta_{rn}$
Notes	$T = N\Delta T$	$\Delta F = 1/T$	$X(f) = \Delta F \sum_{n=-\infty}^{\infty} A_n \delta(f - n/T)$

Relationship between DFT and FIT

A fundamental result relating DFT and FIT, established in this section, is consistent with the definitions stated in the earlier section, "Fourier Transform Fundamentals."

In view of equation (3.1), the frequency-spectrum values $X_m = X(m \cdot \Delta F)$, $m = 0, \pm 1, \pm 2, \ldots$ sampled at the discrete frequency points of sample step ΔF, are given by

$$X_m = \int_{-\infty}^{\infty} x(t) \exp(-j2\pi m \Delta F t)\, dt$$

$$= \sum_{k=-\infty}^{\infty} \int_{kT}^{(k+1)T} x(t) \exp(-j2\pi mt/T)$$

in which $T = 1/\Delta F$. On implementing the change of variable $t \to t + kT$ (that is, let $t' = t - kT$ and drop the prime) and interchanging the summation and the integration operations:

$$X_m = \int_0^T \tilde{x}(t) \exp(-j2\pi mt/T)\, dt \tag{3.12}$$

in which

$$\tilde{x}(t) = \sum_{k=-\infty}^{\infty} x(t + kT) \tag{3.13}$$

The fact that $\exp(-j2\pi mk) = 1$ for integer m, k was used in obtaining equation (3.12). For a finite-duration time history $\tilde{x}(t)$ [see figure 3-2(a)], the nature of $\tilde{x}(t)$ is shown in figure 3-2(b). Since $\tilde{x}(t)$ is periodic, having the period T, it has an FSE given by

$$\tilde{x}(t) = \frac{1}{T} \sum_{n=-\infty}^{\infty} X_n \exp(j2\pi nt/T) \tag{3.14}$$

which follows from equation (3.5). The Fourier series coefficients can be identified by comparing equation (3.9) with equation (3.12). The sampled values $\tilde{x}_m = \tilde{x}(m \cdot \Delta T)$, $m = 0, \pm 1, \pm 2, \ldots$ at sample steps of ΔT, are given by

$$\tilde{x}_m = \frac{1}{T} \sum_{n=-\infty}^{\infty} X_n \exp(j2\pi nm/N)$$

$$= \frac{1}{T} \sum_{k=-\infty}^{\infty} \sum_{n=kN}^{(k+1)N-1} X_n \exp(j2\pi nm/N)$$

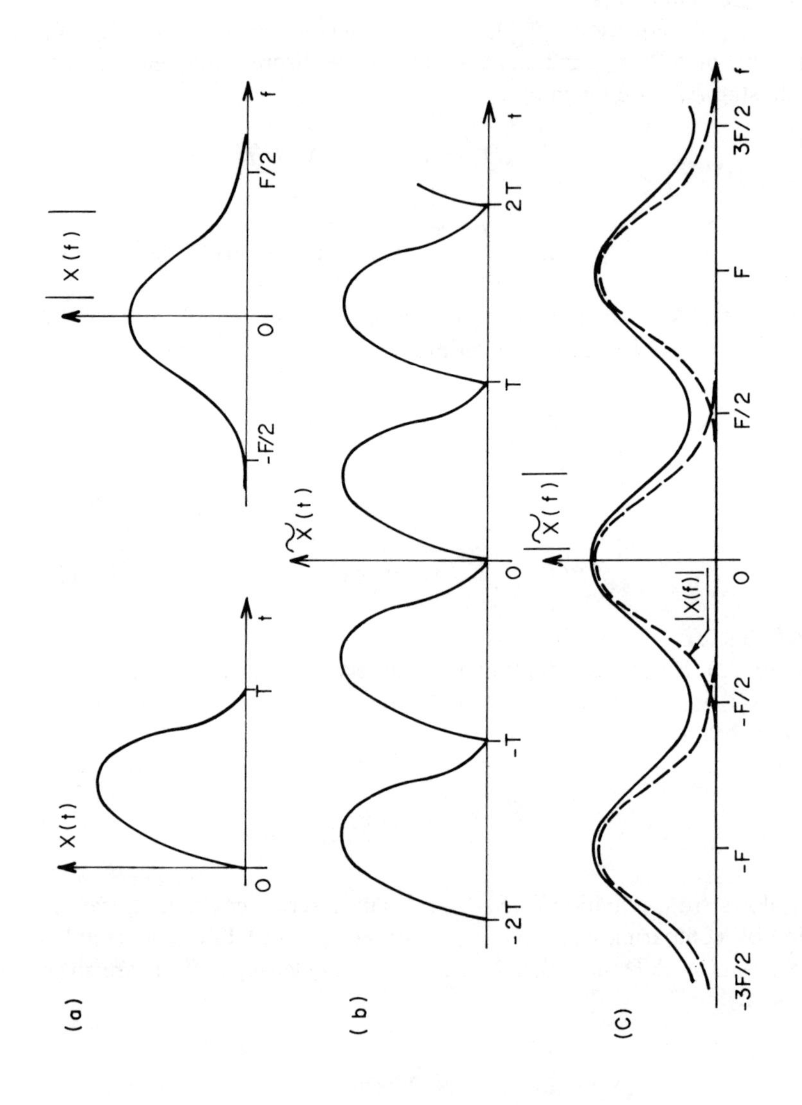
(a)
X(t)
0
T
t
|X(f)|
-F/2
0
F/2
f
(b)
X̃(t)
-2T
-T
0
T
2T
t
(C)
|X̃(f)|
|X(f)|
-3F/2
-F
-F/2
0
F/2
F
3F/2
f

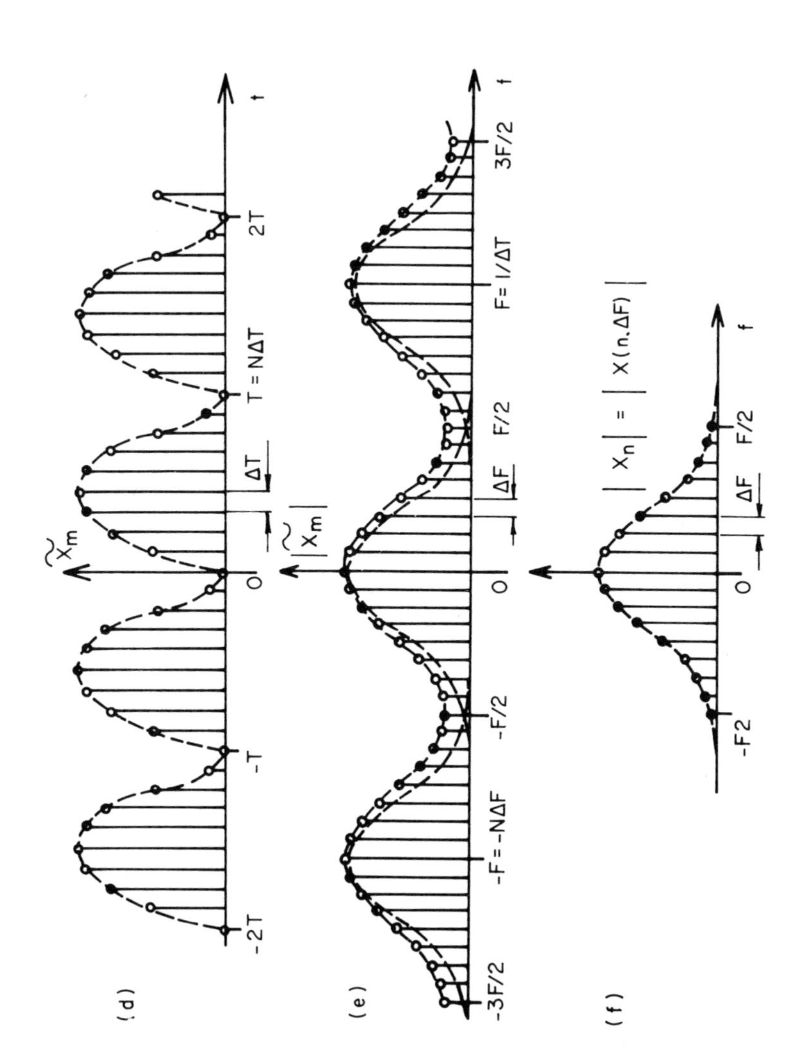

Figure 3-2. Demonstration of the Relationship between FIT and DFT (Aliasing Distortion in the Frequency Domain)

in which $\Delta T = T/N = 1/(N \cdot \Delta F)$. In a manner analogous to the procedure for obtaining equation (3.12), the change of variable $n \rightarrow n + kN$ (that is, let $n' = n - kN$ and drop the prime) is implemented, and the summation operations are interchanged. This results in

$$\tilde{x}_m = \frac{1}{T} \sum_{n=0}^{N-1} \tilde{X}_n \exp(j2\pi nm/N) \tag{3.15}$$

in which

$$\tilde{X}(f) = \sum_{k=-\infty}^{\infty} X(f + kF) \tag{3.16}$$

Note that $\tilde{X}_n = \tilde{X}(n \cdot \Delta F)$, $n = 0, \pm 1, \pm 2, \ldots$ are the sampled values of the periodic function $\tilde{X}(f)$, having the period F. The frequency parameter $F = N \cdot \Delta F = 1/\Delta T = N/T$ represents the number of samples in a record of unity time duration. The discussion in the later section on aliasing distortion will show that it is not possible to extract any information about the frequency spectrum for frequencies $f > F/2 = f_c$ from time-response data sampled at steps of ΔT. The parameter $f_c = 1/(2\Delta T)$ is known as the Nyquist frequency.

It is evident by comparing equation (3.15) with equation (3.11) that the sequence $\{\tilde{X}_n\} = [\tilde{X}_0, \tilde{X}_1, \ldots, \tilde{X}_{n-1}]$ represents the DFT of the sequence $\{\tilde{x}_m\} = [\tilde{x}_0, \tilde{x}_1, \ldots, \tilde{x}_{N-1}]$. From equation (3.10), the forward transform is given by

$$\tilde{X}_n = \frac{1}{F} \sum_{m=0}^{N-1} \tilde{x}_m \exp(-j2\pi nm/N) \tag{3.17}$$

In summary, if $X(f)$ is the FIT of $x(t)$, then the N-element sequence $\{\tilde{X}_n\}$ is the DFT of the N-element sequence $\{\tilde{x}_m\}$. The periodic functions $\tilde{x}(t)$ and $\tilde{X}(f)$ are related to $x(t)$ and $X(f)$, respectively, through equations (3.13) and (3.16), $\{\tilde{x}_m\}$ and $\{\tilde{X}_n\}$ being their individual sampled data (see figure 3-2).

Relationship between DFT and FSE

A fundamental result relating DFT and FSE will be established in this section. From equation (3.5), it follows that, for a periodic function $x(t)$ of period T, the sampled values $x_m = x(m \cdot \Delta T)$, $m = 0, \pm 1, \pm 2, \ldots$ are given by

$$x_m = \frac{1}{T} \sum_{n=-\infty}^{\infty} A_n \exp(j2\pi nm\Delta T/T)$$

$$= \frac{1}{T} \sum_{k=-\infty}^{\infty} \sum_{n=kN}^{(k+1)N-1} A_n \exp(j2\pi nm/N)$$

By definition, the sequence $\{x_m\}$ is periodic with N-element periodicity where $N = T/\Delta T$. The procedure for obtaining equation (3.15) is now adopted to obtain

$$x_m = \frac{1}{T} \sum_{n=0}^{N-1} \widetilde{A}_n \exp(j2\pi nm/N) \tag{3.18}$$

in which

$$\widetilde{A}_n = \sum_{k=-\infty}^{\infty} A_{n+kN} \tag{3.19}$$

This situation is illustrated in figure 3-3.

The sequence $\{\widetilde{A}_n\}$ is periodic, with N-element periodicity. By comparing equation (3.18) with equation (3.11), it becomes clear that the N-element sequence $\{\widetilde{A}_n\} = [\widetilde{A}_0, \widetilde{A}_1, \ldots, \widetilde{A}_{N-1}]$ is the DFT of the N-element sequence $\{x_m\} = [x_0, x_1, \ldots, x_{N-1}]$. From equation (3.10), the forward transform follows; that is,

$$\widetilde{A}_n = \Delta T \sum_{m=0}^{N-1} x_m \exp(-j2\pi nm/N) \tag{3.20}$$

In summary, if $\{A_n\}$ are the coefficients of the FSE of a periodic function $x(t)$, then the N-element sequence $\{\widetilde{A}_n\}$ is the DFT of the N-element sequence $\{x_m\}$, where $x_m = x(m \cdot \Delta T)$ and $\{\widetilde{A}_m\}$ is given by equation (3.19). The foregoing relationships are summarized in table 3-2.

Aliasing Distortion in the Frequency Domain

A signal $x(t)$ of finite duration $(0, T)$ is shown in figure 3-2(a). It is desired to obtain its FIT $X(f)$ as accurately as possible, using a digital computer.

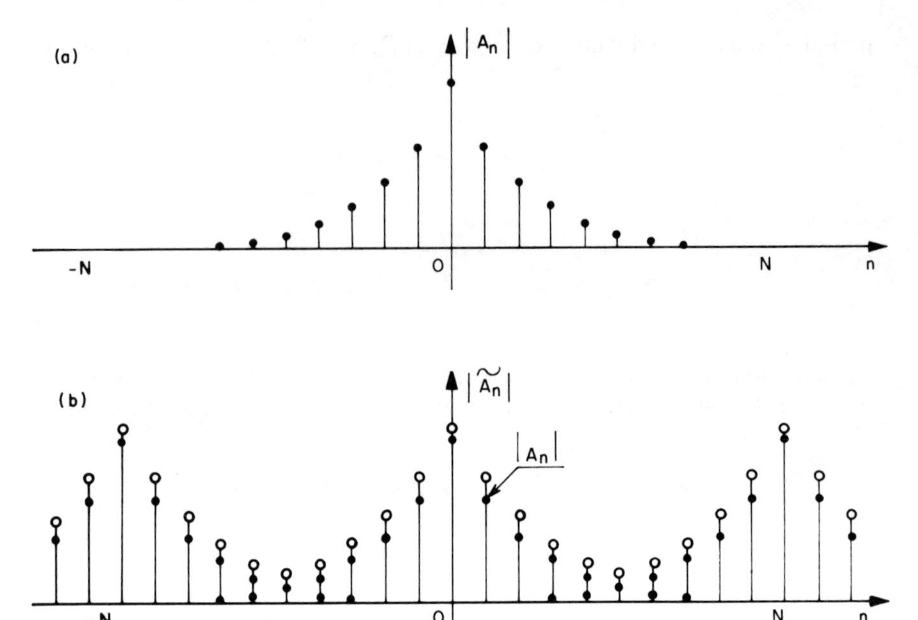

Figure 3-3. Demonstration of the Relationship between FSE and DFT

It should be recalled that $X(f)$ is a generally complex function and, as such, it cannot be displayed as a single curve in a two-dimensional coordinate system. Both magnitude and phase-angle variations with frequency f are needed. For brevity, only the magnitude $|X(f)|$ is shown in figure 3-2(a). Nevertheless, the argument presented in this section applies equaly to the phase angle $\angle X(f)$.

The periodic function $\tilde{x}(t)$ is formed according to equation (3.13), as shown in figure 3-2(b). The periodic function $\tilde{X}(f)$ is formed according to equation (3.16). This is illustrated in figure 3-2(c). It should be remembered that $\tilde{X}(f)$ is not the FIT of $\tilde{x}(t)$. From earlier discussion, it follows that, if $\tilde{x}(t)$ is sampled at steps of ΔT, the resulting periodic sequence $\{\tilde{x}_m = \tilde{x}(m \cdot \Delta T)\}$ [see figure 3-2(d)] has the DFT given by the periodic sequence $\{\tilde{X}_n = \tilde{X}(n \cdot \Delta F)\}$ [see Figure 3-2(e)], where $F = 1/\Delta T$ and $\Delta F = 1/T$. It is obvious that, in the time interval $(0, T)$, $x(t) = \tilde{x}(t)$ and $x_m = \tilde{x}_m$. However, over the frequency interval $(0, F)$, $\tilde{X}(f)$ is not identical to the required $X(f)$. This is known as the aliasing distortion in the frequency domain. If ΔT is sufficiently small (that is, F is sufficiently large), then $\tilde{X}(f)$ will approximate $X(f)$ in the frequency range $(0, F/2)$, as is clear from figure 3-2(c). Also, because of the F-periodicity of $\tilde{X}(f)$, its values in the frequency range $(F/2, F)$ will approximate $X(f)$ in the frequency range $(-F/2, 0)$.

It is clear from the preceding discussion that, if a time history $x(t)$ is

Table 3-2
Relationships among FIT, DFT, and FSE

Description	*Relationship*	
	DFT and FIT	*DFT and FSE*
Given	$x(t) \xrightarrow{\text{FIT}} X(f)$	$x(t) \xrightarrow{\text{FSE}} \{A_n\}$
Define	$\tilde{x}(t) = \sum_{k=-\infty}^{\infty} x(t + kT)$	$\tilde{A}_n = \sum_{k=-\infty}^{\infty} A_{n} + kN$
	$\tilde{X}(f) = \sum_{k=-\infty}^{\infty} X(f + kF)$	
Then	$\{\tilde{x}_m\} \xrightarrow{\text{DFT}} \{\tilde{X}_n\}$	$\{x_m\} \xrightarrow{\text{DFT}} \{\tilde{A}_n\}$
Where	$\tilde{x}_m = \tilde{x}(m \cdot \Delta T), \tilde{X}_n = \tilde{X}(n \cdot \Delta F)$	$x_m = x(m \cdot \Delta T)$
	$F = 1/\Delta T,\ T = 1/\Delta F$	$N = T/\Delta T$

sampled at equal steps of ΔT, no information regarding its frequency spectrum $\tilde{X}(f)$ can be obtained for frequencies higher than $f_c = 1/(2\Delta T)$. This limiting (cut-off) frequency is also known as the Nyquist frequency.

The discussion given in this section suggests the following in regard to DFT applications:

1. A sufficiently small sample step ΔT in the time domain should be picked in order to reduce the aliasing distortion in the frequency domain.
2. The highest frequency for which the Fourier transform (frequency-spectrum) information is obtainable is the Nyquist frequency $f_c = 1/(2\Delta T)$.
3. DFT results in the frequency range $(f_c, 2f_c)$ merely approximate the frequency spectrum in the negative frequency range $(-f_c, 0)$.

Aliasing Distortion in the Time Domain

The inverse problem is considered in this section. Specifically, given a band-limited Fourier integral transform (frequency-spectrum function) $X(f)$, it is desired to obtain the inverse transform (time response) $x(t)$ as accurately as possible, using a digital computer (see figure 3-4). Both $X(f)$ and $x(t)$ are complex in theory. In real situations, however, such as the seismic response of equipment, $x(t)$ is a real function. Only the magnitudes of the complex

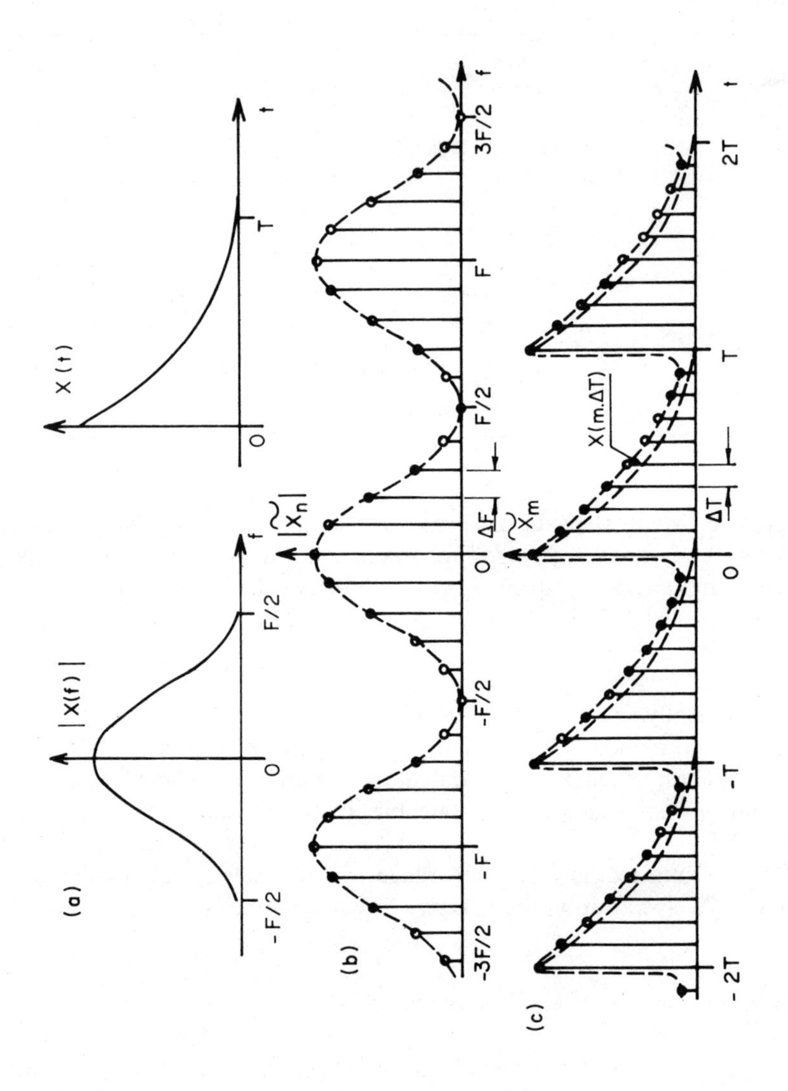

Figure 3-4. Demonstration of Aliasing Distortion in the Time Domain

functions are shown in the figure. As is generally true, however, the discussion also holds for the phase angle.

First, $\widetilde{X}(f)$ is constructed using equation (3.16), and $\widetilde{x}(t)$ is constructed using equation (3.13). Next, $\widetilde{X}(f)$ is sampled at steps of ΔF to form $\{\widetilde{X}_n\}$, and $\widetilde{x}(t)$ is sampled at steps of ΔT to form $\{\widetilde{x}_m\}$, where $\Delta T = 1/F$ and $\Delta F = 1/T$. From earlier discussion, it follows that $\{\widetilde{x}_m\}$ is the inverse DFT of $\{\widetilde{X}_n\}$. The sequences $\{\widetilde{X}_n\}$ and $\{\widetilde{x}_m\}$ are shown (schematically in the case of complex function) in Figures 3-4(b) and (c), respectively. It is noted that, for the frequencies in the range $(-F/2, F/2)$, the sampled sequence $\{\widetilde{X}_n\}$ is identical to the sampled $X(f)$. However, in the time interval $(0, T)$, the inverse DFT sequence $\{\widetilde{x}_m\}$ is only an approximation of the desired discrete values $\{x(m \cdot \Delta T)\}$. The resulting error is termed the aliasing distortion in the time domain. In order to reduce this error, the chosen sample step ΔF in the frequency domain should be sufficiently small (that is, T should be sufficiently large).

Fast Fourier Transform (FFT)

The direct computation of the discrete Fourier transform (DFT) using equation (3.10) is not recommended, particularly in real-time applications, because of the inefficiency of this procedure. For a sequence of N sampled data points, N^2 complex multiplications and $N(N-1)$ complex additions are necessary in the direct evaluation of the DFT equation (3.10), assuming that the complex exponential factors $\exp(-j2\pi mn/N)$ are already computed. Many of these arithmetic operations are redundant.

The Cooley and Tukey algorithm, commonly known as the radix-two fast Fourier transform (FFT) algorithm is an efficient procedure for computing DFT. Efficiency of the algorithm is achieved by dividing the numerical procedure into several stages so that the redundant computations are avoided.

Development of the Radix-Two FFT Algorithm

The multiplicative constant ΔT in the DFT equation (3.10) is a parameter that was introduced to maintain the consistency with the FIT equation (3.1). This constant may be treated as a scaling factor for the final results; or, equivalently, it could be combined with the input data sequence $\{x_n\}$. In any event, the primary computational effort in equation (3.10) is directed toward computing the sequence $[A(0), A(1), \ldots, A(N-1)]$ from the data sequence $[a(0), a(1), \ldots, a(N-1)]$, using the relationship

$$A(m) = \sum_{n=0}^{N-1} a(n) W^{mn} \qquad m = 0, 1, \ldots, N-1 \qquad (3.21)$$

in which

$$W = \exp(-j2\pi/N) \tag{3.22}$$

The FFT algorithm requires that N be highly composite (that is, factorizable into many nonunity integers). In particular, for the radix-two algorithm, it is required that $N = 2^r$, in which r is a positive integer. If the given data sequence does not satisfy this condition, it must be augmented by a sufficient number of trailing zeros.

A systematic development of the radix-two FFT algorithm is presented in this section. The integers m and n are expressed in the binary-number-system expansion. Recalling that $0 \le m \le N-1$ and $0 \le n \le N-1$,

$$\begin{aligned} m &= m_{r-1}2^{r-1} + m_{r-2}2^{r-2} + \cdots + m_0 \\ n &= n_{r-1}2^{r-1} + n_{r-2}2^{r-2} + \cdots + n_0 \end{aligned} \tag{3.23}$$

in which m_i and n_j take values 0 or 1. Equivalently,

$$\begin{aligned} m &= \text{binary}(m_{r-1}, m_{r-2}, \ldots, m_0) \\ n &= \text{binary}(n_{r-1}, n_{r-2}, \ldots, n_0) \end{aligned}$$

Next, the indices of the elements $A(\cdot)$ and $a(\cdot)$ in equation (3.21) are expressed by their binary counterparts;

$$A(m_{r-1}, \ldots, m_0) = \sum_{n_0=0}^{1} \sum_{n_1=0}^{1} \cdots \sum_{n_{r-1}=0}^{1} a(n_{r-1}, \ldots, n_0) W^{mn} \tag{3.24}$$

in which

$$\begin{aligned} W^{mn} &= W^{(m_{r-1}2^{r-1} + \cdots + m_0)(n_{r-1}2^{r-1} + \cdots + n_0)} \\ &= W^{n_{r-1}2^{r-1}m_0} W^{n_{r-2}2^{r-2}(2m_1 + m_0)} W^{n_{r-3}2^{r-3}(2^2m_2 + 2m_1 + m_0)} \\ &\quad \ldots W^{n_0(m_{r-1}2^{r-1} + \cdots + m_0)} \end{aligned}$$

The fact that $W^{2^r} = W^N = 1$ has been used in the foregoing expansion. By defining the intermediate set of sequences $\{A_1(\cdot)\}$, $\{A_2(\cdot)\}, \ldots, \{A_r(\cdot)\}$, equation (3.24) can be expressed as the set of equations:

$$A_1(m_0, n_{r-2}, \ldots, n_0) = \sum_{n_{r-1}=0}^{1} a(n_{r-1}, \ldots, n_0) W^{n_{r-1}2^{r-1}m_0}$$

$$A_2(m_0, m_1, n_{r-3}, \ldots, n_0) = \sum_{n_{r-2}=0}^{1} A_1(m_0, n_{r-2}, \ldots, n_0) W^{n_{r-2}2^{r-2}(2m_1+m_0)}$$

$$\vdots \tag{3.25}$$

$$A_r(m_0, m_1, \ldots, m_{r-1}) = \sum_{n_0=0}^{1} A_{r-1}(m_0, \ldots, m_{r-2}, n_0) \times W^{n_0(m_{r-1}2^{r-1} + \cdots + m_0)}$$

with

$$A(m_{r-1}, \ldots, m_0) = A_r(m_0, \ldots, m_{r-1}) \tag{3.26}$$

Consequently, the single set of computations given by equation (3.21) for the N-element sequence $\{A(\cdot)\}$ has been replaced by r stages of computations. In each stage, an N-element sequence $\{A_i(\cdot)\}$ must be computed from the immediately preceding N-element sequence $\{A_{i-1}(\cdot)\}$. It soon will be apparent that, as a result of this r-stage factorization, the number of arithmetic operations required has been considerably reduced.

It should be noted that each relationship given in equation (3.25) corresponds to a set of N separate relationships, because the index within the parenthesis of $A_i(\cdot)$ runs from 0 to $N-1$. In order to observe some important characteristics of the FFT algorithm, the ith relationship of equation (3.25) is examined. Specifically,

$$A_i(m_0, m_1, \ldots, m_{i-1}, n_{r-i-1}, \ldots, n_0) = \sum_{n_{r-i}=0}^{1} A_{i-1}(m_0, m_1, \ldots, m_{i-2}, n_{r-i}, \ldots, n_0) \times W^{n_{r-i}2^{r-i}(2^{i-1}m_{i-1} + \cdots + m_0)} \quad i = 1, \ldots, r \tag{3.27}$$

in which $A_0(\cdot) = a(\cdot)$. The summation on the right-hand side of equation (3.27) is expanded as

$$A_i(m_0, m_1, \ldots, m_{i-1}, n_{r-i-1}, \ldots, n_0) = A_{i-1}(m_0, \ldots, m_{i-2}, 0, n_{r-i-1}, \ldots, n_0) + A_{i-1}(m_0, \ldots, m_{i-2}, 1, n_{r-i-1}, \ldots, n_0) W^{2^{r-i}(2^{i-1}m_{i-1} + \cdots + m_0)} \tag{3.28}$$

which involves only one complex multiplication and one complex addition, assuming that the complex exponential terms W^p are precomputed. It is noted that the variable m_{i-1}, which is the binary coefficient of 2^{r-i} in the index of the left-hand-side term $A_i(\cdot)$, does not appear in the binary index of the right-hand-side terms $A_{i-1}(\cdot)$. Consequently, the values of $A_{i-1}(\cdot)$ terms on the right-hand side remain unchanged as m_{i-1} switches from 0 to 1 in the left-hand-side index. This switch corresponds to a jump in the index of $A_i(\cdot)$ through a value of 2^{r-i}. Accordingly, the computation of, for example, the kth term $A_i(k)$ and the $(k+2^{r-i})$th term $A_i(k+2^{r-i})$ of the sequence $\{A_i(\cdot)\}$ in the ith stage involves the same two terms $A_{i-1}(k)$ and $A_{i-1}(k+2^{r-i})$ of the previous $(i-1)$th sequence $A_{i-1}(\cdot)$. It follows that equation (3.28) takes the more familiar decimal format:

$$\begin{aligned} A_i(k) &= A_{i-1}(k) + A_{i-1}(k+2^{r-i})W^p \\ A_i(k+2^{r-i}) &= A_{i-1}(k) + A_{i-1}(k+2^{r-i})W^{\tilde{p}} \end{aligned} \tag{3.29}$$

$$p = 2^{r-i}(2^{i-2}m_{i-2} + \cdots + m_0) \tag{3.30}$$

$$\tilde{p} = 2^{r-i}(2^{i-1} + 2^{i-2}m_{i-2} + \cdots + m_0) \tag{3.31}$$

Equation (3.30) results when $m_{i-1} = 0$, and equation (3.31) results when $m_{i-1} = 1$. On closer examination, it is evident that $\tilde{p} = N/2 + p$, which follows from $2^{r-i}2^{i-1} = 2^{r-1} = 2^r/2$. Hence,

$$W^{\tilde{p}} = W^{N/2}W^p$$

From the definition of W [equation (3.22)], however, $W^{N/2} = -1$. Consequently,

$$W^{\tilde{p}} = -W^p \tag{3.32}$$

By substituting equation (3.32) in equation (3.29), we have

$$\begin{aligned} A_i(k) &= A_{i-1}(k) + A_{i-1}(k+2^{r-i})W^p \\ A_i(k+2^{r-i}) &= A_{i-1}(k) - A_{i-1}(k+2^{r-i})W^p \end{aligned} \tag{3.33}$$

for $i = 1, \ldots, r$ and $k = 0, 1, \ldots, 2^{r-i} - 1, 2^{r-i+1}, \ldots$, in which p is given by equation (3.30).

The in-place simultaneous computation of the so-called dual terms $A_i(k)$ and $A_i(k+2^{r-i})$ in the N-term sequence $\{A\cdot(\cdot)\}$ involves just one complex multiplication and two complex additions. As a result, the number of multiplications required has been further reduced by a factor of two. At each stage, the computation of the N-term sequence requires $N/2$ complex multiplications and N complex additions. Since there are r stages, the radix-two FFT requires a total of $rN/2$ complex multiplications and rN complex additions, in which $r = \ln_2 N$. In otherwords, the number of multiplications required has been reduced by a factor of $(2N/\ln_2 N)$ and the number of additions has been reduced by a factor of $(N/\ln_2 N)$. For large N, these ratios correspond to a sizable reduction in the computer time required for a DFT. This is a significant breakthrough in real-time digital Fourier analysis.

An insignificant shortcoming of the Cooley-Tukey FFT procedure is evident from equation (3.26). The final sequence $\{A_r(\cdot)\}$ is a scrambled version of the desired transform $\{A(\cdot)\}$. To unscramble the result, it is merely required to interchange the term in the binary location $(m_0, \ldots, m_{r-1})$ with that in the binary location $(m_{r-1}, \ldots, m_0)$. It should be remembered not to duplicate any interchanges while proceeding down the array during the unscrambling procedure. Since an in-place interchange of the elements is performed, there is no necessity for defining a new array. There is an associated saving in computer memory requirements.

The Radix-Two FFT Procedure

The basic steps of the radix-two FFT algorithm are as follows: $N = 2^r$ elements of the data sequence $\{A(\cdot)\}$ are available.

Step 1 Initialize variables. Stage number $i = 1$. Sequence element number $k = 0$.

Step 2 Determin p, as follows: From equation (3.30), $p = \text{binary}(m_{i-1}, \ldots, m_0, 0, \ldots, 0)_{N\text{ bits}}$. From equation (3.27), $k = \text{binary}(m_0, m_1, \ldots, m_{i-1}, n_{r-i-1}, \ldots, n_0)$. Shift k register through $(r - i)$ bits to the right, and augment the vacancies by leading zeros. This gives binary $(0, \ldots, 0, m_0, \ldots, m_{i-1})_{N\text{ bits}}$. Reverse the bits to obtain p.

Step 3 Compute in place the dual terms $A_i(k)$ and $A_i(k + 2^{r-i})$, using equation (3.33). *Note*: Since $A_{i-1}(k)$ and $A_{i-1}(k + 2^{r-i})$ are not needed in the subsequent computations, they are destroyed by storing $A_i(k)$ and $A_i(k + 2^{r-i})$ in those locations. As a result, only one array of N elements is needed in the computer memory.

Step 4 Increment $k = k + 1$. If an already-computed dual element is encountered, skip k through 2^{r-i} (that is, $k = k + 2^{r-i}$). If $k \geq N$, increment $i = i + 1$. If $i > r$, go to Step 5. Otherwise, go to Step 2.

Step 5 Unscramble the sequence, using equation (3.26), and stop.

Illustrative Example

The elementary problem of four sampled data points $[a(0), a(1), a(2), a(3)]$ is considered an illustrative example. It is noted that $N = 2^r = 4$ in this example. The objective is to determine the DFT sequence $[A(0), A(1), A(2), A(3)]$ for the given sequence. Letting $A_0(\cdot) = a(\cdot)$, the radix-two FFT consists of the following computations:

Stage 1 ($i = 1$): From equation (3.28), the binary index version is

$$A_1(0, 0) = A_0(0, 0) + A_0(1, 0)W^0$$

$$A_1(0, 1) = A_0(0, 1) + A_0(1, 1)W^0$$

$$A_1(1, 0) = A_0(0, 0) - A_0(1, 0)W^0$$

$$A_1(1, 1) = A_0(0, 1) - A_0(1, 1)W^0$$

Equation (3.32) has been used in obtaining the last two equations. Even though $W^0 = 1$, this substitution is not made in order to retain the format of the algorithm. The dual jump for the first stage is $2^{r-i} = 2$. Accordingly, the third equation is the dual of the first equation, and they possess a common multiplication operation. Similarly, the fourth equation is the dual of the second equation. Using Step 2 from the previous section, $p =$ binary(0, 0) $= 0$ for $k = 0$ and 1.

Alternatively, using the decimal index equation (3.33),

$$A_1(0) = A_0(0) + A_0(2)W^0$$

$$A_1(1) = A_0(1) + A_0(3)W^0$$

$$A_1(2) = A_0(0) - A_0(2)W^0$$

$$A_1(3) = A_0(1) - A_0(3)W^0$$

Stage 2 ($i = 2$): From equation (3.28), the binary index version is

$$A_2(0, 0) = A_1(0, 0) + A_1(0, 1)W^0$$

$$A_2(0, 1) = A_1(0, 0) - A_1(0, 1)W^0$$

$$A_2(1, 0) = A_1(1, 0) + A_1(1, 1)W^1$$

$$A_2(1, 1) = A_1(1, 0) - A_1(1, 1)W^1$$

The decimal index version of equation (3.33) is

$$A_2(0) = A_1(0) + A_1(1)W^0$$

$$A_2(1) = A_1(0) - A_1(1)W^0$$

$$A_2(2) = A_1(2) + A_1(3)W^1$$

$$A_2(3) = A_1(2) - A_1(3)W^1$$

The dual jump for the second stage is $2^{r-i} = 1$. It follows that the second equation is the dual of the first equation and that the fourth equation is the dual of the third equation. The dual-equation pairs have common multiplication operations. Also, for example, the value of p for the third equation [for $A_2(2)$] is obtained using Step 2 of the previous section, $k = 2 = \text{binary}(1, 0)$. Shifting bits through $r - i = 0$ (that is, no shifts) to the right and finally reversing the bits, we have $p = \text{binary}(0, 1) = 1$.

Unscrambling: Finally, the results are unscrambled. In binary index notation,

$$A(0, 0) = A_2(0, 0)$$

$$A(0, 1) = A_2(1, 0)$$

$$A(1, 0) = A_2(0, 1)$$

$$A(1, 1) = A_2(1, 1)$$

In decimal index notation

$$A(0) = A_2(0)$$

$$A(1) = A_2(2)$$

$$A(2) = A_2(1)$$

$$A(3) = A_2(3)$$

The two primary sources of numerical error associated with the FFT algorithm are quantization error, and round-off accumulation.

The first type of error is caused by the analog-to-digital conversion of the time history. This error propagates through each stage of the FFT and finally appears in the transformed result. The second type of error is caused by the accumulation of the round-off error in each arithmetic operation.

Some Useful Fourier Transform Results

Fourier transformation is useful in computing time integrals such as correlation and convolution. These integrals in the time domain are transformed into algebraic products in the frequency (Fourier) domain. Analogous results exist for the discrete case (DFT). In this section, we shall examine some useful theorems that use Fourier transformation.

Correlation Theorem

Three forms of the mean lagged product (also known as the correlation function) are commonly used in engineering literature, depending on the specific application.

For a pair of rapidly decaying (aperiodic) deterministic signals $x(t)$ and $y(t)$, the cross-correlation function is given by

$$\phi_{xy}(\tau) = \int_{-\infty}^{\infty} x(t)y(t+\tau)\,dt \tag{3.34}$$

For a pair of periodic and deterministic signals $x(t)$ and $y(t)$, each having the same period T, the cross-correlation function is given by

$$\phi_{xy}(\tau) = \frac{1}{T}\int_{0}^{T} x(t)y(t+\tau)\,dt \tag{3.35}$$

Finally, for a pair of ergodic random (stochastic) signals $x(t)$ and $y(t)$, the cross-correlation function is given by

$$\phi_{xy}(\tau) = \lim_{T\to\infty}\left[\frac{1}{T}\int_{0}^{T} x(t)y(t+\tau)\,dt\right] \tag{3.36}$$

When $x(t) = y(t)$, the autocorrelation function $\phi_{xx}(\tau)$ results. The FIT of equation (3.36) gives the power spectral density (psd) of the ergodic signal $x(t)$ in this case. Consequently, the evaluation of equation (3.36) is necessary in the spectral analysis of stochastic processes, and this has numerous applications. These terms are defined and the related concepts are discussed in chapter 5.

By using the definition of the inverse FIT equation (3.3) in equation (3.34) and following straightforward mathematical manipulation, it may be shown that

$$\Phi_{xy}(f) = [X(f)]^* Y(f) \tag{3.37}$$

in which the cross-spectral density $\Phi_{xy}(f)$ is the FIT of $\phi_{xy}(\tau)$, as given by

$$\Phi_{xy}(f) = \int_{-\infty}^{\infty} \phi_{xy}(\tau) \exp(-j2\pi f\tau) d\tau \tag{3.38}$$

and $[\ ]^*$ denotes the complex conjugation operation. This result, which is known as the correlation theorem (see table 3-3) has applications in the evaluation of the correlation functions and psd's of finite-record-length data.

The mean lagged product defined by equation (3.36) cannot be evaluated exactly in the case of random signals. The signals $x(t)$ and $y(t)$ are truncated at T_1 and T_2, respectively, and an approximation to $\phi_{xy}(\tau)$ is obtained, using

$$z(\tau) = \frac{1}{T} \int_0^T x(t) y(t+\tau) dt \tag{3.39}$$

The sampled data are formed according to

$$\begin{aligned} x_m &= x(m \cdot \Delta T) \quad \text{for} \quad m = 0, \ldots, M-1 \\ &= 0 \quad \text{otherwise} \\ y_k &= y(k \cdot \Delta T) \quad \text{for} \quad k = 0, \ldots, K-1 \\ &= 0 \quad \text{otherwise} \end{aligned}$$

Using the trapezoidal rule, the sequence $\{z_n\}$ that approximates the sampled values $z(n \cdot \Delta T)$ may be computed, using

$$z(n \cdot \Delta T) \simeq z_n = \frac{1}{N} \sum_{r=0}^{N-1} x_r y_{r+n} \tag{3.40}$$

In which $N > \max(M, K)$. It is noted that, in the summation, an upper limit greater than $\min(M-1, K-1-n)$ is redundant. Since the divisor is the

Table 3-3
Some Useful Fourier Transform Results

Description		*Continuous*	*Discrete*
Correlation theorem	If	$z(\tau) = \int_{-\infty}^{\infty} x(t)y(t+\tau)dt$	$z_m = \Delta T \sum_{r=0}^{N-1} x_r y_{r+m}$
	Then	$Z(f) = [X(f)]^* Y(2\ f)$	$Z_n = [X_n]^* Y_n$
Parseval's Theorem	If	$y(t) \xrightarrow{\text{FIT}} Y(f)$	$\{y_m\} \xrightarrow{\text{DFT}} \{y_n\}$
	Then	$\int_{-\infty}^{\infty} y^2(t)dt = \int_{-\infty}^{\infty} \mid Y(f) \mid^2 df$	$\Delta T \sum_{m=0}^{N-1} y_m^2 = \Delta F \sum_{n-0}^{N-1} \mid Y_n \mid^2$
Convolution theorem	If	$y(t) = \int_{-\infty}^{\infty} h(\tau)u(t-\tau)d\tau$	$y_m = \Delta T \sum_{r=0}^{N-1} h_r u_{m-r}$
		$= \int_{-\infty}^{\infty} h(t-\tau)u(\tau)d\tau$	$= \Delta T \sum_{r=0}^{N-1} h_{m-r} u_r$
	Then	$Y(f) = H(f)U(f)$	$Y_n = H_n U_n$

constant value N rather than the actual number of terms in the summation, equation (3.40) represents a biased estimate of the mean lagged product. Nevertheless, it is convenient to use equation (3.40) in this analysis.

Discrete Correlation Theorem. A result analogous to equation (3.37) is established for discrete data. The inverse DFT equation (3.11) is used in equation (3.40) in conjunction with the fact that $x_m = [x_m]^*$ for real x_m:

$$z_n = \frac{1}{N} \sum_{r=0}^{N-1} \frac{1}{N\Delta T} \sum_{m=0}^{N-1} [X_m]^* \exp(-j2\pi mr/N)$$

$$\times \frac{1}{N\Delta T} \sum_{k=0}^{N-1} Y_k \exp[j2\pi k(n+r)/N] = \frac{1}{N^3 \Delta T} \sum_{m=0}^{N-1} \sum_{k=0}^{N-1}$$

$$\times [X_m]^* Y_k \exp(j2\pi kn/N) \sum_{r=0}^{N-1} \exp[j2\pi r(k-m)/N]$$

The orthogonality condition (see table 3-1) is used in the last summation term. Consequently,

$$Tz_n = \frac{1}{N\Delta T} \sum_{m=0}^{N-1} [X_m]^* Y_m \exp(j2\pi mn/N) \qquad (3.41)$$

in which $T = N \cdot \Delta T$. It follows that $T\{z_n\}$ is the inverse DFT of $\{[X_m]^* Y_m\}$ (see table 3-1). The discrete correlation theorem is summarized in table 3-3.

Parseval's Theorem

The inverse FIT relation corresponding to equation (3.38) is

$$\phi_{xy}(\tau) = \int_{-\infty}^{\infty} \Phi_{xy}(f) \exp(j2\pi f\tau) df \qquad (3.42)$$

From equation (3.37),

$$\phi_{xy}(\tau) = \int_{-\infty}^{\infty} [X(f)]^* Y(f) \exp(j2\pi f\tau) df \qquad (3.43)$$

If we set $\tau = 0$ and $x = y$ in equation (3.43),

$$\phi_{xy}(0) = \int_{-\infty}^{\infty} |Y(f)|^2 df \qquad (3.44)$$

Similarly, from equation (3.34),

$$\phi_{xy}(0) = \int_{-\infty}^{\infty} y^2(t) dt \qquad (3.45)$$

By comparing equations (3.44) and (3.45), we obtain Parseval's theorem:

$$\int_{-\infty}^{\infty} y^2(t) dt = \int_{-\infty}^{\infty} |Y(f)|^2 df \qquad (3.46)$$

By using the discrete correlation theorem analogously, we can establish the discrete version of equation (3.46):

$$\Delta T \sum_{m=0}^{N-1} y_m^2 = \Delta F \sum_{n=0}^{N-1} |Y_n|^2 \qquad (3.47)$$

Convolution Theorem

The convolution of the two functions $h(t)$ and $u(t)$ has been discussed with reference to system response in chapter 2. Consider

$$\begin{aligned} y(t) &= \int_{-\infty}^{\infty} h(\tau)u(t-\tau)d\tau \\ &= \int_{-\infty}^{\infty} h(t-\tau)u(\tau)d\tau \end{aligned} \tag{3.48}$$

The convolution theorem is conveniently denoted by $h(t) * u(t)$.

By substituting the inverse FIT relations [equations (3.3)] in equation (3.48) and following straightforward mathematical manipulation, using the definition of the Dirac delta function (see chapter 2), we obtain

$$y(t) = \int_{-\infty}^{\infty} H(f)U(f) \exp(j2\pi ft)df \tag{3.49}$$

Equivalently, the FIT of $y(t)$ is

$$Y(f) = H(f)U(f) \tag{3.50}$$

Equation (3.50) is commonly known as the convolution theorem. Because of the similiarity of FIT and inverse FIT, it can be shown by following a procedure analogous to the foregoing that $h(t)u(t)$ is the inverse FIT of the frequency-domain convolution $H(f) * U(f)$ given by

$$\begin{aligned} H(f) * U(f) &= \int_{-\infty}^{\infty} H(F)U(f-F)dF \\ &= \int_{-\infty}^{\infty} H(f-F)U(F)dF \end{aligned} \tag{3.51}$$

In other words, when considering time and frequency domains, convolution in one domain corresponds to multiplication in the other domain. The convolution theorem is summarized in table 3-3.

Discrete Convolution Theorem. The convolution theorem equation (3.48) of two signals $u(t)$ and $h(t)$, defined over the finite durations $(0, T_1)$ and $(0, T_2)$, respectively, may be determined using a digital computer. First, the sample step ΔT is chosen and the two sequences $\{u_m\}$ and $\{h_k\}$ of sampled data are formed according to

$$\begin{aligned} u_m &= u(m \cdot \Delta T) \quad \text{for} \quad m = 0, \dots, M-1 \\ &= 0 \quad \text{otherwise} \\ h_k &= h(k \cdot \Delta T) \quad \text{for} \quad k = 0, \dots, K-1 \\ &= 0 \quad \text{otherwise} \end{aligned} \tag{3.52}$$

in which $M = \text{integer}(T_1/\Delta T)$ and $K = \text{integer}(T_2/\Delta T)$. In order to eliminate the wraparound error, it is required that, for the number of samples of $y(t)$, $N = M + K - 1$.

The direct digital computation of equation (3.48) may be performed using the trapezoidal rule:

$$y(n \cdot \Delta T) = y_n = \Delta T \sum_{m=0}^{N-1} u_m h_{n-m}$$
$$= \Delta T \sum_{m=0}^{N-1} u_{n-m} h_m \qquad (3.53)$$
$$n = 0, 1, \ldots, N-1$$

In view of the zero terms in the two sequences $\{u_m\}$ and $\{h_k\}$, as given by equation (3.52), it is equally correct to make the lower and upper limits of the first summation $\max(0, n-k+1)$ and $\min(n, M-1)$, respectively. Similarly, the two limits in the second summation could be $\max(0, n-M+1)$ and $\min(n, K-1)$. In any event, by direct counting through summation of series, it can be shown that the computation of equation (3.53) needs KM real multiplications and KM-N real additions. Alternatively, the discrete convolution result that is analogous to the continuous counterpart equation (3.49) may be used to evaluate equation (3.53) indirectly.

By substituting the inverse DFT equation (3.11) in equation (3.53), we have

$$y_n = \Delta T \sum_{m=0}^{N-1} \frac{1}{N\Delta T} \sum_{r=0}^{N-1} U_r \exp(j2\pi rm/N)$$
$$\times \frac{1}{N\Delta T} \sum_{k=0}^{N-1} H_k \exp[j2\pi k(n-m)/N]$$
$$= \frac{1}{N^2 \Delta T} \sum_{r=0}^{N-1} \sum_{k=0}^{N-1} U_r H_k \exp(j2\pi kn/N) \sum_{m=0}^{N-1} \exp[j2\pi(r-k)m/N]$$

The orthogonality condition (see table 3-1) is used in the last summation. Consequently,

$$y_n = \frac{1}{N \cdot \Delta T} \sum_{r=0}^{N-1} U_r H_r \exp(j2\pi rn/N) \qquad (3.54)$$

Equation (3.54) should be compared with equation (3.49). Also see table 3-3 for a summary of the results.

Digital Fourier Analysis Procedures

From the developments in the earlier section "Unification of the Three Fourier Transform Types," it is clear that proper interpretation of the DFT results is extremely important in digital Fourier analysis. Only the first $N/2 + 1$ points of the DFT array, for example, approximate the Fourier transform of the data signal. The remaining $N/2 - 1$ points correspond to the negative-frequency spectrum and should be interpreted accordingly. The error caused by interpreting all N points in the DFT array as the positive frequency spectrum corresponding to the data signal is so great that the analysis would become worthless. In this section, the common DFT procedures are outlined. Emphasis is placed on correct interpretation of the results. Some ways to reduce computation time and memory requirements in real-time applications are described.

Fourier Transform Using DFT

Given an analog signal (continous-time) $x(t)$, the major steps for obtaining a suitable approximation to its Fourier transform $X(f)$, using digital Fourier analysis, are as follows:

1. Pick the sample step ΔT. Theoretically, $\Delta T = 1/(2 \times \text{highest frequency of interest})$. This value should be sufficiently small in order to reduce the aliasing distortion in the frequency domain.
2. Sample the signal up to time T, where $T = N \cdot \Delta T$ and $N = 2^r$. The duration $(0, T)$ of the sampled record must be sufficiently long in order to reduce the truncation error (leakage), as will be discussed later.
3. Obtain the DFT $\{\tilde{X}_n\}$ of the sampled data sequence $\{x_m\}$ using FFT.
4. A discrete approximation to the Fourier transform $X(f)$ is constructed from $\{\tilde{X}_n\}$ according to $X(n \cdot \Delta F) \simeq \tilde{X}_n$ for $n = 0, 1, \ldots, N/2$, and $X(-n \cdot \Delta F) \simeq \tilde{X}_{N-n}$ for $n = 1, \ldots, N/2 - 1$, in which $\Delta F = 1/T$.

Inverse DFT Using DFT

The inverse DFT is given by equation (3.11). This can be written

$$[x_n]^* = \frac{1}{N\Delta T} \sum_{m=0}^{N-1} [X_m]^* \exp(-j2\pi mn/N) \tag{3.55}$$

in which [ ]* denotes the complex conjugation operation. It is observed that equation (3.55) is identical to the forward DFT equation (3.10) except for a scaling factor. Consequently, the forward DFT algorithm may be used in the computation of the inverse DFT. The sampled data should be reorganized and complex-conjugated, however, before using DFT. Finally, the scaling factor should be accounted for, so that the final results have the proper units.

Given the complex function $X(f)$, which is the FIT of a real signal $x(t)$ with $x(t) = 0$ for $t < 0$, the main steps of determining a good approximation to the signal using digital Fourier analysis are as follows:

1. Let F be the highest frequency of interest in $X(f)$ and let $(0, T)$ be the interval over which real signal $x(t)$ is required. The sample step $\Delta F = 1/T$. It is required that ΔF be sufficiently small (T sufficiently large) to reduce aliasing distortion in the time domain. Also, F should be sufficiently large to reduce truncation error. Furthermore, the number of samples $F/\Delta F = N = 2^r$, if radix-two FFT is used.
2. Sample $X(f)$ at intervals ΔF over the frequency interval $(-F/2, F/2)$ according to $X_n = X(n \cdot \Delta F)$ for $n = -N/2, \ldots, 0, \ldots, N/2$ and properly scale the data.
3. Form the sequence $\{\widetilde{X}_n\}$ according to

$$\begin{aligned} \widetilde{X}_n &= X_n \quad \text{for} \quad n = 0, 1, \ldots, N/2 \\ &= X_{n-N} \quad \text{for} \quad n = N/2 + 1, \ldots, N-1 \end{aligned}$$

4. Form the complex conjugate sequence $\{[\widetilde{X}_n]^*\}$.
5. Obtain the DFT of $\{[\widetilde{X}_n]^*\}$ using FFT. This results in $\{[\widetilde{x}_m]^*\}$, which has complex elements with negligible imaginary parts.
6. Construct

$$x(m \cdot \Delta T) \simeq \text{real}[\widetilde{x}_m]^* \qquad m = 0, 1, \ldots, N-1$$

Simultaneous DFT of Two Real Data Records

Considerable computational advantages can be realized when the DFTs $\{Y_m\}$ and $\{Z_m\}$ of two real sequences $\{y_n\}$ and $\{z_n\}$ are required simultaneously. The procedure given in this section achieves this by using only a single DFT rather than two separate DFTs.

It is recalled that $\{x_n\}$ in equation (3.10) is generally a complex sequence. When a real sequence is used, half the storage requirements is wasted. Instead, the DFT of the complex sequence

$$\{x_n\} = \{y_n\} + j\{z_n\}$$

is obtained using FFT. This results in $\{X_m\}$. It is evident from equation (3.10) that

$$X_{N-m} = \Delta T \sum_{n=0}^{N-1} x_n \exp(j2\pi mn/N)$$

recalling that $\exp(-j2\pi n) = 1$. Consequently,

$$[X_{N-m}]^* = \Delta T \sum_{n=0}^{N-1} [x_n]^* \exp(-j2\pi mn/N) \qquad (3.56)$$

Since $[x_n]^* = y_n - jz_n$ it is straightforward to observe from equations (3.10) and (3.56) that

$$Y_m = \frac{1}{2} (X_m + \{X_{N-m}]^*) \qquad (3.57)$$

and

$$Z_m = \frac{1}{2j} (X_m - [X_{N-m}]^*) \qquad (3.58)$$

From the complex sequence $\{X_m\}$, the required complex sequences $\{Y_m\}$ and $\{Z_m\}$ are constructed according to equations (3.57) and (3.58).

Reduction of Computation Time for a Real Data Record

The DFT of a $2N$-element real sequence $[x_0, x_1, \ldots, x_{2N-1}]$ can be accomplished by means of a single DFT of an N-element complex sequence, using the concept discussed in the preceding section. From equation (3.10),

$$X_m = \Delta T \sum_{n=0}^{2N-1} x_n \exp[-j2\pi mn/(2N)]$$

$$= \Delta T \sum_{n=0}^{N-1} x_{2n} \exp[-j2\pi m(2n)/(2N)]$$

$$+ \Delta T \sum_{n=0}^{N-1} x_{2n+1} \exp[-j2\pi m(2n+1)/(2N)]$$

Consequently,

$$X_m = \Delta T \sum_{n=0}^{N-1} x_{2n} \exp(-j2\pi mn/N)$$

$$+ \exp(-j\pi m/N)\Delta T \sum_{n=0}^{N-1} x_{2n+1} \exp(-j2\pi mn/N) \qquad (3.59)$$

Two real sequences, each having N elements, are defined by separating the even and the odd terms of the given sequence $\{x_n\}$ according to

$$y_n = x_{2n}$$

$$z_n = x_{2n+1} \qquad n = 0, 1, \ldots, N-1$$

The DFT sequences $\{Y_m\}$ and $\{Z_m\}$ of the two real sequences $\{y_n\}$ and $\{z_n\}$ are obtained using the procedure given in the preceding section. Finally, the required DFT sequence is obtained using equation (3.59):

$$X_m = Y_m + \exp(-j\pi m/N)Z_m \qquad m = 0, 1, \ldots, N-1 \quad (3.60)$$

It should be noted that only the first N terms of the transformed sequence are obtained by this method. This is not a drawback, however, because, as is clear from the earlier discussion in the section on aliasing distortion in the frequency domain, the remaining terms correspond to the negative frequencies of $X(f)$.

Convolution of Finite Duration Signals Using DFT

Direct computation of the convolution is possible using the trapezoidal rule. Also, from equation (3.54), it is clear that the required sequence $\{y_n\}$ is the inverse DFT of $\{U_r H_r\}$, in which $\{U_r\}$ and $\{H_r\}$ are the DFTs of the N-point sequences $\{u_r\}$ and $\{h_r\}$, respectively. This gives rise to the following procedure for evaluating the convolution:

1. Determine $\{U_r\}$ and $\{H_r\}$ by the DFT of the N-point sequences $\{u_r\}$ and $\{h_r\}$, respectivley.
2. Evaluate $\{y_n\}$ from the inverse DFT of $\{H_r U_r\}$.

If the slow DFT, as given by equation (3.10), is used, the foregoing procedure requires $3N^2 + N$ complex multiplications and $3N(N-1)$

complex additions. If the FFT is employed, however, only $1.5N \ln_2 N + N$ complex multiplications and $3N \ln_2 N$ complex additions are necessary. For large N, this can amount to a considerable reduction in computer time. Figure 3-5 shows the number of real multiplications needed to evaluate the discrete convolution, using each of the three procedures outlined here. One complex multiplication is taken as equivalent to four real multiplications. It can be shown that the trapezoidal rule is the most economical method for $N < 200$ (approximately). For larger values of N, the FFT method is recommended.

Wraparound Error. A direct consequence of the definition of the DFT equation (3.10) is the N-term periodicity of the sequence $\{X_m\}$:

$$X_m = X_{m+iN} \qquad \text{for } i = \pm 1, \pm 2, \ldots \tag{3.61}$$

Similarly, from equation (3.11), it follows that the sequence $\{x_n\}$ has the N-term periodicity

$$x_n = x_{n+iN} \qquad \text{for } i = \pm 1, \pm 2, \ldots \tag{3.62}$$

Accordingly, whenever a particular problem allows variation of the indices of X_m or x_n beyond their fundamental period $(0, N-1)$, the periodicity of the sequences should be properly accounted for, and the indiscriminate use of DFT should be avoided under such circumstances. An example for such a situation is the evaluation of the discrete convolution equation (3.53) using DFT.

The direct evaluation of equation (3.53) using the trapezoidal rule does note cause any discrepancy, because the correct values as given by equation (3.52) are used in this case. When the DFT method is used, however, the N-term periodicity is assumed for the sequences $\{u_m\}$ and $\{h_k\}$. Since this is not true according to equation (3.52), the use of DFT can introduce a technical error into computation. It is shown that, unless $N \geq M + K - 1$, the first $M + K - 1 - N$ terms in the N-point sequence $\{y_n\}$ do not represent the correct discrete convolution results.

In the first relation of equation (3.53), as m varies from 0 to $N - 1$, the highest value of m for which $u_m \neq 0$ is $M - 1$ [see figure 3-6(a)]. The corresponding index of h is $n - M + 1$. Because of the N-term periodicity assumed in DFT, the terms in the sequence $\{h_k\}$ with indices ranging from $(-N)$ to $(-N+K-1)$ are also nonzero [see figure 3-6(b)], but if they are included in the discrete convolution, they lead to incorrect results, because, in the correct sequence (equation 3.52), these terms are zero. This is known as the wraparound error. It follows that, in order to avoid the discrepancy, one must require $n - M + 1 > -N + K - 1$. In other words,

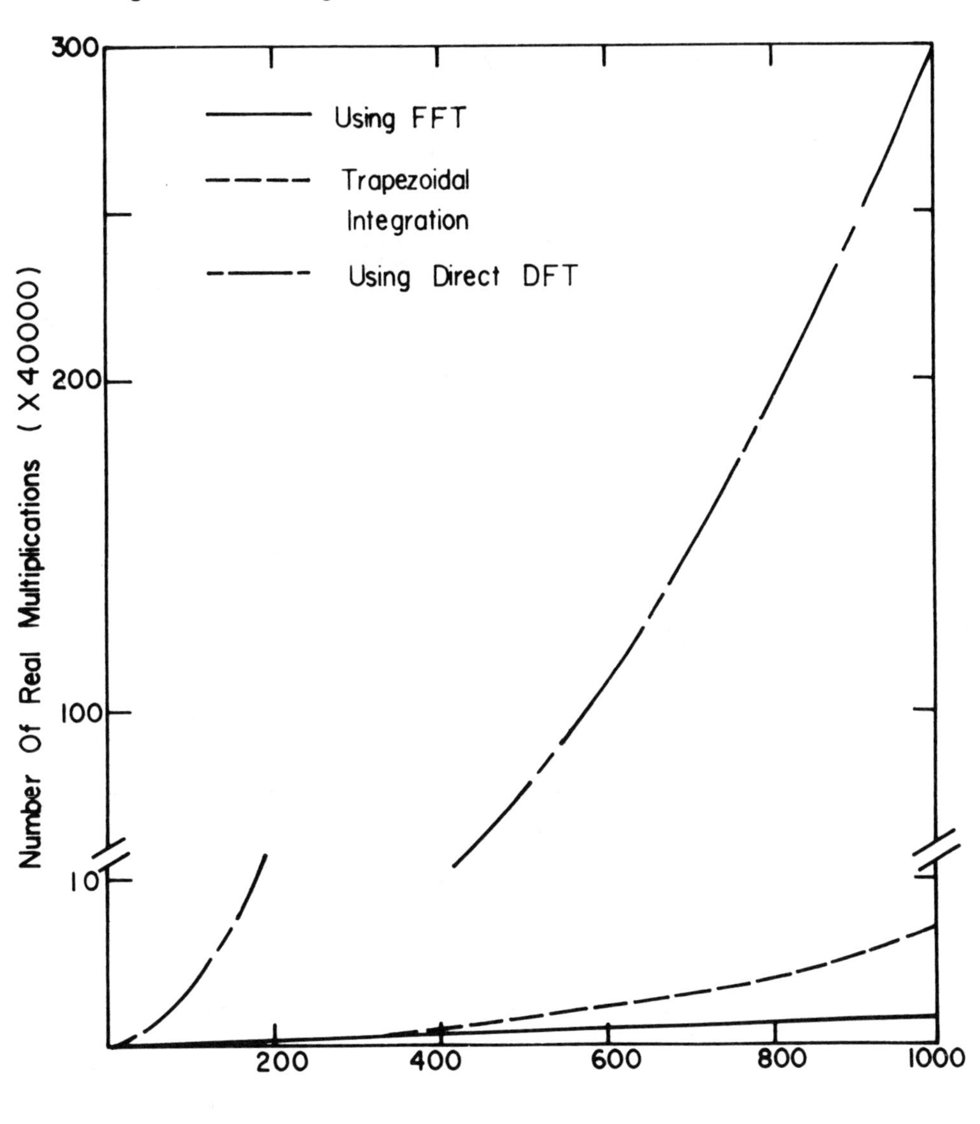

Figure 3-5. Efficiency of Several Methods of Computing Discrete Convolution

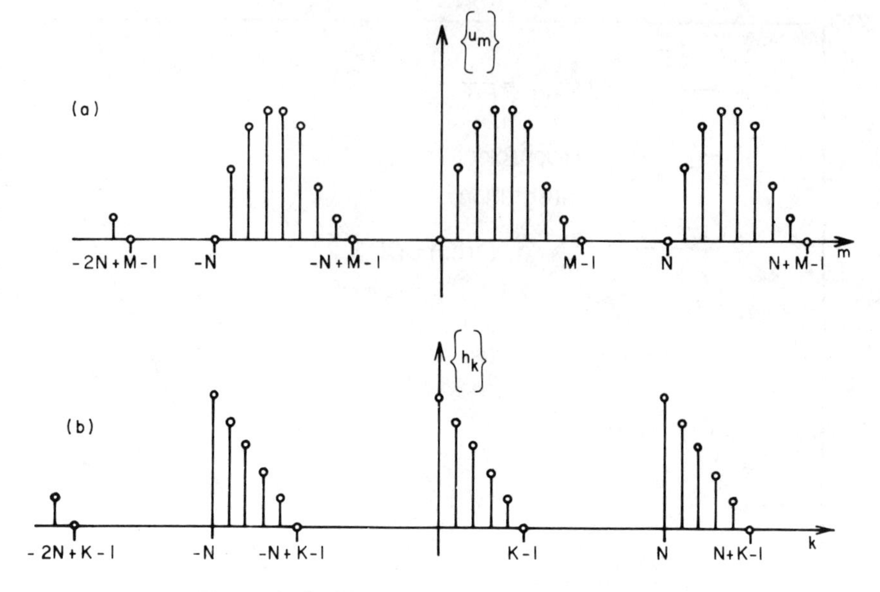

Figure 3-6. Illustration of Wraparound Error

$n > M + K - 2 - N$ is necessary to avoid the discrepancy. Since n ranges from 0 to N, the condition is satisfied if and only if $M + K - 2 - N \leq -1$. Consequently, it is required that $N \geq M + K - 1$ in order to avoid the wraparound error.

Data-Record Sectioning in Convolution. The result

$$\int_{-\infty}^{\infty} u(\tau + t_1)h(t - \tau + t_2)d\tau = \int_{-\infty}^{\infty} u(\tau')h(t + t_1 + t_2 - \tau')d\tau'$$

is obtained using the change of variable $\tau' = \tau + t_1$. In view of equation (3.48),

$$\int_{-\infty}^{\infty} u(\tau + t_1)h(t - \tau + t_2)d\tau = y(t + t_1 + t_2) \qquad (3.63)$$

From equation (3.63), it follows that, if the two convolving functions are shifted to the left through t_1 and t_2, their convolution shifts to the left through $t_1 + t_2$.

Suppose that the time history $u(t)$ is of short duration and that the nonnegligible portion of $h(t)$ represents a relatively long record. If proper sampling of $h(t)$ can exceed the memory of the digital computer used, the function $h(t)$ is sectioned into several portions of equal length T_2 (figure 3-7), and the convolution integral is computed for each section. Finally, the total convolution integral is obtained using these individual results. The philosophy behind this procedure is as follows:

$$h(t) = \sum_i h_i(t)$$

On substituting in equation (3.48),

$$y(t) = \sum_i y_i(t) \tag{3.64}$$

where

$$y_i(t) = \int_{-\infty}^{\infty} u(\tau) h_i(t - \tau) d\tau \tag{3.65}$$

However, $h_i(t) = 0$ over $0 \leq t < iT_2$. Because of these trailing zeros, the use of the DFT method becomes extremely inefficient for large i. To overcome this, each segment $h_i(t)$ is shifted to the left through iT_2, which results in a set

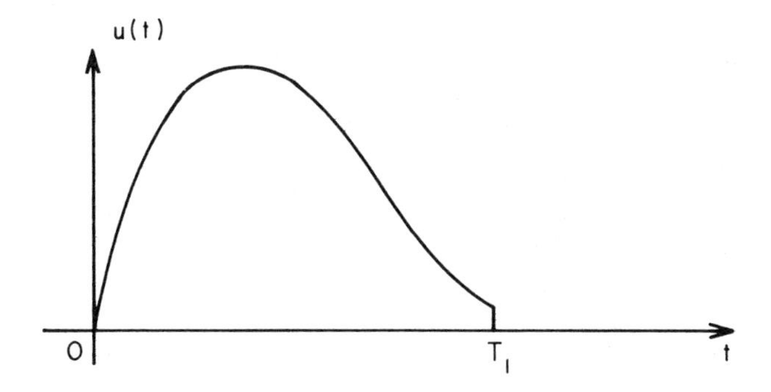

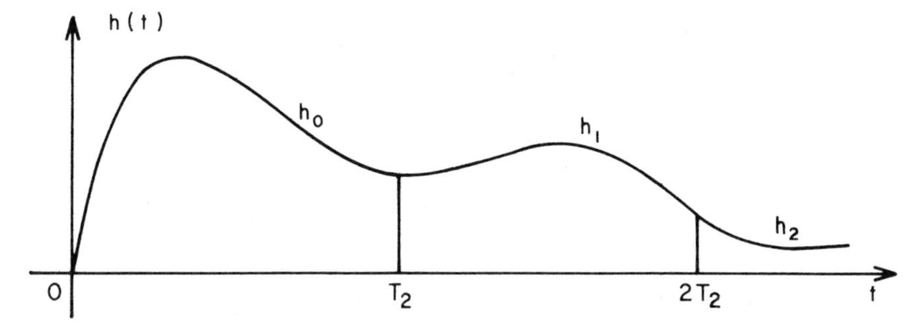

Figure 3-7. Sectioning of Long Data Records in Discrete Convolution

of modified functions $h_i(t + iT_2)$ that do not contain the trailing zeros. The corresponding convolutions,

$$y_i(t + iT_2) = \int_{-\infty}^{\infty} u(\tau)h_i(t - \tau + iT_2)d\tau \tag{3.66}$$

may be evaluated very efficiently using FFT in the usual manner. Subsequently, the functions $y_i(t + iT_2)$ are shifted to the right through iT_2 to obtain $y_i(t)$. Finally, $y(t)$ is constructed by superposition [equation (3.64)]. It should be noted that the DFT or FFT evaluation of equation (3.66) is performed as described earlier in this section. The major steps of the procedure are as follows:

1. Choose the sample step ΔT in the usual manner. Choose T_2 based on computer memory limitations. Section $h(t)$ at periods of T_2. Move each section to the origin and sample each section. An external storage device may be used to store the sectioned and sampled data sequences $\{h_k\}_i$.
2. Sample $u(t)$ at ΔT. This results in the sequence $\{u_m\}$.
3. Using $N = (T_1 + T_2)/\Delta T$ as the period, obtain the discrete convolution $\{y_n\}_i$, of each pair $\{u_m\}$ and $\{h_k\}$.
4. Shift each sequence $\{y_n\}_i$ to the right through $iK = iT_2/\Delta T$ elements and superpose (add the overlapping elements).

Leakage or Truncation Error. It is obvious that signals of infinite duration are not amenable to the digital computer. It is frequently necessary to truncate excessively long time histories, such as the accelerometer outputs from a qualification test, before digital processing. Computer memory limitations, speed and cost of processing, desired accuracy, and the nature of the signal (for example, whether it is fast-decaying, slow-decaying, periodic, or random) must be taken into consideration in selecting the truncation point. A signal $x(t)$ of long duration and its FIT $X(f)$ are shown in Figure 3-8(a). Given a sampled record of $x(t)$, it is desired to determine $X(f)$ with sufficient accuracy using digital Fourier analysis.

The signal is truncated at $t = T$. This is equivalent to multiplication of the signal by the unit box-car function $w(t)$, defined as

$$\begin{aligned} w(t) &= 1 \quad \text{for} \quad -0 \leq t < T \\ &= 0 \quad \text{otherwise} \end{aligned} \tag{3.67}$$

The FIT of $w(t)$ is [see figure 3-8(b)]

$$W(f) = \frac{1}{j2\pi f}[1 - \cos 2\pi fT + j \sin 2\pi fT] \tag{3.68}$$

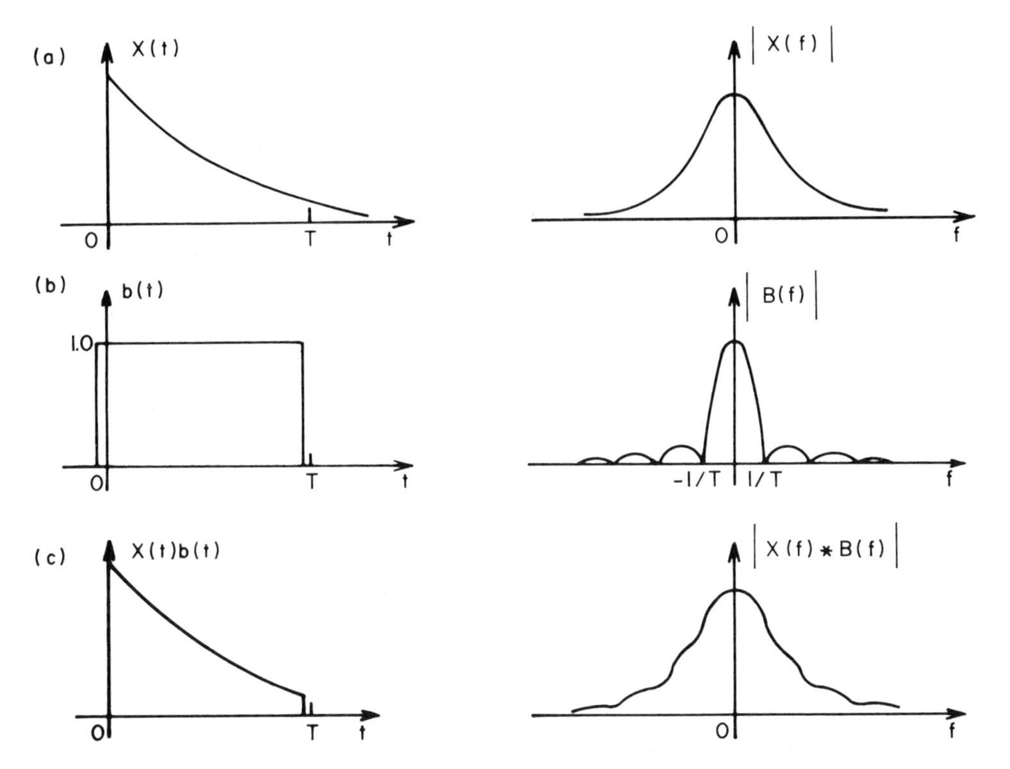

Figure 3-8. Illustration of Truncation Error (Box-Car Window Function)

From the discusion in the earlier section on the convolution theorem, it is clear that, for FIT, the product in one domain is equivalent to the convolution in the other domain. Accordingly, the FIT of the truncated signal $x(t)w(t)$ is the frequency convolution

$$X(f) * W(f) = \int_{-\infty}^{\infty} X(f')W(f-f')df' \qquad (3.69)$$

This procedure introduces ripples (or leakage components) into the FIT, as shown in figure 3-8(c). The resulting error $|X(f) - X(f) * W(f)|$ in the frequency domain is known as leakage or truncation error. Similar leakage effects arise in the time domain during inverse Fourier transformation, as a result of truncation in the frequency domain. It should be clear that, in addition to leakage error caused by truncation, the final results will also contain aliasing effects as a result of sampling of the truncated signal.

Truncation error may be reduced by suppressing the side lobes of $W(f)$.

In other words, the truncation function or window $w(t)$ must be modified from the box-car shape to a more desirable shape.

Window Functions. In spectral analysis of time histories, it is often required to segment the time history into several parts, then perform spectral analysis on the individual segments, and finally average out the individual results or compare individual results to observe the time development of the spectrum. If segmenting is done by simple truncation [multiplication by the box-car window, given by equation (3.67)], the process would introduce rapidly fluctuating side lobes into spectral results. Window functions, or smoothing functions other than the box-car function, are widely used to suppress the side lobes (leakage error). Some common smoothing functions are defined in table 3-4.

A graphical comparison of these four window types is given in figure 3-9. Hanning windows are very popular in practical applications. A related window is the Hamming window, which is simply a hanning window with rectangular cutoffs at the two ends. A hamming window will have characteristics similar to those of a hanning window, except that the side-lobe fall-off rate at higher frequencies is less in the hamming window.

From Figure 3-9(b), we observe that the frequency-domain weight of each window varies with the frequency range of interest. Obviously, the box-car window is the worst. In practical applications, the performance of any window could be improved by simply increasing the window length T.

Correlation Function and Power Spectral Density (psd)

Equation (3.41) suggests a method for evaluating the discrete correlation equation (3.40), through FFT. Also, from equation (3.38), it is observed that the cross psd $\Phi_{xy}(f)$ is given by the FIT of the mean lagged product $\phi_{xy}(\tau)$. It follows that a discrete approximation for cross psd is given by the sequence $\{[X_m]^* Y_m/T\}$. The results for the special case $x(t) \equiv y(t)$ follow. Computational time-saving that results by using FFT in the discrete mean lagged product evaluation is comparable to that for the discrete convolution.

As in the case of discrete convolution, it is necessary to avoid wraparound error when equation (3.40) is evaluated using DFT. It is necessary to assure that the index of y in equation (3.40) does not correspond to a nonzero term in a secondary period. Accordingly, $r + n \leq N-1$ and $r + n \geq -N + K$. However, $\max(r; x_r \neq 0) = M-1$ and $\min(r; x_r \neq 0) = 0$. In view of this, wraparound error is avoided when $M-1 + n \leq N-1$ and $n \geq -N + K$. This corresponds to $-(N-K) \leq n \leq N-M$. This indicates that the values z_n, as computed using DFT, are valid only in range

Table 3-4
Some Common Window Functions

Function Name	*Time-Domain Representation [w(t)]*	*Frequency-Domain Representation [W(f)]*
Box-Car	$= 1$ for $-0 \leq t < T$	$\frac{1}{j2\pi f}[1 - \cos 2\pi fT + j \sin 2\pi fT]$
	$= 0$ otherwise	
Hanning	$= \frac{1}{2} + \frac{1}{2}\cos\frac{\pi t}{T}$ for $\lvert t\rvert < T$	$\frac{T \sin 2\pi fT}{2\pi fT[1 - (2fT)^2]}$
	$= 0$ otherwise	
Parzen	$= 1 - 6\left[\frac{t}{T}\right]^2 + 6\left[\frac{\lvert t\rvert}{T}\right]^3$ for $\lvert t\rvert < \frac{T}{2}$	$\frac{3}{4}T\left[\frac{\sin 1/2\pi fT}{\frac{1}{2}\pi fT}\right]^4$
	$= 2\left[1 - \frac{\lvert t\rvert}{T}\right]^3$ for $\frac{T}{2} < \lvert t\rvert \leq T$	
	$= 0$ otherwise	
Bartlett	$= 1 - \frac{\lvert t\rvert}{T}$ *for* $\lvert t\rvert \leq T$	$T\left[\frac{\sin \pi fT}{\pi fT}\right]^4$
	$= 0$ otherwise	

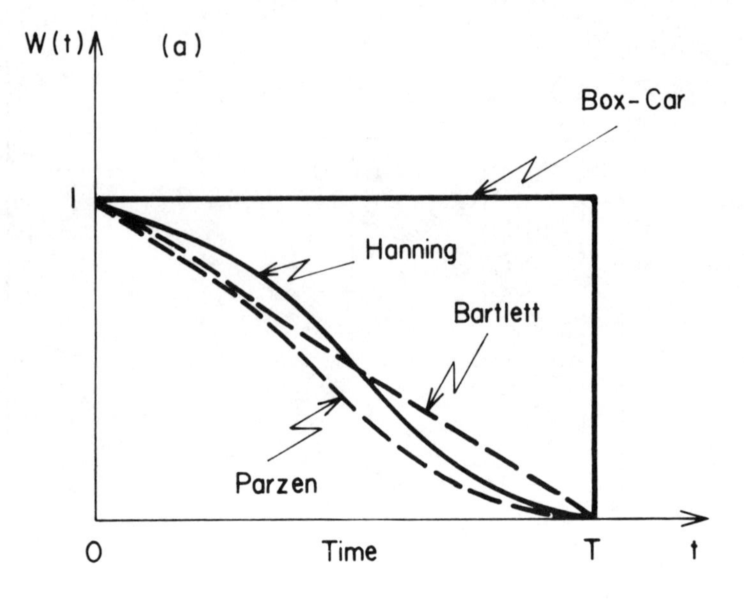

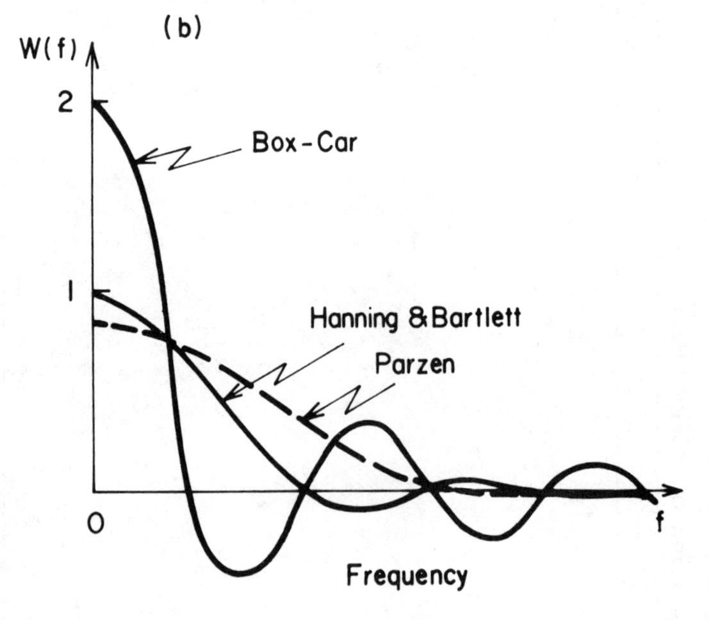

Figure 3-9. Some Common Window Functions: (a) Time-Domain Representation, (b) Frequency-Domain Representation

$n = -(N-K), \ldots, 0, \ldots, N-M$. If the DFT method results in the N-term sequence $\{\tilde{z}_n\}$ (with N-term periodicity as a rule), the correct sequence $\{z_n\}$ is formed according to

$$\begin{aligned} z_{-n} &= \tilde{z}_{N-n} \quad \text{for} \quad n = (N-K), \ldots, 1 \\ z_n &= \tilde{z}_n \quad \text{for} \quad n = 0, 1, \ldots, N-M \end{aligned} \tag{3.70}$$

The procedure gives $2N-M-K+1$ terms in $\{z_n\}$. If $N = M+K-1$ is chosen, as for the case of discrete correlation, a total of N terms in the sequence $\{z_n\}$ are obtained.

Fourier Spectrum Comparison

In dynamic testing, it is assumed that component degradation in the test object is caused by dynamic excitations. Accordingly, a certain dynamic excitation is associated with test-object malfunction.or failure. In this sense, continuous monitoring during testing of mechanical deterioration in various critical components of the test object is of prime importance. This usually cannot be done by simple visual observation, unless malfunction is registered by operability monitoring. Since mechanical degradation is always associated with a change in vibration level, however, by continuously monitoring the development of Fourier spectra with time (during testing) at various critical locations of the test object, it is possible to detect conveniently any mechanical deterioration and impending failure. In this respect, real-time Fourier analysis is very useful in dynamic testing. Special-purpose real-time analyzers with the capability of spectrum comparison (often done by an external command) are available for this purpose.

Various mechanical deteriorations manifest themselves at specific frequency values. A change in spectrum level at a particular frequency (and its multiples) would indicate a certain type of mechanical degradation or component failure. An example is given in figure 3-10, which compares the Fourier spectrum at a monitoring location of a test object at the start of test with the Fourier spectrum after some mechanical degradation has taken place because of test excitations. To facilitate spectrum comparison within a narrow-frequency band, it is customary to plot such Fourier spectra on a linear frequency axis. It is seen that the overall spectrum levels have increased as a result of mechanical degradation. Also, a significant change has occured near 30 Hz. This information is useful in diagnosing the cause of degradation or malfunction. Figure 3-10 might indicate, for example, impending failure of a component having resonance frequency close to 30 Hz.

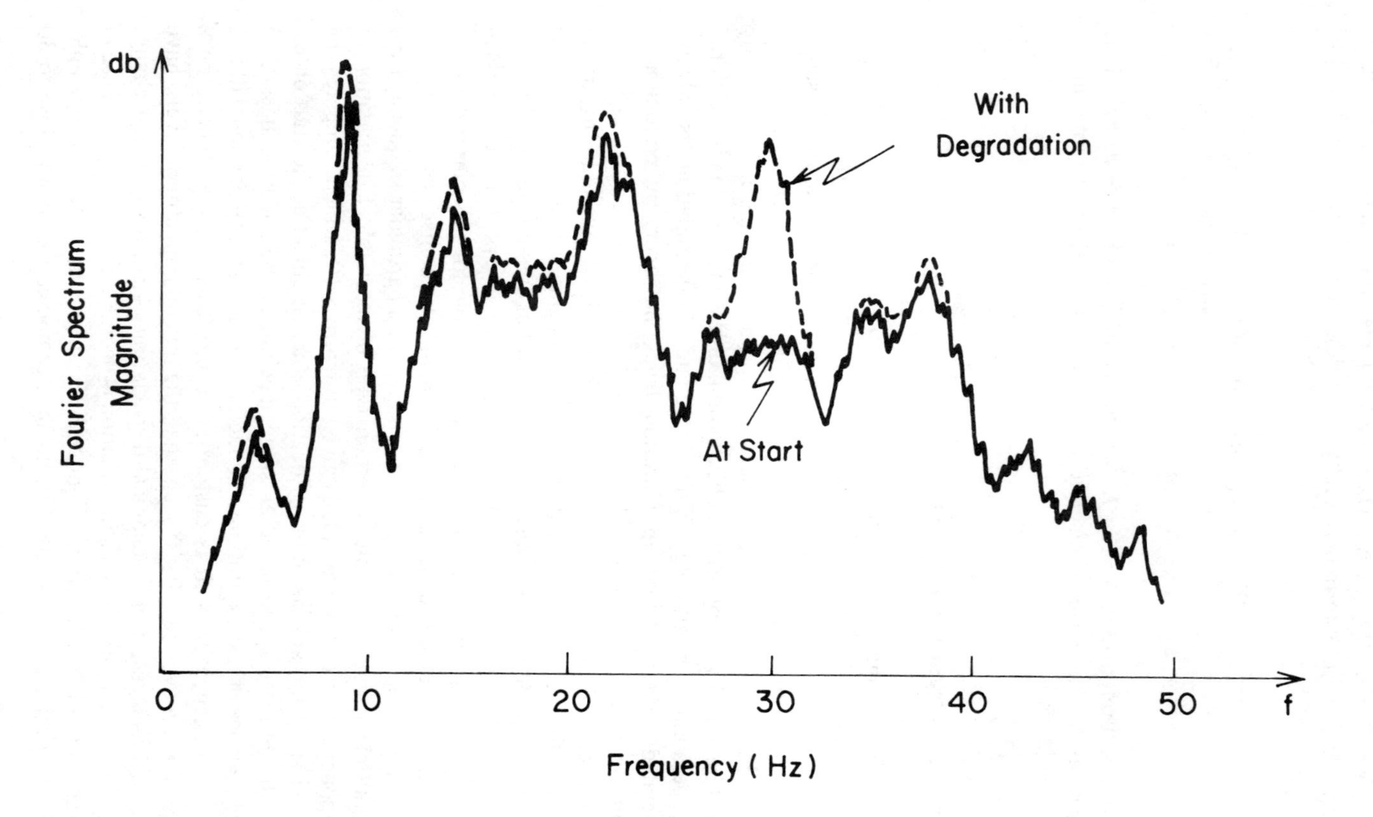

Figure 3-10. Effect of Mechanical Degradation on Fourier Spectrum

Cepstrum

A function known as cepstrum is sometimes used to facilitate the analysis of the Fourier spectrum in detecting mechanical degradation. Cepstrum (complex) $C(\tau)$ of Fourier spectrum $Y(f)$ is defined by[2]

$$C(\tau) = \mathscr{F}^{-1} \log Y(f) \quad (3.71)$$

The independent variable τ is known as quefrency, and it has the units of time.

An immediate advantage of cepstrum arises from the fact that the logarithm of the Fourier spectrum is taken. From equation (2.36), it is clear that, for a system having frequency-transfer function $H(f)$ excited by a signal having Fourier spectrum $U(f)$, the response Fourier spectrum $Y(f)$ could be expressed in the logarithmic form:

$$\log Y(f) = \log H(f) + \log U(f) \quad (3.72)$$

Since the right-hand-side terms are added rather than multiplied, any variation in $H(f)$ at a particular frequency will be less affected by a possible low spectrum level in $U(f)$ at that frequency, when considering $\log Y(f)$ instead of $Y(f)$. Consequently, any degradation will be more conspicuous in the cepstrum than in the Fourier spectrum. Another advantage of cepstrum is that it is more capable of detecting variations in phenomena that manifest themselves as periodic components in the Fourier spectrum (for example, harmonics and sidebands). Such phenomena which appear as repeated peaks in the Fourier spectrum occur as a single peak in the cepstrum, and so any variations could be detected more easily.

Notes

1. J.W. Cooley and J.W. Tukey., "An Algorithm for the Machine Computation of Complex Fourier Series," *Mathematics of Computation* 19(April 1965): 297–301.

2. R.B. Randall, *Application of B&K Equipment to Frequency Analysis*, 2nd ed. (Naerum, Denmark: Brüel and Kjaer, 1977).

4 Damping

Damping is the phenomenon by which mechanical energy is dissipated (usually converted into internal thermal energy) in dynamic systems. A knowledge of the level of damping in a dynamic system is important in dynamic analysis and testing of the system. A device having natural frequencies within the seismic range (that is, less than 33 Hz) and having relatively low damping, for example, could produce damaging motions under resonance conditions when subjected to a seismic disturbance. Also, the device motions could be further magnified by low-frequency support structures and panels having low damping. This indicates that a knowledge of damping in constituent devices, components, and support structure is particularly useful in seismic qualification of a dynamic system by analysis. The nature and the level of component damping should be known in order to develop a dynamic model (see chapter 2) of the system and its peripherals. A knowledge of damping in a system is also important in imposing dynamic environmental limitations on the system (that is, the maximum dynamic excitation the system could withstand) under in-service conditions. Furthermore, in order to make design modifications in a system that has failed the acceptance test, a knowledge of its damping could be useful. The significance of a knowledge of damping level in a test object, for the development of test excitation (input), often is overemphasized, however. Specifically, if the response-spectrum method (chapter 5) is used to represent the required excitation in a dynamic test, it is not necessary that the damping value used in the development of the required response spectrum be equal to actual damping in the test object. It is only necessary that the damping used in the required response spectrum be equal to that used in the test-response spectrum. The degree of dynamic interaction between test object and shaker table, however, will depend on damping in these systems. Furthermore, when testing near test-object resonance frequency, it is desirable to have a knowledge of damping in the test object.

In characterizing damping in a dynamic system, it is important, first, to understand the major mechanisms associated with mechanical-energy dissipation in the system. Then a suitable damping model should be chosen. Finally, damping values (model parameters) are determined, for example, by testing the system or a representative physical model, by monitoring system response under transient conditions during normal operation, or by employing already available data.

Types of Damping

There is some form of mechanical-energy dissipation in any dynamic system. In the modeling of systems, damping can be neglected if the mechanical energy dissipated during the time duration of interest is small in comparison to the initial total mechanical energy of excitation in the system. Even for highly damped systems, it is useful to perform an analysis with the damping terms neglected, in order to study several crucial dynamic characteristics.

Three primary mechanisms of damping are important to us in the study of mechanical systems: material (internal) damping, structural damping, and fluid damping. Material damping results from mechanical-energy dissipation within the material. Structural damping is caused by mechanical-energy dissipation resulting from relative motions between members in a structure that has common points of contact or joints. Fluid damping arises from the mechanical-energy dissipation resulting from drag forces when a mechanical member moves in a fluid.

Material (Internal) Damping

Internal damping of materials originates from the energy dissipation associated with microstructural defects, such as grain boundaries and impurities; thermoelastic effects caused by local temperature gradients resulting from nonuniform stresses, as in vibrating beams; eddy-current effects in ferromagnetic materials; dislocation motion in metals; and chain motion in polymers. Several models have been employed to represent energy dissipation caused by the internal damping mechanism. This variability is primarily a result of the vast range of engineering materials; no single model can satisfactorily represent the internal damping characteristics of all materials. Nevertheless, two general types of internal damping can be identified: viscoelastic damping and hysteretic damping. The latter term is actually a misnomer, because all types of internal damping are associated with hysteresis-loop effects. The stress (σ) and strain (ε) relationship at a point in a vibrating continuum possesses a hysteresis loop, such as the one shown in figure 4-1. The area of the hysteresis loop gives the energy dissipation per unit volume of the material, per stress cycle. This is termed specific damping capacity, and is denoted by d. It is clear that d is given by the cyclic integral:

$$d = \oint \sigma \, d\varepsilon \tag{4.1}$$

For a linear viscoelastic material, the stress-strain relationship is given by a

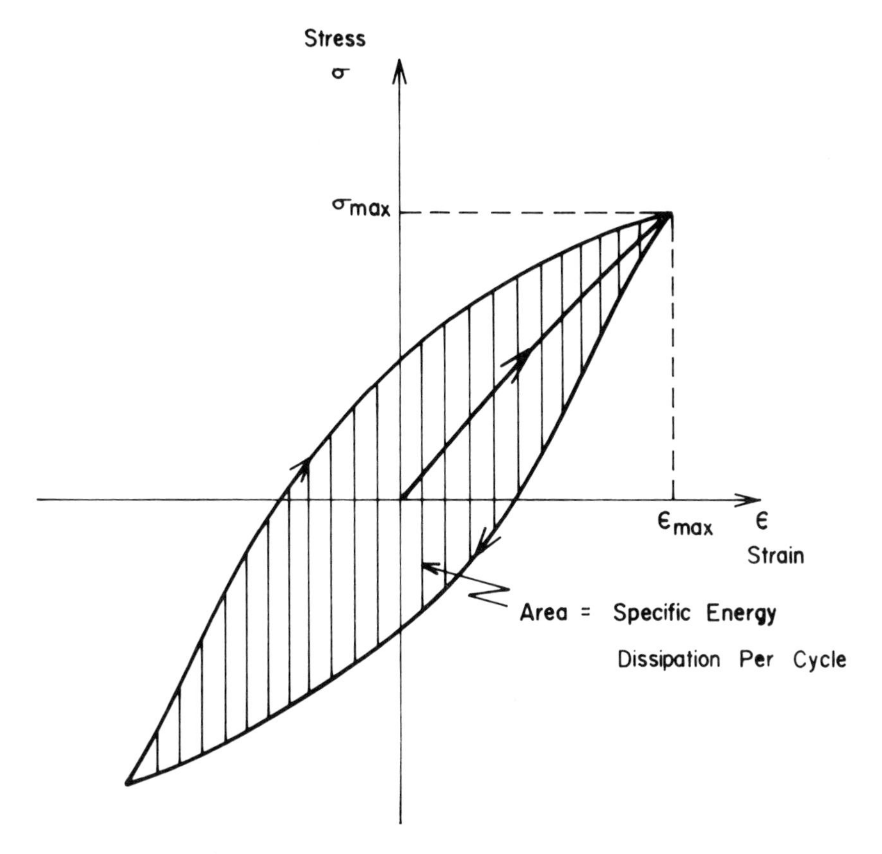

Figure 4-1. A Typical Hysteresis Loop for Mechanical Damping

linear differential equation with respect to time, having constant coefficients. A commonly employed relationship is

$$\sigma = E\varepsilon + E^* \frac{d\varepsilon}{dt} \tag{4.2}$$

which is known as the Kelvin-Voigt model. In equation (4.2), E is Young's modulus and E^* is a viscoelastic parameter that is assumed to be time-independent. The elastic term $E\varepsilon$ does not contribute to damping, and, mathematically, its cyclic integral vanishes. Consequently, for the Kelvin-Voigt model, specific damping capacity is

$$d_v = E^* \oint \frac{d\varepsilon}{dt}\, d\varepsilon \tag{4.3}$$

For a harmonically excited material at steady state:

$$\varepsilon = \varepsilon_{max} \cos \omega t \tag{4.4}$$

When equation (4.4) is substituted in equation (4.3), we obtain

$$d_v = \frac{\pi E^* \varepsilon_{max}^2}{\omega} \tag{4.5}$$

or

$$d_v = \frac{R}{\omega} \sigma_{max}^2 \tag{4.6}$$

These expressions for d_v depend on the frequency of excitation, ω.

Specific damping capacity d_h for hysteretic damping is independent of the frequency of excitation and can be approximated by

$$d_h = J\sigma_{max}^n \tag{4.7}$$

A simple model that satisfies equation (4.7) is given by[1]

$$\sigma = E\varepsilon + \frac{F}{\omega} \frac{d\varepsilon}{dt} \tag{4.8}$$

which is equivalent to using a viscoelastic parameter E^* that depends on the excitation frequency in equation (4.2). For this model, $n = 2$.

By combining equations (4.2) and (4.8), a simple model for combined viscoelastic and hysteretic damping is obtained:

$$\sigma = E\varepsilon + \left(E^* + \frac{F}{\omega} \right) \frac{d\varepsilon}{dt} \tag{4.9}$$

The equation of motion for a system whose damping is represented by equation (4.9) can be deduced from the pure elastic equation of motion by simply substituting E by the operator

$$E + \left(E^* + \frac{F}{\omega} \right) \frac{\partial}{\partial t}$$

in the time domain. The Bernoulli-Euler equation for transverse vibration of slender beams [equation (2.131)], for instance, becomes

$$\frac{\partial^2}{\partial x^2} EI \frac{\partial^2 v}{\partial x^2} + \frac{\partial^2}{\partial x^2}\left(E^* + \frac{F}{\omega}\right) I \frac{\partial^3 v}{\partial t \partial x^2} + \rho A \frac{\partial^2 v}{\partial t^2} = f(x, t) \tag{4.10}$$

in which ω is the frequency of the external excitation $f(x, t)$ in the case of steady forced vibration. In the case of free vibration, however, ω represents the frequency of free-vibration decay. Consequently, when analyzing the modal decay of free vibrations, ω in equation (4.10) should be replaced by the appropriate frequency (ω_i) of modal vibration in each modal equation. The resulting damped vibratory system possesses the same normal mode shapes as the undamped system. The analysis of the damped case is very similar to that given in chapter 2 for the undamped system.

Structural Damping

Structural damping is a result of the mechanical-energy dissipation caused by rubbing friction resulting from relative motion of the members and by impacting or intermittent contact at the joints in a structure. Energy-dissipation behavior depends on the details of the particular structure in this case. Consequently, it is extremely difficult to develop a generalized analytical model that would satisfactorily describe structural damping. Energy dissipation caused by rubbing is usually represented by a coulomb-friction model. Energy dissipation caused by impacting, however, should be determined from the coefficient of restitution of the two members that are in contact.

The common method of estimating structural damping is by measurement. The measured values, however, represent the overall damping in the structure. The structural damping component is obtained by subtracting the values corresponding to other types of damping, such as material damping present in the structure (estimated by environment-controlled experiments, previous data, and so forth), from the overall damping value. Experimental methods of determining damping in systems are discussed in the later section, "Measurement of Damping."

Usually, internal damping is negligible compared to structural damping. A large portion of mechanical-energy dissipation in tall buildings, bridges, vehicle guideways, and many other civil engineering structures takes place through the structural-damping mechanism. A major form of structural damping is the slip damping that results from the energy dissipation by interface shear at a structural joint.[2] The degree of slip damping that is directly caused by coulomb (dry) friction depends on such factors as joint forces (for example, bolt tensions), surface properties, and the nature of the materials of the mating surfaces. This is associated with wear, corrosion, and general deterioration of the structural joint. In this sense, slip damping is time-dependent. Damping layers are placed at joints to reduce undesirable deterioration of the joints. Sliding will cause shear distortions in the damping layers, causing energy dissipation by internal damping and thereby considerably reducing energy dissipation by coulomb friction. In this way, a high level of structural damping can be maintained without causing excessive joint deterioration. These damping layers should have a high stiffness (as well as a high specific-damping capacity) in order to take the structural loads at the joint.

For structural damping at a joint, force varies as slip occurs at the joint. This is primarily caused by local deformations at the joint, which occur with slipping. A typical hysteresis loop for this case is shown in figure 4-2(a). An approximate damping model corresponding to this is

$$F = c_1 v + c_2 \operatorname{sgn}(\dot{v}) \tag{4.11}$$

in which F is the damping force, v is the relative displacement at the joint, and c_1 and c_2 are constant parameters. For idealized coulomb friction, the frictional force (F) remains constant in each direction of relative motion. An idealized hysteresis loop for structural coulomb damping is shown in figure 4-2(b). The corresponding constitutive relation is

$$F = c \operatorname{sgn}(\dot{v}) \tag{4.12}$$

A simplified model for structural damping caused by local deformation may be given by

$$F = c|v| \operatorname{sgn}(\dot{v}) \tag{4.13}$$

The corresponding hysteresis loop is shown in figure 4-2(c). Note that the signum function is defined by

$$\begin{aligned} \operatorname{sgn}(x) &= 1 \quad \text{for} \quad x \geq 0 \\ &= -1 \quad \text{for} \quad x < 0 \end{aligned} \tag{4.14}$$

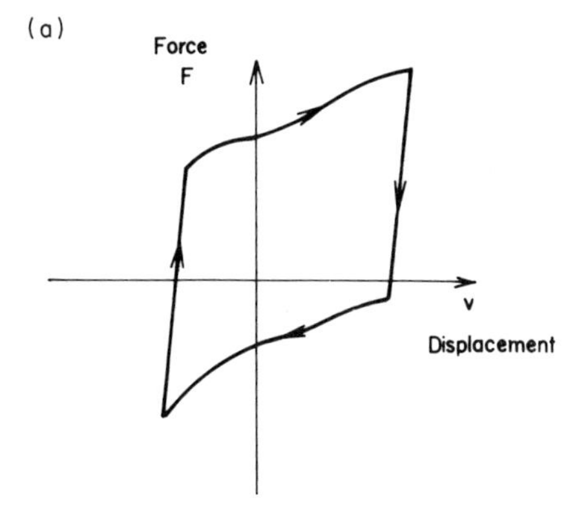

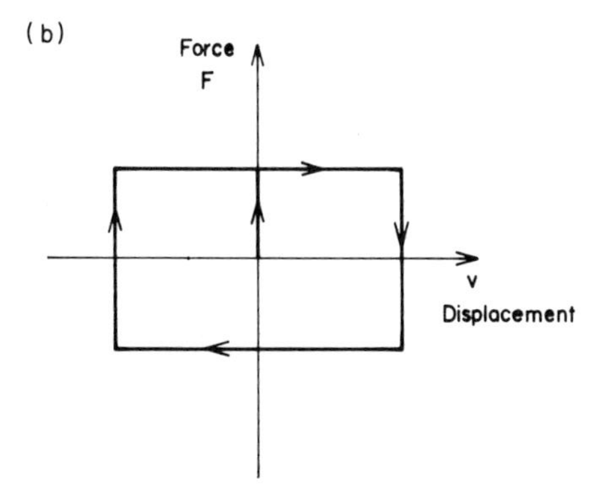

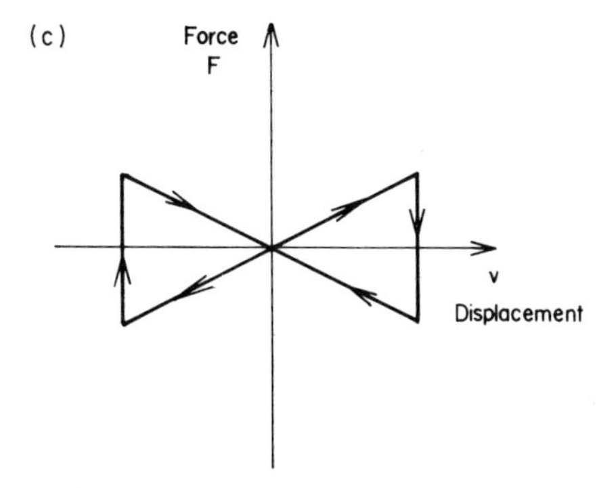

Figure 4-2. Some Representative Hysteresis Loops: (a) Typical Structural Damping, (b) Coulomb Friction Damping, (c) Simplified Structural Damping

Fluid Damping

Consider a structural element moving in a fluid space. The direction of relative motion is shown parallel to the *y*-axis in figure 4-3. Local displacement of the element relative to the surrounding fluid is denoted by $v(x, y, t)$. The resulting drag force per unit area of projection on the *x*-*z* plane is denoted by f_d. This resistance force is the cause of mechanical-energy dissipation in fluid damping. It is usually expressed as

$$f_d = \frac{1}{2} C_d \rho \dot{v}^2 \operatorname{sgn}(\dot{v}) \tag{4.15}$$

in which $\dot{v} = \partial v(x, z, t)/\partial t$ is the relative velocity. The drag coefficient C_d is a function of the Reynolds number and the geometry of the structural cross section. Net damping effect is generated by viscous drag produced by the boundary-layer effects at the fluid-structure interface, and by pressure drag produced by the turbulent effects resulting from flow separation at the wake. The two effects are illustrated in figure 4-4. Fluid density is ρ.

For fluid damping, specific-damping capacity associated with the configuration shown in figure 4-3 is given by

$$d_f = \frac{\oint \int_0^{L_x} \int_0^{L_z} f_d dz\, dx\, dv(x, z, t)}{L_x L_z v_0} \tag{4.16}$$

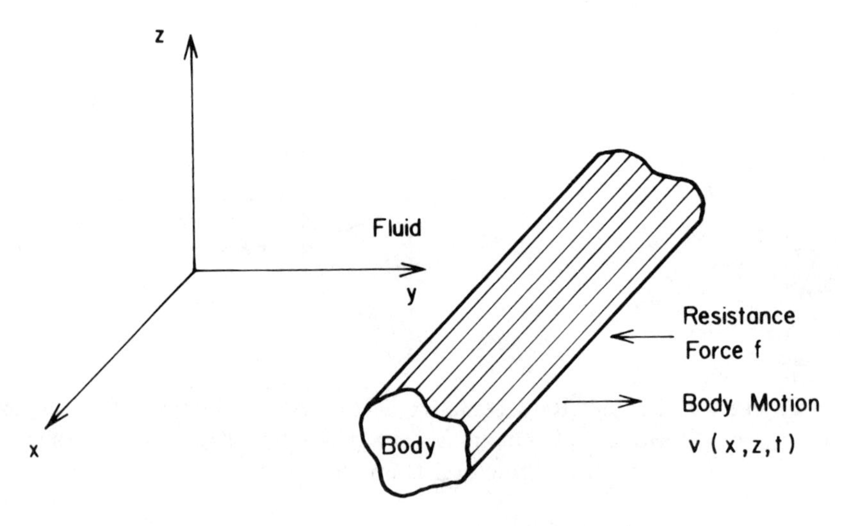

Figure 4-3. Fluid Damping System Nomenclature

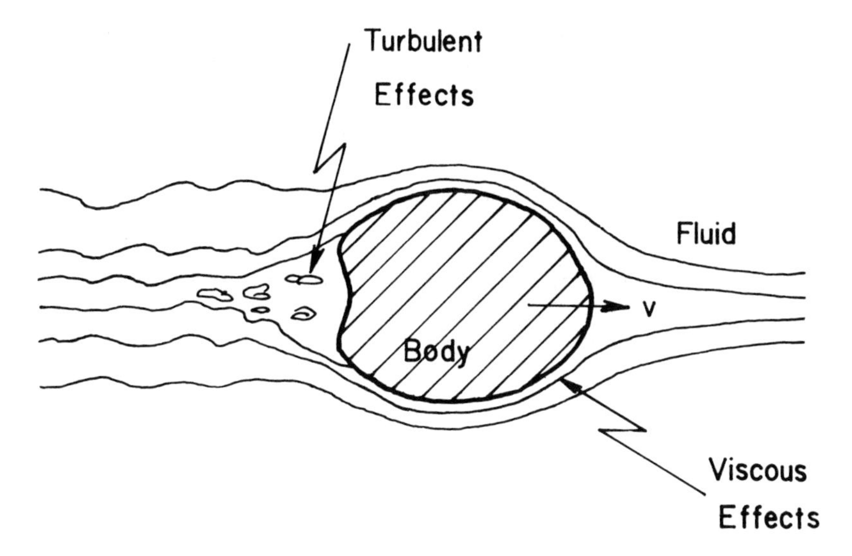

Figure 4-4. Illustration of Fluid Damping Mechanics

in which, L_x and L_z are cross-sectional dimensions of the element in x and y directions, respectively, and v_0 is a normalizing amplitude parameter for relative displacement.

As an example, consider a uniform beam of length L that is undergoing transverse vibration in a stationary fluid. For this case,

$$d_f = \frac{\oint \int_0^L f_d dx \, dv(x, t)}{L v_0}$$

or

$$d_f = \frac{\int_0^T \int_0^L f_d \dot{v}(x, t) dx \, dt}{L v_0} \tag{4.17}$$

in which T is the period of the oscillations. Assuming constant C_d, we substitute equation (4.15) into equation (4.17):

$$d_f = \frac{1}{2} \frac{C_d \rho}{L v_0} \int_0^L \int_0^T |\dot{v}|^3 \, dt \, dx \tag{4.18}$$

For steady-excited harmonic vibration at frequency ω (or for free-modal vibration at mode frequency ω) we have, from chapter 2,

$$v(x, t) = v_{max} V(x) \sin \omega t$$

In this case, equation (4.18) becomes

$$d_f = 2 C_d \rho \frac{v_{max}^3}{L v_0} \int_0^L |V(x)|^3 \, dx \, \omega^2 \int_0^{\pi/2} \cos^3 \theta \, d\theta$$

or

$$d_f = \frac{4}{3} C_d \rho \, v_{max}^3 \omega^2 \frac{\int_0^L |V(x)|^3 dx}{L v_0}$$

If the normalizing parameter is defined as

$$v_0 = \frac{1}{L} v_{max} \int_0^L |V(x)|^3 \, dx$$

then

$$d_f = \frac{4}{3} C_d \rho v_{max}^2 \omega^2 \tag{4.19}$$

Analysis of Lumped-Parameter Damped Systems

Consider a general n-degree-of-freedom mechanical system. Its motion can be represented by the vector $\boldsymbol{x}$ of n generalized coordinates x_i, representing the independent motions of the inertia elements. For small displacements, linear spring elements can be assumed. The corresponding equations of motion may be expressed in the vector-matrix form:

$$\boldsymbol{M}\ddot{\boldsymbol{x}} + \boldsymbol{d} + \boldsymbol{K}\boldsymbol{x} = \boldsymbol{f}(t) \tag{4.20}$$

in which $\boldsymbol{M}$ is the mass (inertia) matrix and $\boldsymbol{K}$ is the stiffness matrix. Both these matrices are generally symmetrical; that is,

$$m_{ij} = m_{ji}, \quad k_{ij} = k_{ji} \qquad i, j = 1, 2, \ldots, n \tag{4.21}$$

The forcing-function vector is $\boldsymbol{f}(t)$. The damping-force vector $\boldsymbol{d}(\boldsymbol{x}, \dot{\boldsymbol{x}})$ is generally a nonlinear function of $\boldsymbol{x}$ and $\dot{\boldsymbol{x}}$. The type of damping used in the system model is classified according to the nature of $\boldsymbol{d}$ that is employed in the system equations. Several commonly used damping models are listed in table 4-1. Only the linear viscous damping term given in table 4-1 is amenable to simplified mathematical analysis. In simplified dynamic models, other types of damping terms are usually replaced by an equivalent viscous damping term. Equivalent viscous damping is chosen so that its energy dissipation per cycle of oscillation is equal to that for the original damping. The resulting equations of motion are expressed by

$$\boldsymbol{M}\ddot{\boldsymbol{x}} + \boldsymbol{C}\dot{\boldsymbol{x}} + \boldsymbol{K}\boldsymbol{x} = \boldsymbol{f}(t) \tag{4.22}$$

This equation can be converted into the linear state-space form (chapter 2), and the analysis can be performed using state-space techniques. Classical methods of analysis, however, employ the second-order, coupled differential equation format given by equation (4.22).

Proportional Damping

Consider the homogeneous, undamped case of equation (4.22):

$$\boldsymbol{M}\ddot{\boldsymbol{x}} + \boldsymbol{K}\boldsymbol{x} = 0 \tag{4.23}$$

Table 4-1
Some Common Damping Models Used in System Equations

Damping Type	*Simplified Model d_i*
Viscous	$\sum_j c_{ij}\dot{x}_j$
Hysteretic	$\sum_j \frac{1}{\omega} c_{ij}\dot{x}_j$
Structural	$\sum_j c_{ij} \lvert x_j \rvert \operatorname{sgn}(\dot{x}_j)$
Structural coulomb	$\sum_j c_{ij} \operatorname{sgn}(\dot{x}_j)$
Fluid	$\sum_j c_{ij} \lvert \dot{x}_j \rvert \dot{x}_j$

A free solution response) of the form

$$\boldsymbol{x} = \boldsymbol{X} \cos \omega t \tag{4.24}$$

is sought. By substituting equation (4.24) in equation (4.23), the eigenvalue problem,

$$[\boldsymbol{K} - \omega^2 \boldsymbol{M}]\boldsymbol{X} = 0 \tag{4.25}$$

results. For nontrivial $\boldsymbol{X}$,

$$\det[\boldsymbol{K} - \omega^2 \boldsymbol{M}] = 0 \tag{4.26}$$

Equation (4.26) is the characteristic equation of the system.

We shall assume that the n roots of equation (4.26), $\omega_1, \omega_2, \ldots, \omega_n$ are positive and distinct (that is, no repeated roots) for an n-degree-of-freedom system. These roots are the natural frequencies of the system. For each natural frequency ω_i, there is a corresponding characteristic vector $\boldsymbol{X}_i$. This gives the mode shape at that frequency and is obtained by solving

$$[\boldsymbol{K} - \omega_i^2 \boldsymbol{M}]\boldsymbol{X}_i = 0 \tag{4.27}$$

for nontrivial $\boldsymbol{X}_i$ up to one unknown parameter.

Orthogonal properties of mode shapes (generalized eigenvectors) are obtained as follows. First, equation (4.27) is premultiplied by $\boldsymbol{X}_j^T$:

$$\boldsymbol{X}_j^T \boldsymbol{K} \boldsymbol{X}_i = \omega_i^2 \boldsymbol{X}_j^T \boldsymbol{M} \boldsymbol{X}_i \tag{4.28}$$

Next, indices i and j are interchanged in equation (4.28), and the matrix transpose is obtained:

$$\boldsymbol{X}_j^T \boldsymbol{K}^T \boldsymbol{X}_i = \omega_j^2 \boldsymbol{X}_j^T \boldsymbol{M}^T \boldsymbol{X}_i \tag{4.29}$$

In view of equation (4.21),

$$\boldsymbol{M}^T = \boldsymbol{M} \quad \text{and} \quad \boldsymbol{K}^T = \boldsymbol{K} \tag{4.30}$$

which are substituted into equation (4.29), and the result is subtraced from equation (4.28). Then, noting that $\omega_i \neq \omega_j$ for $i \neq j$, the first orthogonality relation is obtained:

$$\boldsymbol{X}_j^T \boldsymbol{M} \boldsymbol{X}_i = 0 \quad \text{for} \quad i \neq j \tag{4.31}$$

The second orthogonality relation is obtained by substituting equation (4.31) in equation (4.28):

$$\boldsymbol{X}_j^T \boldsymbol{K} \boldsymbol{X}_i = 0 \quad \text{for} \quad i \neq j \tag{4.32}$$

Modal decomposition of the system equations is achieved by transforming it using the modal matrix $\boldsymbol{X}$, which is the matrix formed by mode-shape vectors:

$$\boldsymbol{X} = [\boldsymbol{X}_1, \boldsymbol{X}_2, \ldots, \boldsymbol{X}_n] \tag{4.33}$$

The coordinate transformation

$$\boldsymbol{x} = \boldsymbol{X}\boldsymbol{q} \tag{4.34}$$

is applied to equation (4.22), followed by a premultiplication by $\boldsymbol{X}^T$:

$$\boldsymbol{X}^T\boldsymbol{M}\boldsymbol{X}\ddot{\boldsymbol{q}} + \boldsymbol{X}^T\boldsymbol{C}\boldsymbol{X}\dot{\boldsymbol{q}} + \boldsymbol{X}^T\boldsymbol{K}\boldsymbol{X}\boldsymbol{q} = \boldsymbol{X}^T\boldsymbol{f}(t) \tag{4.35}$$

In view of equations (4.31) and (4.32), the matrices $\boldsymbol{X}^T\boldsymbol{M}\boldsymbol{X}$ and $\boldsymbol{X}^T\boldsymbol{K}\boldsymbol{X}$ are diagonal matrices. If $\boldsymbol{X}^T\boldsymbol{C}\boldsymbol{X}$ is also a diagonal matrix, the viscous damping used in the system equations is termed proportional damping. In this case, equation (4.35) will represent a set of uncoupled second-order differential equations of the form

$$\ddot{q}_i + 2\zeta_i\omega_i\dot{q}_i + \omega_i^2 q_i = \omega_i^2 u_i(t) \qquad i = 1, 2, \ldots, n \tag{4.36}$$

which can be solved for q_i independently and finally combined, using equation (4.34), to get **x**. This results in considerable analytical simplification. Unfortunately, when a system is modeled by an interconnected set of mass, spring, and viscous damping elements, there is no guarantee that the resulting damping matrix $\boldsymbol{C}$ is of proportional type. Instead, an overall $\boldsymbol{C}$ matrix has to be specified globally in order to guarantee that $\boldsymbol{X}^T\boldsymbol{C}\boldsymbol{X}$ is diagonal. This procedure is often adopted in modeling, mainly for analytical convenience. Care should be exercised, however, to use equivalent proportional damping, which has the same energy dissipation per cycle as the actual damping present in the system.

One way (but not the only way) to guarantee proportional damping is to pick a $\boldsymbol{C}$ matrix that satisfies

$$\boldsymbol{C} = c_m\boldsymbol{M} + c_k\boldsymbol{K} \tag{4.37}$$

The first term on the right-hand side of equation (4.37) is known as the inertial damping matrix. The corresponding damping force on each concentrated mass is proportional to its momentum. It represents the energy loss associated with change in momentum (for example, during an impact). The second term is known as the stiffness damping matrix. The corresponding damping force is proportional to the rate of change of the local deformation forces at joints near the concentrated mass elements. Consequently, it represents a simplified form of linear structural damping.

If damping is of the proportional type, it follows from equation (4.36) that the damped motion can be uncoupled into individual modes. This means that the damped system (as well as the undamped system) possesses classical normal modes if the damping is of the proportional type.

Example. The equations of motion for the system modeled as in figure 4-5 are

$$\begin{bmatrix} m & 0 \\ 0 & m \end{bmatrix} \begin{bmatrix} \ddot{x}_1 \\ \ddot{x}_2 \end{bmatrix} + \begin{bmatrix} c_1 + c_2 & -c_2 \\ -c_2 & c_2 \end{bmatrix} \begin{bmatrix} \dot{x}_1 \\ \dot{x}_2 \end{bmatrix} + \begin{bmatrix} 2k & -k \\ -k & 2k \end{bmatrix} \begin{bmatrix} x_1 \\ x_2 \end{bmatrix} = \begin{bmatrix} 0 \\ f(t) \end{bmatrix}$$

The undamped natural frequencies are given by the positive roots of equation (4.26):

$$\det \begin{bmatrix} 2k - \omega^2 m & -k \\ -k & 2k - \omega^2 m \end{bmatrix} = 0$$

which are

$$\omega_1 = \sqrt{k/m} \quad \text{and} \quad \omega_2 = \sqrt{3k/m}$$

The corresponding mode-shape vectors are given by the nontrivial solution of equation (4.27):

$$\begin{bmatrix} 2k - \omega_i^2 m & -k \\ -k & 2k - \omega_i^2 m \end{bmatrix} \begin{bmatrix} X_1 \\ X_2 \end{bmatrix}_i = \begin{bmatrix} 0 \\ 0 \end{bmatrix} \qquad i = 1, 2$$

They are

$$X_1 = \begin{bmatrix} 1 \\ 1 \end{bmatrix} \quad \text{and} \quad X_2 = \begin{bmatrix} 1 \\ -1 \end{bmatrix}$$

The modal matrix equation (4.33) is

$$X = \begin{bmatrix} 1 & 1 \\ 1 & -1 \end{bmatrix}$$

Consequently,

$$X^T M X = \begin{bmatrix} 2m & 0 \\ 0 & 2m \end{bmatrix}$$

$$X^T K X = \begin{bmatrix} 2k & 0 \\ 0 & 6k \end{bmatrix}$$

$$X^T C X = \begin{bmatrix} c_1 & c_1 \\ c_1 & c_1 + 4c_2 \end{bmatrix}$$

$$X^T f = \begin{bmatrix} f(t) \\ -f(t) \end{bmatrix}$$

It is noted that proportional damping is realized if and only if $c_1 = 0$. In this case, the decoupled modal equations are

$$2m\ddot{q}_i + 2kq_i = f(t)$$
$$2m\ddot{q}_i + 4c_i\dot{q}_i + 6kq_i = -f(t)$$

It follows that the first mode is always undamped in this case of proportional damping.

Equivalent Viscous Damping

Consider a linear, single-degree-of-freedom system with viscous damping, subjected to an external excitation. The equation of motion is given by

$$\ddot{x} + 2\zeta\omega_n\dot{x} + \omega_n^2 x = \omega_n^2 u(t) \tag{4.38}$$

If the excitation is harmonic, with frequency ω:

$$u(t) = u_0 \cos \omega t \tag{4.39}$$

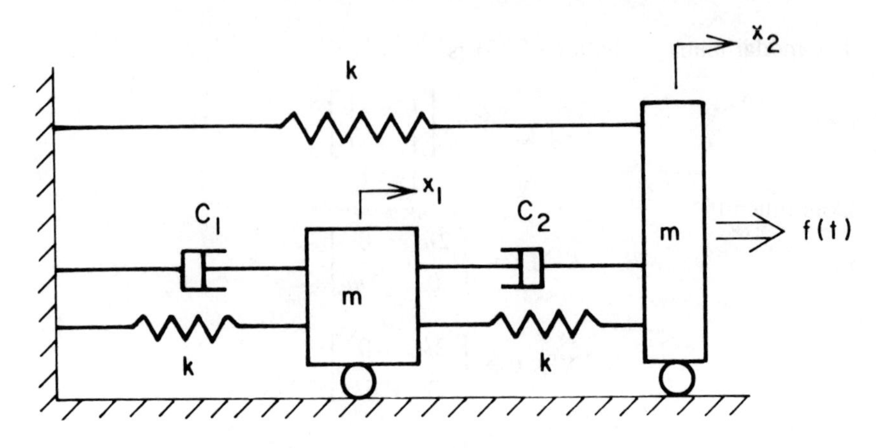

Figure 4-5. A Damped System Model.

Then, the response at steady state is given by

$$x = x_0 \cos(\omega t + \phi) \tag{4.40}$$

in which

$$x_0 = u_0 \frac{\omega_n^2}{[(\omega_n^2 - \omega^2)^2 + 4\zeta^2 \omega_n^2 \omega^2]^{1/2}} \tag{4.41}$$

and

$$\phi = -\tan^{-1} \frac{2\zeta\omega_n\omega}{(\omega_n^2 - \omega^2)} \tag{4.42}$$

The energy dissipation D in one cycle is given by the net work done by the damping force F_d:

$$D = \oint F_d \, dx = \int_{-\phi/\omega}^{(2\pi-\phi)\omega} F_d \dot{x} \, dt \tag{4.43}$$

Since the viscous damping force normalized with respect to mass is given by

$$F_d = 2\zeta\omega_n \dot{x} \tag{4.44}$$

the energy dissipation D_v in one cycle for viscous damping can be obtained from

$$D_v = 2\zeta\omega_n \int_0^{2\pi/\omega} \dot{x}^2\, dt$$

Finally, by using equation (4.40),

$$D_v = 2\pi x_0^2 \omega_n \omega \zeta \tag{4.45}$$

For any other type of damping (see table 4-1), the equation of motion becomes

$$\ddot{x} + d(x, \dot{x}) + \omega_n^2 x = \omega_n^2 u(t) \tag{4.46}$$

The energy dissipation in one cycle [equation (4.43)] is given by

$$D = \int_{-\phi/\omega}^{(2\pi-\phi)/w} d(x, \dot{x})\dot{x}\, dt \tag{4.47}$$

Various damping force expressions $d(x, \dot{x})$, normalized with respect to mass, are given in table 4-2. For fluid damping, for example,

$$D_f = \int_{-\phi/\omega}^{(2\pi-\phi)/\omega} c|\dot{x}|\dot{x}^2\, dt \tag{4.48}$$

Table 4-2
Equivalent Damping-Ratio Expressions for Some Common Types of Damping

Damping Type	*Damping Force* $d(x, \dot{x})$	*Equivalent Damping Ratio* ζ_{eq}
Viscous	$2\zeta\omega_n\dot{x}$	ζ
Hysteretic	$\dfrac{c}{\omega}\dot{x}$	$\dfrac{c}{2\omega_n\omega}$
Structural	$c\lvert x\rvert \operatorname{sgn}(\dot{x})$	$\dfrac{c}{\pi\omega_n\omega}$
Structural coulomb	$c \operatorname{sgn}(\dot{x})$	$\dfrac{2c}{\pi x_0\omega_n\omega}$
Fluid	$c\lvert\dot{x}\rvert\dot{x}$	$\dfrac{4}{3\pi}\left(\dfrac{\omega}{\omega_n}\right)x_0 c$

By substituting equation (4.40) in equation (4.48), for steady, harmonic motion, we obtain

$$D_f = \frac{8}{3} c x_0^3 \omega^2 \tag{4.49}$$

By comparing equation (4.49) with equation (4.45), the equivalent damping ratio for fluid damping is obtained as

$$\zeta_f = \frac{4}{3\pi} \left(\frac{\omega}{\omega_n} \right) x_0 c \tag{4.50}$$

in which x_0 is the amplitude of steady-state vibrations, as given by equation (4.41). For other types of damping listed in table 4-1, expressions for the equivalent damping ratio could be obtained in a similar manner. The corresponding equivalent damping-ratio expressions are given in table 4-2. It should be noted that, for nonviscous damping types, ζ is generally a function of the frequency of oscillation ω and the amplitude of excitation u_0. It should be noted that the expressions given in table 4-2 are derived assuming harmonic excitation. Engineering judgment should be exercised when empolying these expressions for nonharmonic excitations.

For multi-degree-of-freedom systems that incorporate proportional damping, the equations of motion can be transformed into a set of one-degree-of-freedom equations of the type given by equation (4.38), as is clear from equation (4.36). In this case, damping ratio and natural frequency correspond to the respective modal values.

Measurement of Damping

Damping may be represented by various parameters (such as specific damping capacity and damping ratio) and models (such as viscous, hysteretic, structural, and fluid). Before attempting to measure damping in a system, we should decide on a representation (model) that will adequately characterize the nature of mechanical-energy dissipation in the system. Next, we should decide on the parameter or parameters to be measured.

It is extremely difficult to develop a realistic model for damping in a complex piece of equipment operating under various mechanical interactions. Even if a damping modal is developed, experimental determination of its parameters could be tedious. A major difficulty arises because it usually is not possible to isolate various types of damping (for example, material, structural, and fluid) from an overall measurement. Damping measurements must be conducted under actual operating conditions for them to be realistic.

If one type of damping (such as fluid damping) is eliminated during the actual measurement, it would not represent the true operating conditions. This would also eliminate possible interacting effects of the eliminated damping type with the other types. Thus, overall damping in a system is not generally equal to the sum of individual damping values when they are acting independently. Another limitation is that, when computing equivalent damping values using experimental data, it is assumed for analytical simplicity that the dynamic system behavior is linear. If the system is highly nonlinear, a significant error could be introduced into the damping estimate. Nevertheless, it is customary to assume linear viscous behavior when estimating damping parameters using experimental data.

There are two general ways by which damping measurements can be made: time-response methods and frequency-response methods. The basic difference between the two types of measurements is that the first type uses a time-response record of the system to estimate damping, whereas the second type uses a frequency-response record.

Logarithmic Decrement Method

This is the most popular time-response method used to measure damping. When a single-degree-of-freedom oscillatory system with viscous damping [equation (4.38)] is excited by an impulse input (or initial condition excitation), its response takes the form of a time decay (see figure 4-6), given by

$$y(t) = y_0 \exp(-\zeta\omega_n t) \sin \omega_d t \tag{4.51}$$

in which the damped natural frequency

$$\omega_d = \sqrt{1 - \zeta^2}\,\omega_n \tag{4.52}$$

If the response at $t = t_i$ is denoted by y_i, and the response at $t = t_i + 2\pi r/\omega_d$ is denoted by y_{i+r}, then, from equation (4.51),

$$\frac{y_{i+r}}{y_i} = \exp\left(-\zeta \frac{\omega_n}{\omega_d} 2\pi r\right) \quad i = 1, 2, \ldots, n$$

In particular, if y_i corresponds to a peak point in the time decay, having magnitude A_i, then y_{i+r} corresponds to the peak-point r cycles later in the time history, and its magnitude is denoted by A_{i+r} (see figure 4-6). Then

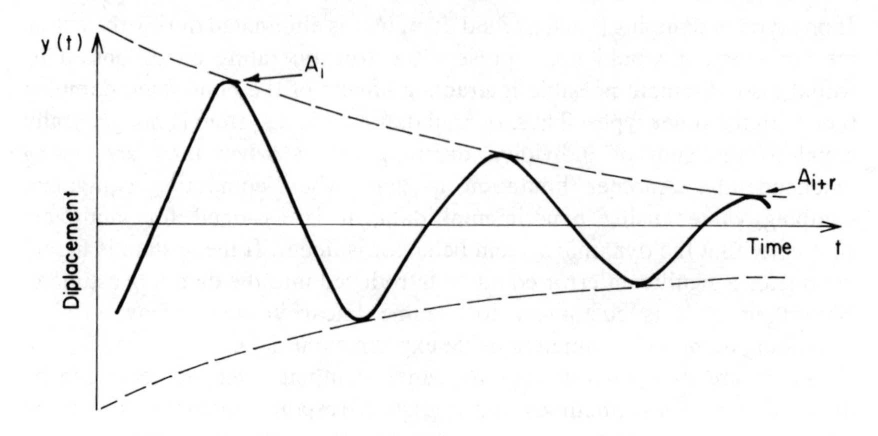

Figure 4-6. A Typical Time-Decay Record

$$\frac{A_{i+r}}{A_i} = \exp(-\zeta \frac{\omega_n}{\omega_d} 2\pi r) \tag{4.53}$$

By using equation (4.52),

$$\ln\left(\frac{A_i}{A_{i+r}}\right) = \frac{\zeta}{\sqrt{1-\zeta^2}} 2\pi r$$

or

$$\zeta = 1/\left[1 + 1/\left\{\frac{1}{2\pi r} \ln\left(\frac{A_i}{A_{i+r}}\right)\right\}^2\right]^{1/2} \tag{4.54}$$

For low damping (typically, $\zeta < 0.1$), $\omega_d \simeq \omega_n$, and

$$\frac{A_{i+r}}{A_i} \simeq \exp(-\zeta 2\pi r) \tag{4.55}$$

or

$$\zeta = \frac{1}{2\pi r} \ln\left(\frac{A_i}{A_{i+r}}\right) \quad \text{for} \quad \zeta < 0.1 \tag{4.56}$$

Damping ratio can be estimated from a free-decay record, using equation (4.56). Specifically, the ratio of the extreme amplitudes in prominent r cycles of decay is determined and substituted into equation (4.56) to get the equivalent damping ratio.

Alternatively, if n cycles of damped oscillation are needed for the amplitude to decay by a factor of two for example, then, from equation (4.56),

$$\zeta = \frac{1}{2\pi n}\ln(2) = \frac{0.11}{n} \quad \text{for} \quad \zeta < 0.1 \tag{4.57}$$

For slow decays (low damping),

$$\ln\left(\frac{A_i}{A_{i+1}}\right) \simeq \frac{2(A_i - A_{i+1})}{(A_i + A_{i+1})} \tag{4.58}$$

Then, from equation (4.56),

$$\zeta = \frac{A_i - A_{i+1}}{\pi(A_i + A_{i+1})} \quad \text{for} \quad \zeta < 0.1 \tag{4.59}$$

Any one of the equations (4.54), (4.56), (4.57), and (4.59) could be employed in computing ζ from test data. It should be cautioned that the results assume single-degree-of-freedom system behavior. For multi-degree-of-freedom systems, the modal damping ratio for each mode can be determined using this method if the initial excitation is such that the decay takes place primarily in one mode of vibration.

Step-Response Method

This is also a time-response method. If a unit-step excitation is applied to the single-degree-of-freedom oscillatory system given by equation (4.38), its time response is given by

$$y(t) = 1 - \frac{1}{\sqrt{1-\zeta^2}}\exp(-\zeta\omega_n t)\sin(\omega_d t + \phi) \tag{4.60}$$

in which $\phi = \cos^{-1}\zeta$. A typical step-response curve is shown in figure 4-7. The time at the first peak (peak time), T_p, is given by

$$T_p = \frac{\pi}{\omega_d} = \frac{\pi}{\sqrt{1-\zeta^2}\,\omega_n} \tag{4.61}$$

The response at peak time (peak value), M_p, is given by

$$M_p = 1 + \exp(-\zeta\omega_n T_p) = 1 + \exp(-\pi\zeta/\sqrt{1-\zeta^2}) \tag{4.62}$$

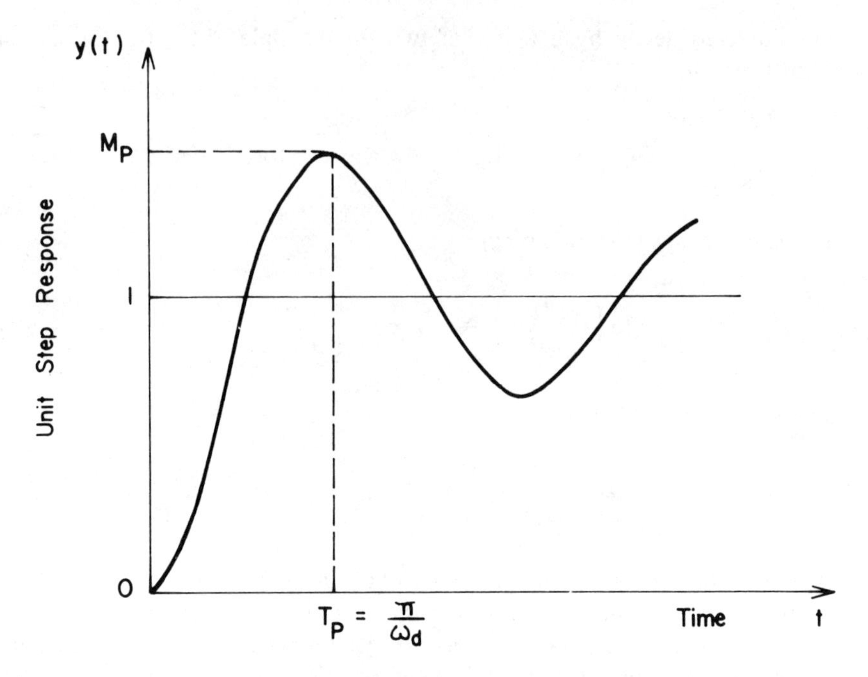

Figure 4-7. A Typical Step Response

The percentage overshoot, *PO*, is given by,

$$PO = (M_p - 1) \times 100\% = 100 \exp(-\pi\zeta/\sqrt{1 - \zeta^2}) \quad (4.63)$$

It follows that, if any one parameter of T_p, M_p, or *PO* is known from a step-response record, the corresponding damping ratio ζ can be computed by using the appropriate relationship from the following:

$$\zeta = \sqrt{1 - \left(\frac{\pi}{T_p \omega_n}\right)^2} \quad (4.64)$$

$$\zeta = 1 \bigg/ \sqrt{1 + 1/\left[\frac{\ln(M_p - 1)}{\pi}\right]^2} \quad (4.65)$$

$$\zeta = 1 \bigg/ \sqrt{1 + 1/\left[\frac{\ln(PO/100)}{\pi}\right]^2} \quad (4.66)$$

It should be noted that, when determining M_p, the response curve should be normalized to unit steady-state value. Furthermore, the results are valid only for single-degree-of-freedom systems and modal excitations in multi-degree-of-freedom systems.

Magnification-Factor Method

This is a frequency-response method. Consider the single-degree-of-freedom oscillatory system with viscous damping. The magnitude of its frequency-response function is

$$|H(\omega)| = \frac{\omega_n^2}{[(\omega_n^2 - \omega^2)^2 + 4\zeta^2\omega_n^2\omega^2]^{1/2}} \tag{4.67}$$

A plot of this expression with respect to ω, the frequency of excitation, is given in Figure 4-8. The peak value of the magnitude occurs when the denominator of the expression is minimum. This corresponds to

$$\frac{d}{d\omega}[(\omega_n^2 - \omega^2)^2 + 4\zeta^2\omega_n^2\omega^2] = 0 \tag{4.68}$$

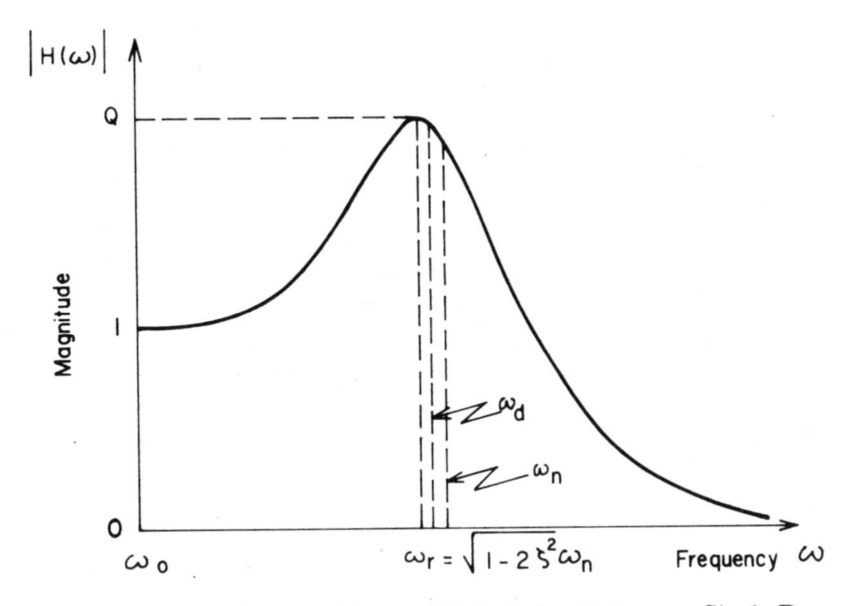

Figure 4-8. Magnification Factor Method Applied to a Single-Degree-of-Freedom System

The resulting solution for ω is termed the resonant frequency ω_r. From equation (4.61), it follows that

$$\omega_r = \sqrt{1 - 2\zeta^2}\,\omega_n \tag{4.69}$$

It is noted that $\omega_r < \omega_d$ [equation (4.52)], but, for low damping ($\zeta < 0.1$), the values of ω_n, ω_d, and ω_r are nearly equal. The amplication factor Q, which is the magnitude of the frequency-response function at resonance frequency, is obtained by substituting equation (4.69) in equation (4.67):

$$Q = \frac{1}{2\zeta\sqrt{1 - \zeta^2}} \tag{4.70}$$

For low damping ($\zeta < 0.1$),

$$Q = \frac{1}{2\zeta} \tag{4.71}$$

In fact, equation (4.71) corresponds to the magnitude of the frequency-response function at $\omega = \omega_n$.

It follows that, if the magnitude curve of the frequency-response function (or a Bode plot) is available, then the system damping ratio ζ can be estimated using equation (4.71). In using this method, it should be remembered to normalize the frequency-response curve so that its magnitude at zero frequency (termed *gain*) is unity.

For a multi-degree-of-freedom system, modal damping values may be estimated from the magnitude Bode plot of its frequency-response function, provided that the modal frequencies are not too closely spaced and the system is lightly damped. Consider the logarithmic (to the base 10) magnitude plot shown in figure 4-9. The magnitude is expressed in decibels (db), which is done by multiplying the $\log_{10}$ (magnitude) by the factor 20. At the ith resonant frequency ω_i, the amplication factor q_i (in db) is obtained by drawing an asymptote to the preceding segment of the curve and measuring the peak value from the asymptote. Then,

$$Q_i = (10)^{q_i/20} \tag{4.72}$$

and the modal damping ratio

$$\zeta_i = \frac{1}{2Q_i} \qquad i = 1, 2, \ldots, n \tag{4.73}$$

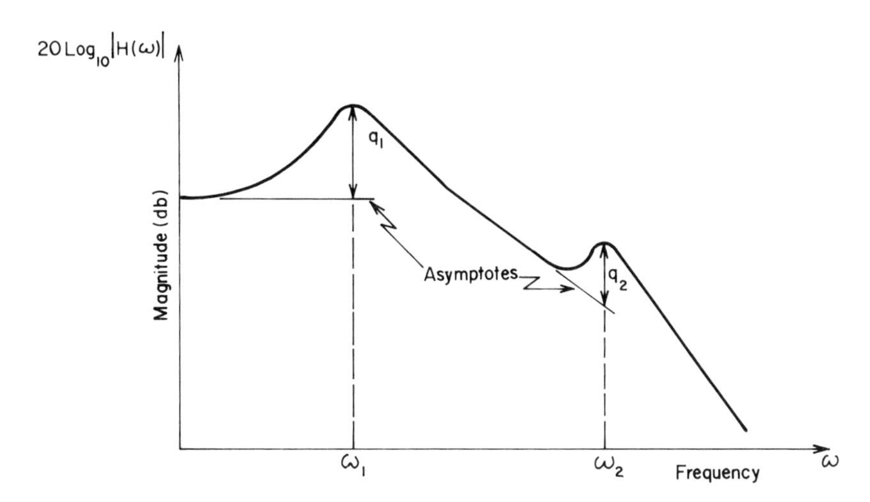

Figure 4-9. Magnification Factor Method Applied to a Multi-Degree-of-Freedom System.

If the significant resonances are closely spaced, curve-fitting to a suitable function might be necessary in order to determine the corresponding modal damping values.

Bandwidth Method

The bandwidth method of damping measurement is also based on frequency response. Consider the frequency-response-function magnitude given by equation (4.67) for a single-degree-of-freedom, oscillatory system with viscous damping. The peak magnitude is given by equation (4.71) for low damping. Bandwidth (half-power) is defined as the width of the frequency-response magnitude curve when the magnitude is $1/\sqrt{2}$ times the peak value. This is denoted by $\Delta\omega$ (see figure 4-10). An expression for $\Delta\omega = \omega_2 - \omega_1$ is obtained, using equation (4.67). By definition, ω_1 and ω_2 are the roots of the equation:

$$\frac{\omega_n^2}{[(\omega_n^2 - \omega^2)^2 + 4\zeta^2\omega_n^2\omega^2]^{1/2}} = \frac{1}{\sqrt{2}\,2\zeta} \tag{4.74}$$

for ω. Equation (4.74) can be expressed in the form

$$\omega^4 - 2(1 - 2\zeta^2)\omega_n^2\omega^2 + (1 - 8\zeta^2)\omega_n^4 = 0 \tag{4.75}$$

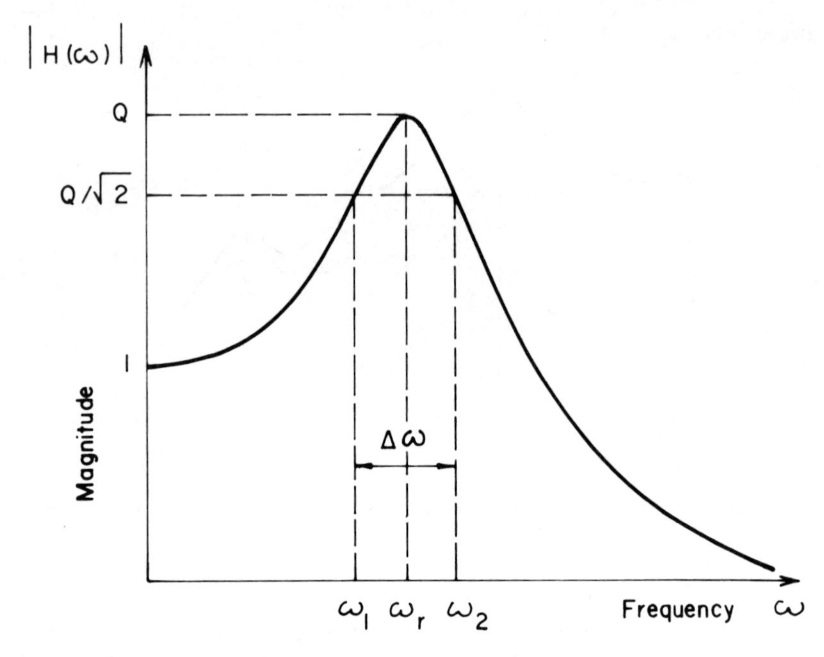

Figure 4-10. Bandwidth Method Applied to a Single-Degree-of-Freedom System

This is a quadratic equation in ω^2, having roots ω_1^2 and ω_2^2, which satisfy

$$(\omega^2 - \omega_1^2)(\omega^2 - \omega_2^2) = \omega^4 - (\omega_1^2 + \omega_2^2)\omega^2 + \omega_1^2\omega_2^2 = 0$$

Consequently,

$$\omega_1^2 + \omega_2^2 = 2(1 - 2\zeta^2)\omega_n^2 \tag{4.76}$$

and

$$\omega_1^2\omega_2^2 = (1 - 8\zeta^2)\omega_n^4 \tag{4.77}$$

It follows that

$$\begin{aligned}(\omega_2 - \omega_1)^2 &= \omega_1^2 + \omega_2^2 - 2\omega_1\omega_2 \\ &= 2(1 - 2\zeta^2)\omega_n^2 - 2\sqrt{1 - 8\zeta^2}\,\omega_n^2\end{aligned}$$

For small ζ (in comparison to 1),

$$\sqrt{1 - 8\zeta^2} \simeq 1 - 4\zeta^2$$

Hence,

$$(\omega_2 - \omega_1)^2 \simeq 4\zeta^2\omega_n^2$$

or, for low damping,

$$\Delta\omega = 2\zeta\omega_n = 2\zeta\omega_r \tag{4.78}$$

From equation (4.78) it follows that the damping ratio can be estimated from bandwidth using the relation

$$\zeta = \frac{1}{2}\frac{\Delta\omega}{\omega_r} \tag{4.79}$$

For a multi-degree-of-freedom system having widely spaced resonances, the foregoing method can be extended to estimate modal damping. Consider the frequency-response magnitude plot (in db) shown in figure 4-11. Since a factor of $\sqrt{2}$ corresponds to 3 db, the bandwidth corresponding to a resonance is given by the width of the magnitude plot at 3 db below that resonance peak. For the *i*th mode, the damping ratio is given by

$$\zeta_i = \frac{1}{2}\frac{\Delta\omega_i}{\omega_i} \tag{4.80}$$

The bandwidth method of damping measurement indicates that the bandwidth at a resonance is a measure of the energy dissipation in the system in the neighborhood of that resonance. The simplified relationship given by equation (4.80) is valid for low damping, however, and is based on linear system analysis.

General Remarks

There are limitations to the use of damping values that are experimentally determined. In time-response methods of determining modal damping of a device for higher modes, for instance, the customary procedure is first to

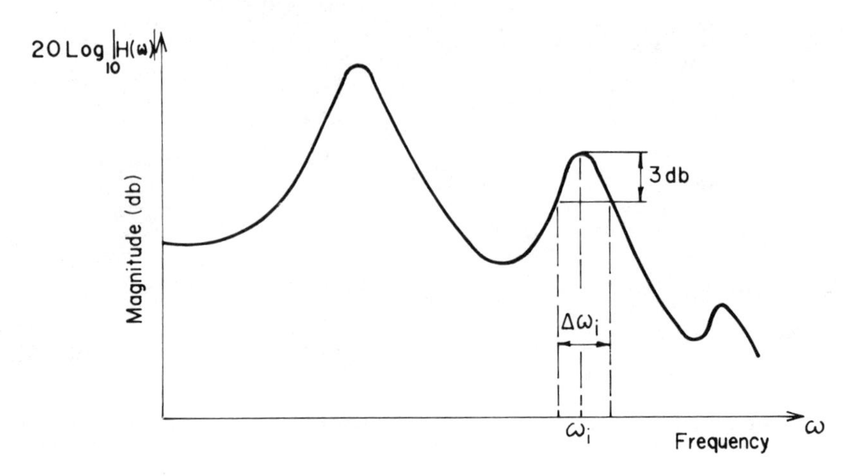

Figure 4-11. Bandwidth Method Applied to a Multi-Degree-of-Freedom System

excite it at the desired resonant frequency, using a harmonic exciter, and then to release the excitation mechanism. In the resulting transient vibration, however, there invariably will be modal interactions, except in the case of proportional damping. In this type of tests, it is tacitly assumed that the device can be excited in that mode. In essence, proportional damping is assumed in modal damping measurements. This introduces a certain amount of error into the measured damping values.

Expressions used in computing damping parameters from test measurements are usually based on linear system theory. All practical devices exhibit some nonlinear behavior, however. If the degree of nonlinearity is high, the measured damping values will not be representative of the actual system behavior. Furthermore, testing to determine damping is usually done at low amplitudes of vibration. The corresponding responses could be at least one order of magnitude lower than, for instance, the amplitudes exhibited under strong-motion earthquakes. Damping in practical devices increases with the amplitude of motion, except for relatively low amplitudes. This nonlinear behavior is illustrated in figure 4-12. Consequently, the damping values determined from experiments should be extrapolated when they are used to study the system behavior under operating-basis earthquake (OBE) or safe-shutdown earthquake (SSE) conditions, for instance. Alternatively, damping could be associated with a stress level under the American Society of Mechanical Engineers (ASME) category of upset for OBE and one under the ASME category of faulted condition for SSE. Different components in a device are subjected to varying levels of stress, however, and it might be difficult to obtain a representative stress value for the device. A recom-

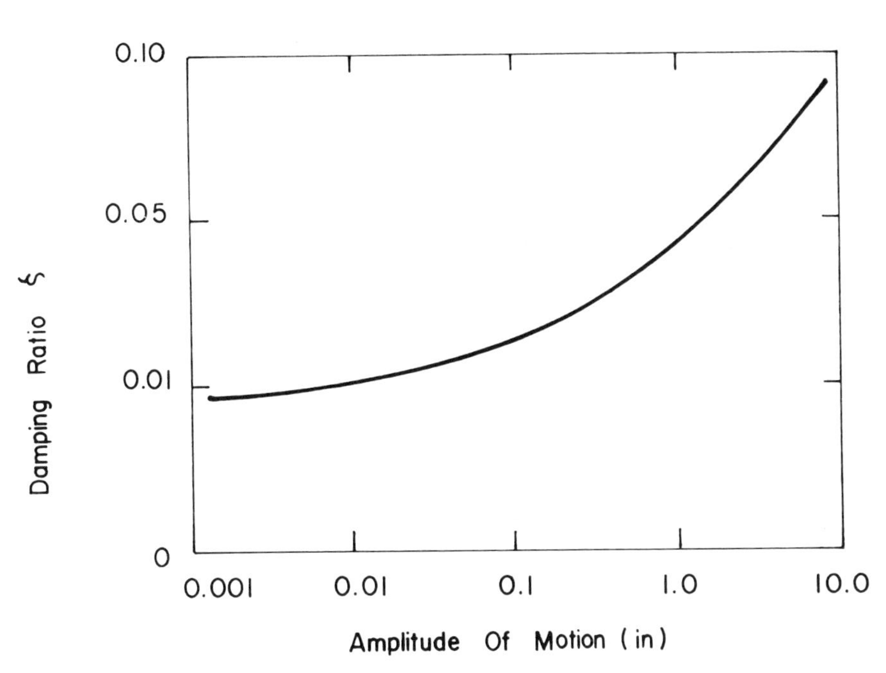

Figure 4-12. Illustration of the Effect of Vibration Amplitude on Damping in Structures

Table 4-3
Typical Damping Values for Seismic Applications

	Damping Ratio (ζ %)	
	OBE	*SSE*
Equipment and large-diameter piping systems[a] (> 12 in. diameter)	2	3
Small-diameter piping systems (≤ 12 in. diameter)	1	2
Welded-steel structures	2	4
Bolted-steel structures	4	7
Prestressed-concrete structures	2	5
Reinforced-concrete structures	4	7

Source: Table N-1230-1, reproduced from Section III—Division 1, Appendices, of the *ASME Boiler and Pressure Vessels Code*, with the permission of the publisher, The American Society of Mechanical Engineers.

[a]Includes both material and structural damping. If the piping system consists of only one or two spans, with little structural damping, use values for small diameter piping.

mended method for estimating damping in structures under seismic disturbances is by analyzing earthquake-response records for structures that are similar to the one being considered. Some typical damping ratios that are applicable under OBE and SSE conditions for a range of items are given in table 4-3.

When damping values are estimated using frequency-response magnitude curves, accuracy becomes poor at very low damping ratios ($<$ 1 percent). The main reason for this is the difficulty in obtaining a sufficient number of points in the magnitude curve near a poorly damped resonance when the frequency-response function is determined experimentally. As a result, the magnitude curve is poorly defined in the neighborhood of a weakly damped resonance. For low damping ($<$ 2 percent), time-response methods are particularly useful. At high damping values, the decay could be so fast that the measurements would contain large errors. Modal interference in closely spaced modes also could affect measured damping results.

Notes

1. M.P. Paidoussis and P.E. Des Trois Maisons, "Free Vibration of a Heavy, Damped, Vertical Cantilever," *Journal of Applied Mechanics* (Transactions of the ASME), June 1971, pp. 524–526.

2. C.M. Harris and C.E. Crede, *Shock and Vibration Handbook,* 2nd ed. (New York: McGraw-Hill, 1976), pg. 36–22.

5 Representation of the Dynamic Environment

The two primary methods of seismic qualifications are qualification by analysis and qualification by testing. Often, a combination of the two methods is used in a seismic-qualification program. No matter what method of qualification is used, a primary requirement is generation of the seismic environment. A disturbance-input motion should be developed, which, when applied to the equipment, produces a response (stresses, displacements, and so on) that is representative of the equipment's response to a severe earthquake.

In qualification by analysis, the first step is to develop an analytical model of the equipment to be qualified, by mathematical formulation, by computer simulation, or by a combination of the two. Next, the disturbance motion, which is represented by a mathematical expression or expressed numerically as digital data, is applied (or simulated) and the critical response of the model is determined. From this output, the true behavior of the equipment can be estimated. These results, in turn, might be used to establish structual integrity and functional operability of the equipment in the defined severe seismic environment. This procedure is shown in figure 5-1.

Qualification by testing is usually accomplished using a shaking table apparatus (see chapter 7). The equipment to be qualified (the test package) is properly mounted on the horizontal table. In-service operating conditions are simulated while the table is actuated by applying suitable input signals to the actuators. More than one signal is often required to simulate the three-dimensional characteristics of seismic motions. The input signal is either stored in an analog magnetic tape or generated in real time by a signal generator. The capability of that equipment or a similar unit to withstand predefined seismic disturbances may be evaluated by monitoring the dynamic-response and functional-operability variables of the test unit.

Essentially, both of the foregoing qualification procedures simulate the seismic environment by initially defining a time history that represents the most severe seismic motion that is expected at the installation site of the equipment during its design life. This is known as the design earthquake for that piece of equipment. Thus, the necessary two steps of generating the seismic environment are

Step 1 Define the design earthquake in terms of the qualification requirements.

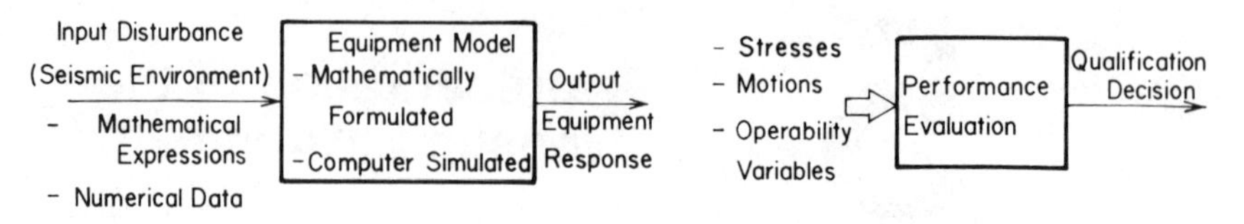

Figure 5-1. Seismic Qualification by Analysis

Step 2 Generate a signal that conservatively satisfies the specifications of Step 1.

The seismic ground motion of a building foundation is known as the ground response. The response of the location of the building at which the equipment is installed is known as the floor response. If the ground response is known, the floor response may be determined, at least theoretically, using the system-simulation concepts described in chapter 2. For this, a dynamic model of the building, relating the input motion (ground response) to the output motion (floor response), is necessary. This is illustrated in figure 5-2. The building acts as a filter and usually produces a smoothing effect on seismic ground motions, assuming, of course, that its structural integrity is maintained during the motion. Note that we have assumed that there is no relative motion between the building foundation and the supporting ground. In other words, soil-structure interaction due to boundary flexibility has been neglected. The validity of this assumption depends on subsoil conditions.

The foregoing discussion indicates that a primary requirement in seismic qualification is the proper representation of seismic ground motions. Using this as the starting point, the response, under a specified seismic environment, of buildings, structures, and the equipment they support can be estimated. Before specifying a seismic environment (the design earthquakes), the particular representation used should be thoroughly understood. The objective of this chapter is to discuss the common ways of representing a seismic environment.

The Nature of Earthquake Motions

A study of the mechanisms that give rise to earthquakes is outside the scope of this book. Our interest lies in studying the effect rather than the cause of seismic motions, even though the two are very much interrelated. In this section, we outline some important characteristics of seismic ground motions and state some common assumptions made in developing simple representations (models) for these motions.

Earthquake ground motions result from the sudden release of strain

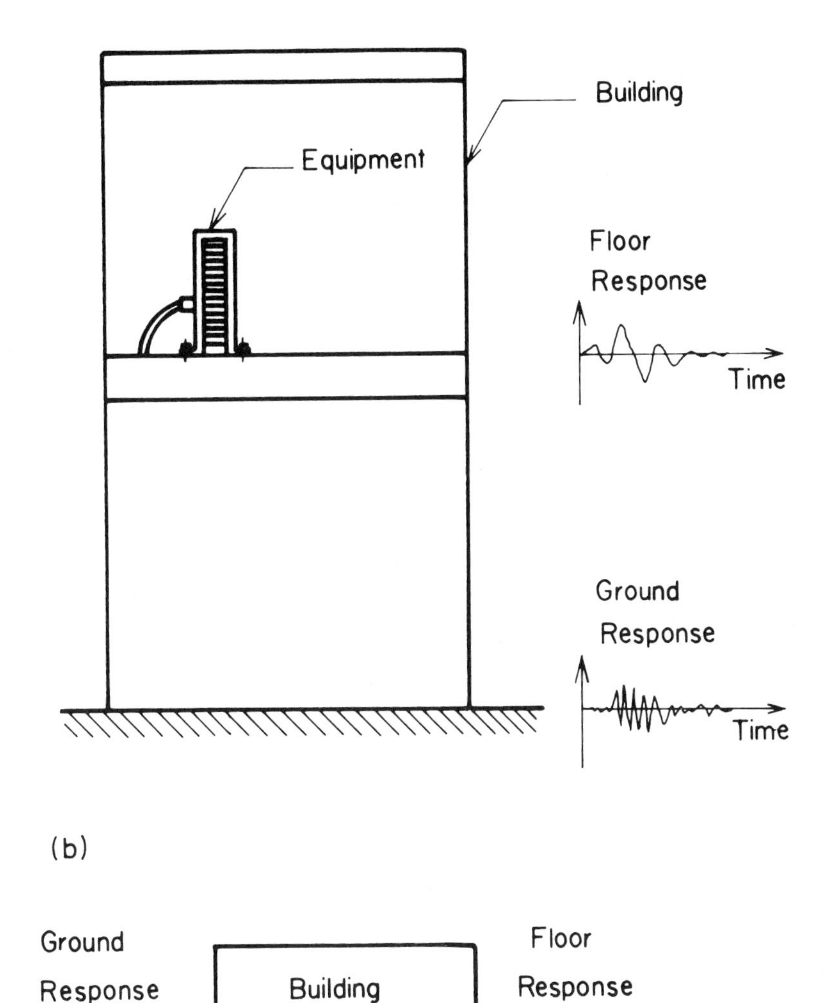

Figure 5-2. Relationship between Ground Response and Floor Response: (a) Physical Representation, (b) Simulation

energy caused by a sequence of ruptures on a fault surface within the earth's crust. The location at which rupture started is known as the focus or hypocenter, and the point on the ground surface vertically above it is called the *epicenter.* The nomenclature is described in figure 5-3. Note that it is not possible to know the location of the focus exactly. It is usually estimated by calculations using earthquake data. If we were to measure the total energy release by the entire rupture, we could determine a location on the rupture surface where that energy could be concentrated to produce an effect that is equivalent to the distributed energy release on the rupture surface. This location of the hypothetical point source of energy is known as the energy center. The three-dimensional displacement waves that start from the rupture locations on the rupture surface reach the ground surface in a random fashion. The random behavior is further complicated by reflection and refraction of the seismic waves. The amount of energy released and the distance from the energy center to the equipment site (see figure 5-3) are two important parameters that determine the intensity of the ground motions at the site. The density of energy released is also important. Various measures of earthquake damage potential are available to us.

Consider a small cubic lump of soil at the ground surface of an equipment site that is subjected to a seismic disturbance. This element feels motions in all directions in the space, and hence the seismic ground motions are three-dimensional. It is not possible to predict these motions exactly, and hence they are random. We assume that the element is sufficiently rigid. Its motion from a point O to an arbitrary point P in the neighborhood, as a result of the

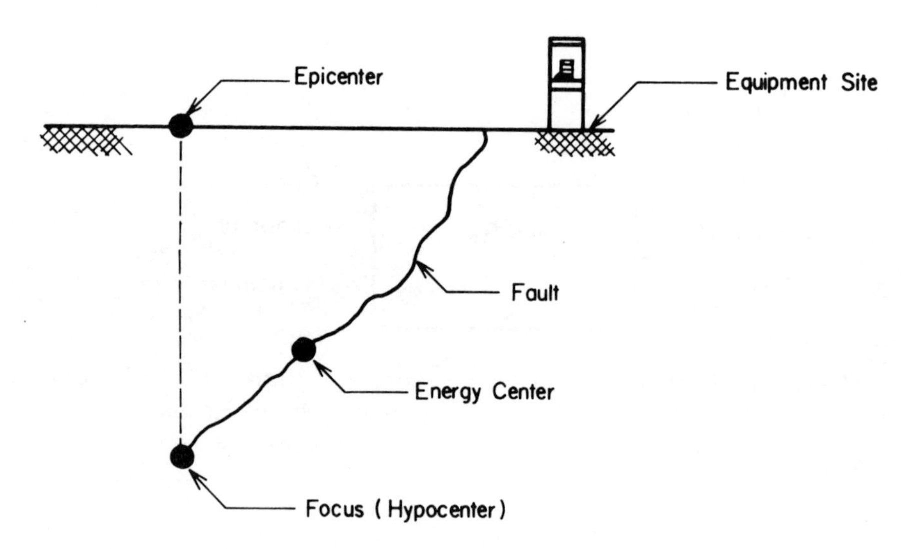

Figure 5-3. Earthquake Distance Nomenclature

seismic motions, can be completely represented by three translations (u_x, u_y, u_z) along the three axes (X, Y, Z) of an orthogonal cartesian coordinate frame ($OXYZ$) and three rotations (θ_x, θ_y, θ_z) about the same axes, as shown in figure 5-4. For convenience let us take the X-axis to be pointing toward the east, the Y-axis to be pointing toward the north, and the Z-axis vertically upwards. These three-dimensional motion components can be recorded by means of seismographic instruments installed in the seismic region, if an earthquake were to occur in that region. The three components of translatory acceleration, as recorded using a seismograph (seismometer), during a strong-motion earthquake, are shown in figure 5-5. The acceleration levels are measured in the units of acceleration due to gravity (g). It is customary to record the accelerations. These records are called accelerograms. The velocities and the displacements may be obtained by successive integration of the acceleration time histories, using an analog or a digital computer. It should be noted that the time histories shown in figure 5-5 are recordings made at only one location on the earth's surface. Different locations generally will have different time histories. Consequently, different points of the building foundation will not be excited in an identical manner during an earthquake. It therefore appears that determining the seismic response of a structure is a very complex task. We can simplify the problem for most purposes, however, by using some common assumptions. Consider the translatory motions. If the foundation of the structure is rigid and is resting over a distributed area of firm ground, the translatory motions acting on the structure will average out. Then the three components of translatory ground motion recorded at one location are assumed to be acting over the entire area of soil-structure interaction. Also, if the plan dimensions of the structure are small in comparison to the distance of the building site from the energy center of the earthquake, the translatory seismic motions for just one location of the

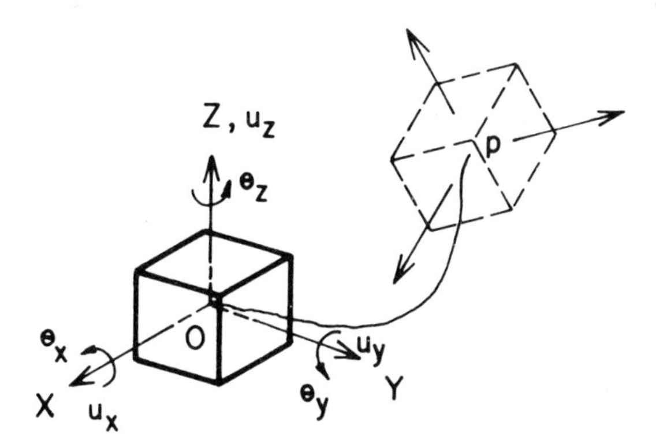

Figure 5-4. Three-Dimensional Seismic Motion of a Rigid Soil Element

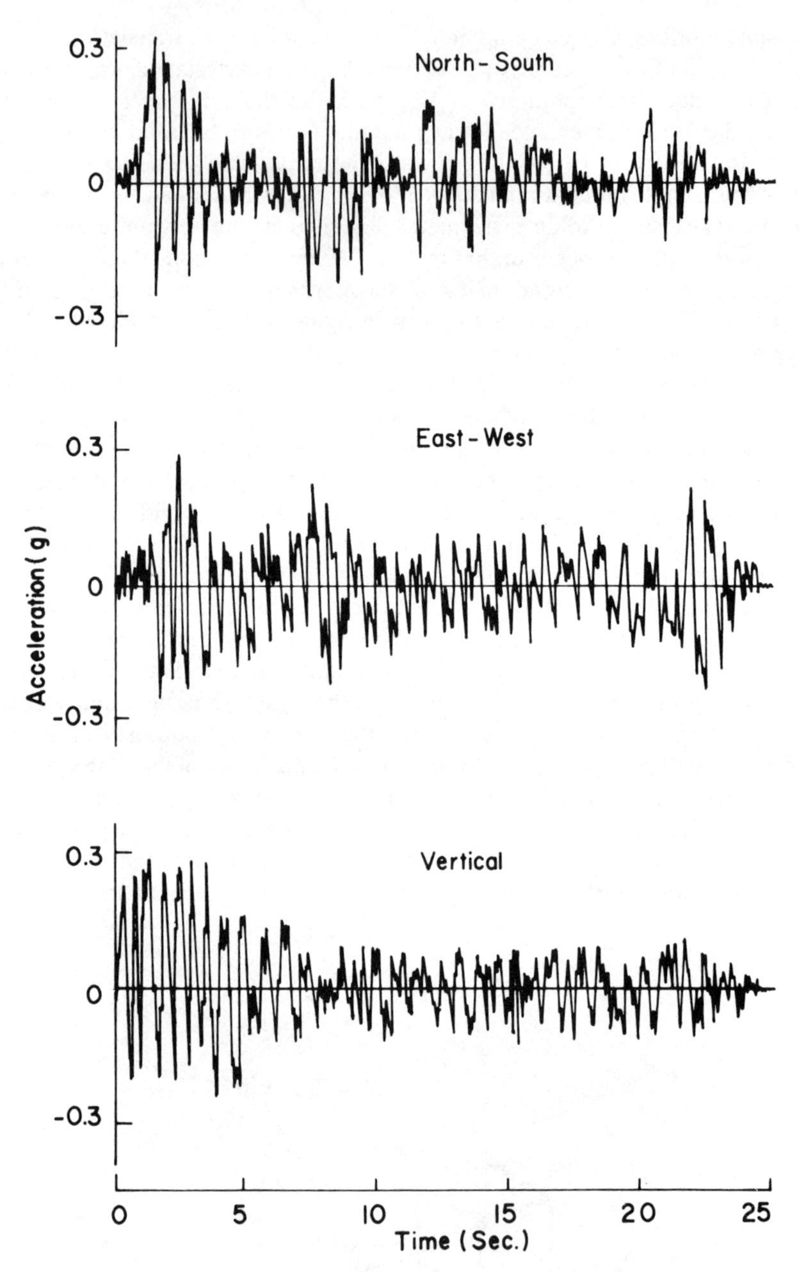

Figure 5-5. A Typical Earthquake Ground-Motion Record

ground may again be used satisfactorily in the analysis of the seismic response of that structure. The surface variation of translatory ground motions is of great concern only if the length of the structure is comparable to the distance of its site from the energy center and if it is supported at only a few discrete locations on the ground (for example, long bridges, dams, tunnels). In this book, we neglect the surface variations of seismic ground motions at a given equipment site.

Not much attention has been given in the earthquake literature to rotational components of seismic ground motions. Rotational motions are important if the foundation dimensions of the structure are small (typically one-quarter or less) in comparison to the wave length of the seismic waves. Newmark estimates the shortest relevant wavelength to be of the order of 600 m.[1] Since many structures have plan dimensions smaller than 150 m, it is noted that the rotational components should be considered for such structures. The rotational components are also important in the analysis of slender, flexible structures. I believe that the rotational components of seismic ground motions are usually neglected not because they are not important but because of the lack of data from field measurements. In seismic qualification, however, the equipment can be subjected to motions of any severity by using translatory disturbances alone. Therefore, in this book, we do not consider the rotational components of seismic ground mctions.

Examination of Seismograms

Figure 5-6 gives a schematic for obtaining a seismic ground-acceleration record. During an earthquake, the seismograph, which is insalled on ground in a seismic region, produces an analog record of the ground-acceleration time history on a magnetic tape. This tape is processed in the laboratory by playing it through a filter system to remove any undesirable noise. If the signal is weak, it may have to be amplified at this stage. The conditioned signal is then plotted on a strip chart, using a brush recorder. The amplitude and time scales should be correctly determined and stated on the chart for it to be of much use. The record thus obtained is a true acceleration time history of an earthquake ground motion (subject to possible equipment error). Unfortunately, the usefulness of such a record is very limited. An analysis is required to extract the information it contains. There are a few characteristic parameters that can be extracted by direct observation. By referring to the acceleration time history shown in figure 5-7, some useful parameters can be identified:

a_p = peak acceleration
T_p = time period near the peak acceleration

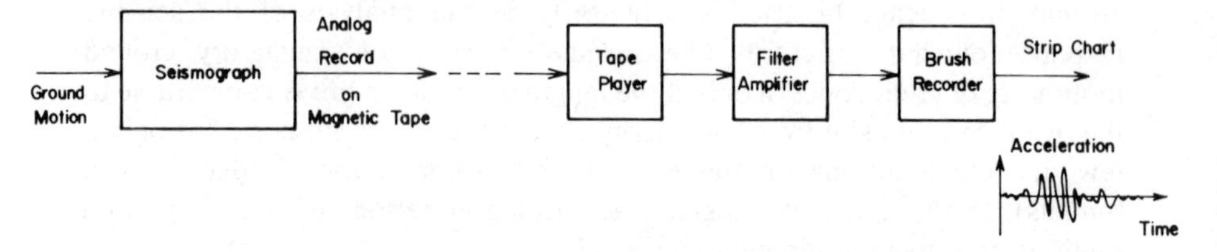

Figure 5-6. Seismic Accelerogram Generation Schematic

T_s = duration of strong motion (time interval beyond which no peaks over $a_{max}/2$ absolute occur)
T_d = total duration of the record
N_z = number of zero-crossings within T_s (this is equal to 14 for the case shown).

Further analysis, often of statistical nature, is required, however, to correlate these parameters to such earthquake parameters as magnitude, distance from energy center, and subsoil conditions.

We cannot perform a meaningful seismic-qualification test by applying this time history to a shaking table. The reason for this is simple. Earthquakes are random phenomena, and their occurrence is not known in advance. Even if we were able to predict the occurrence, it is virtually impossible to know the exact time history of a future seismic ground motion at a particular location. As a result, it is not wise to qualify a piece of equipment on the basis of only one past record of a seismic ground motion. In order to develop a design-basis time history that can be used in seismic qualification, a large collection of such records is necessary; preferably recorded at locations that are close to or have geological characteristics

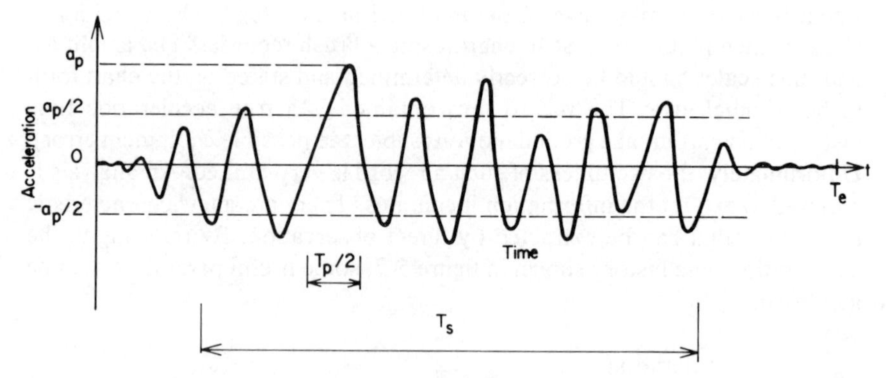

Figure 5-7. Parameters Directly Observable from a Seismic Time History

similar to those of the equipment site. Additional information is also required, such as

Geographical location of the equipment site with respect to regions of seismic activity

Geological characteristics, such as subsoil condition (for example, rocky or soft alluvial ground)

Equipment characteristics (for example, design life, functional importance, dynamic properties)

Supporting-structure characteristics (for example, rigid or flexible supports, distributed or discrete supports)

The time history can be either stochastic or deterministic and need not have the irregular shape of seismic time history records.

Dynamic Environment

A complete knowledge of the dynamic environment in which equipment would be operating is not available to the test engineer or the qualification analyst. The primary reason for this is that the operating environment is governed by many random factors, which are not completely known ahead of their occurrence. In this sense, the dynamic environment is a random process. When performing a dynamic test, however, either a deterministic or a random input excitation could be employed to meet the test requirements. This is known as the test environment. Similarly, in qualification by analysis, random as well as deterministic excitation inputs may be employed.

Based on the dynamic-testing specifications or qualification requirements, the test environment should be developed to have the required characteristics of (1) intensity (amplitude), (2) frequency content (effect on system resonances and the like), (3) decay rate (damping), and (4) phasing (dynamic interactions). Usually, these parameters are chosen to conservatively represent the worst possible dynamic environment reasonably expected during the design life of the test object. So long as this requirement is satisfied, it is not necessary for the test environment to be identical to the operating dynamic environment.

In dynamic testing (and also in dynamic analysis), the excitation input (test environment) can be represented in several ways. The common representations are (1) by time history, (2) by response spectrum, (3) by Fourier spectrum, and (4) by power spectral density functions.

Once the required environment is specified by one of these forms, the test should be conducted either by employing them directly or by using a more conservative excitation when the required environment cannot be exactly reproduced.

Time History

Dynamic testing employs both random and deterministic time histories as test excitation inputs. Regardless of its nature, the test input should conservatively meet the specified requirements for that test.

Stochastic versus Deterministic Time Histories

We refer again to the time-history record obtained using the setup shown in figure 5-6. This ground-motion record is not stochastic. It is true that the earthquakes are random phenomena and the process by which the time history was produced is a random process. Once the time history is recorded, however, it is known completely as a curve of magnitude versus time (a deterministic function of time). Therefore, it is a deterministic set of information. It is also a sample function of the stochastic process (earthquake) by which it was generated. Hence, very valuable information about the stochastic process itself can be determined by analyzing this sample function on the basis of the ergodic hypothesis (see the later section on stochastic representation). Many fail to understand this fact. Many also think that an irregular time-history record is a random signal. It should be remembered that some random processes produce very smooth signals. As an example, consider the sine wave given by $a \sin(\omega t + \phi)$. Let us assume that the amplitude a and the frequency ω are deterministic quantities and the phase angle ϕ is a random variable. This is a random process. Every time this particular random process is activated, a sine wave is generated that has the same amplitude and frequency but generally a different phase angle. Nevertheless, the sine wave will always look as smooth as it can be.

In a seismic-qualification program, if we use a time history (analog or digital) recorded on a magnetic tape or stored in the computer memory, it is a deterministic signal, even if it was originally produced by a random phenomenon such as an earthquake. Also, if we use a mathematical expression for the time history in terms of completely known (deterministic) parameters, it is again a deterministic signal. If the signal is generated by some random mechanism (computer simulation or physical) in real time, however, and if that signal is used in the qualification at the same time it is being generated, then we have a true random signal. Also, if we use a mathematical expression for the time history and some of its parameters are not known numerically and the values are assigned to them during the qualification (test or analysis) in a random manner, we again have a true random signal.

Deterministic Time-History Representation

In dynamic testing, time histories that are completely predefined can be used as test excitation inputs. They should be capable, however, of subjecting the test object to specified levels of intensity, frequency, decay rate, and phasing (in the case of simultaneous multiple excitations).

Deterministic time histories used in dynamic testing are divided into two broad categories: single-frequency time histories and multifrequency time histories.

Single-Frequency Time Histories. Single-frequency time histories have only one predominant frequency component at a given time. For the entire duration, however, the frequency range covered is representative of the frequency content of the dynamic environment. For seismic-qualification purposes, this range should be at least 1 to 33 Hz. Some typical single-frequency time histories that are used as excitation inputs in dynamic testing of equipment are shown in figure 5-8. The time histories shown in the figure can be expressed by simple mathematical expressions. This is not a requirement, however. It is equally satisfactory to store a very complex time history on a magnetic tape (analog or digital) and subsequently use it in the qualification procedure. In picking a particular time history, we should give proper consideration to its ease of reproduction and the accuracy with which it satisfies the test specifications.

Let us describe mathematically the acceleration time histories shown in figure 5-8.

Sine Sweep. We obtain a sine sweep by continuously varying the frequency of a sine wave. Mathematically,

$$u(t) = a \sin[\omega(t)\, t + \phi] \tag{5.1}$$

The amplitude a and the phase angle ϕ are usually constants, and the frequency $\omega(t)$ is a function of time. Both linear and exponential variations over the duration of the test are in common usage, but exponential variations are more common. For the linear variation (see figure 5-9):

$$\omega(t) = \omega_{\min} + (\omega_{\max} - \omega_{\min}) \frac{t}{T_d} \tag{5.2}$$

in which

$\omega_{\min}$ = lowest frequency in the sweep
$\omega_{\max}$ = highest frequency in the sweep
T_d = duration of the sweep

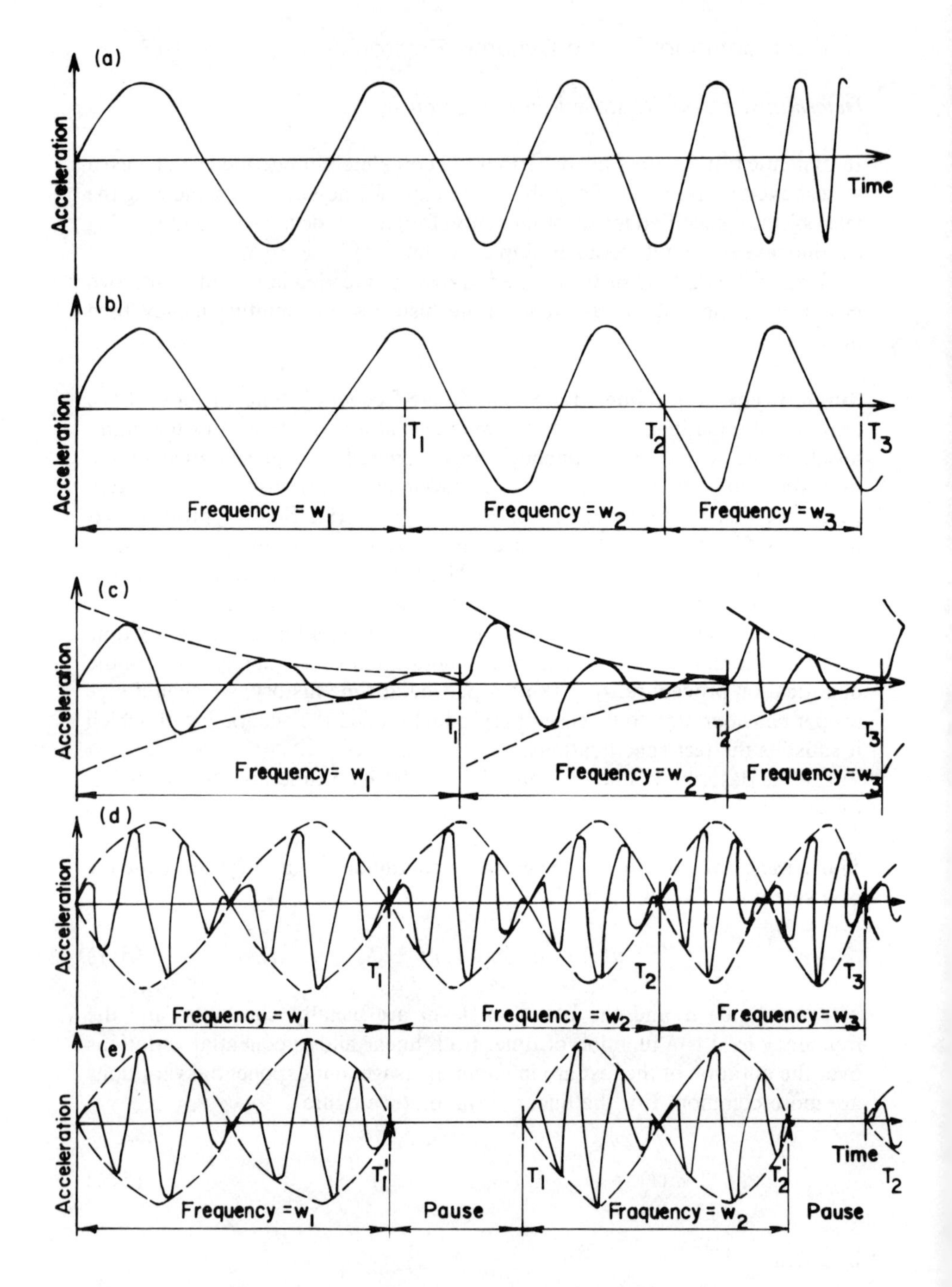

Figure 5-8. Typical Single-Frequency Time Histories: (a) Sine Sweep, (b) Sine Dwell, (c) Sine Decay, (d) Sine Beat, (e) Sine Beat with Pause

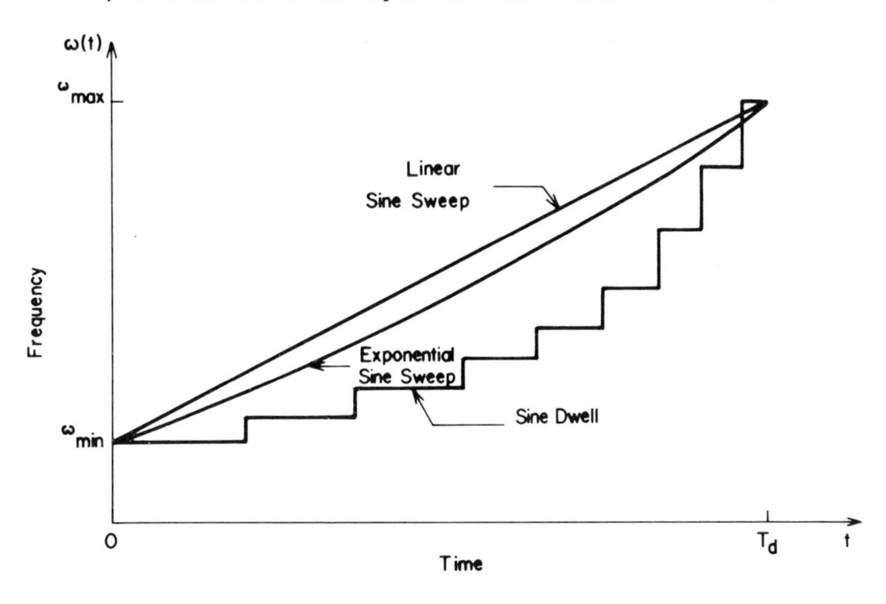

Figure 5-9. Frequency Variation in Some Single-Frequency Test Inputs

For the exponential variation (see Figure 5-9):

$$\log\left[\frac{\omega(t)}{\omega_{\min}}\right] = \frac{t}{T_d}\log\left[\frac{\omega_{\max}}{\omega_{\min}}\right] \tag{5.3}$$

or

$$\omega(t) = \omega_{\min}\left[\frac{\omega_{\max}}{\omega_{\min}}\right]^{t/T_d} \tag{5.4}$$

This variation is sometimes incorrectly called the logarithmic variation. This confusion arises because of its definition using equation (5.3) instead of equation (5.4). It is actually an inverse logarithmic variation. Note that the logarithm in equation (5.3) can be taken to any arbitrary base. If base 10 is used, the frequency increments are measured in decades (multiples of 10); if base 2 is used, the frequency increments are measured in octaves (multiples of 2). Thus, the number of decades in the frequency range ω_1 to ω_2 is given by $\log_{10}(\omega_2/\omega_1)$; for example, with $\omega_1 = 1$ rad/s and $\omega_2 = 100$ rad/s, we have $\log_{10}(\omega_2/\omega_1) = 2$, which corresponds to two decades. Similarly, the number of octaves in the range ω_1 to ω_2 is given by $\log_2(\omega_2/\omega_1)$. Then, with $\omega_1 = 2$ rad/s and $\omega_2 = 32$ rad/s, we have $\log_2(\omega_2/\omega_1) = 4$, a range of four octaves. Note that these quantities are ratios and have no physical units. The foregoing definitions can be extended for smaller units; for instance, one-third octave represents increments of $2^{1/3}$. Thus, if we start with 1 rad/s, and increment the frequency successively by one-third octaves, we get 1, $2^{1/3}$, $2^{2/3}$, 2, $2^{4/3}$, $2^{5/3}$, 2^2, and so on. It is clear, for example, that there are four one-third octaves in the frequency range $2^{2/3}$ to 2^2. Note that ω is known as the angular frequency (or radian frequency) and is usually measured in the units of radians per second (rad/s). The more commonly used frequency is the cyclic frequency, which is denoted by f and is measured in Hertz (Hz), which is identical to cycles per second (cps). It is clear that

$$f = \frac{\omega}{2\pi} \tag{5.5}$$

because there are 2π radians in one cycle.

So that all important vibration frequencies of the test object (or its model) are properly excited, the sine sweep rate should be as slow as is feasible. Typically, one octave per minute or slower rates are employed.

Sine Dwell. Sine-dwell time history is the descrete version of the sine sweep. The frequency is not varied continuously but is incremented by discrete amounts at discrete time points. This is shown graphically in figure 5-9. Mathematically, for the rth time interval, the dwell time history is

$$u(t) = a \sin(\omega_r t + \phi_r) \qquad T_{r-1} \leq t \leq T_r \tag{5.6}$$

in which ω_r, a, and ϕ_r are constants during the time interval (T_{r-1}, T_r). The frequency may be incremented by a constant increment or the frequency increments could be made bigger with time (exponential-type increment). The latter procedure is more common. Also, the dwelling-time interval is usually made smaller as the frequency is increased. This is logical because, as the frequency increases, the number of cycles during a given time also increases. Consequently, steady-state conditions may be achieved in a shorter time.

Sine-dwell time histories can be specified using either a graphical form (see figure 5-9) or a tabular form, giving the dwell frequencies and corresponding dwelling-time intervals. The amplitude is usually kept constant for the entire duration $(0, T_d)$, but the phase angle ϕ_r may have to be changed with each frequency increment in order to maintain the continuity of the signal.

Decaying Sine. Actual seismic ground motions decay with time as the wave energy is dissipated by some means. This decay characteristic is not present, however, in sine-sweep and sine-dwell time histories. Sine-decay representation is a sine dwell with decay (see figure 5-8). For an exponential decay, the counterpart of equation (5.6) may be written as

$$u(t) = a \exp(-\lambda_r t) \sin(\omega_r t + \phi_r) \qquad T_{r-1} \leq t \leq T_r \tag{5.7}$$

The damping parameter (inverse of time constant) λ_r is typically increased with each frequency increment in order to represent the increased decay rates of the dynamic environment at higher frequencies.

Sine Beat. When two sine waves having the same amplitude but different frequencies are mixed (added or subtracted) together, a sine beat is obtained. This is considered a sine wave having the average frequency of the two original waves, which is amplitude-modulated by a sine wave of frequency equal to half the difference of the frequencies of the two original waves. The amplitude modulation produces a transient effect that is similar to that caused by the damping term in the sine-decay equation (5.7). The sharpness of the peaks becomes more prominent when the frequency difference of the two frequencies gets smaller.

Consider two cosine waves having frequencies $(\omega_r + \Delta\omega_r)$ and $(\omega_r - \Delta\omega_r)$ and the same amplitude $a/2$. If the first is subtracted from the second (that is, added with a 180° phase change for the first wave), we obtain

$$u(t) = \frac{a}{2}[\cos(\omega_r - \Delta\omega_r)t - \cos(\omega_r + \Delta\omega_r)t] \tag{5.8}$$

By straightforward use of trigonometric identities, we obtain

$$u(t) = a(\sin \omega_r t)(\sin \Delta\omega_r t) \qquad T_{r-1} \leq t \leq T_r \tag{5.9}$$

This is a sine wave of amplitude a and frequency ω_r modulated by a sine wave of frequency $\Delta\omega_r$. Sine-beat time histories are commonly used as test excitation inputs in dynamic testing. Usually, the ratio $\omega_r/\Delta\omega_r$ is kept constant. A typical value used is 20, in which case we get 10 cycles per beat. Here, the cycles refer to the cycles at the higher frequency ω_r, and a beat corresponds to half a cycle at the smaller frequency $\Delta\omega_r$. Thus, a beat is identified by a peak of amplitude a in the modulated wave.

As in the case of sine dwell, the frequency ω_r of a sine-beat input is incremented at discrete time points T_r so as to cover the entire frequency interval of interest $(\omega_{\min}, \omega_{\max})$. It is a common practice to increase the size of the frequency increment and decrease the time duration at each frequency with each frequency increment, just as is done for the sine dwell. The reasoning for this is identical to that given for sine dwell. The number of beats for each duration is usually kept constant (typically at a value over 7). A sine-beat time history is shown in figure 5-8(d).

Sine Beat with Pauses. If we include pauses between sine-beat durations, we obtain a sine-beat time history with pauses. Mathematically, we have

$$
\begin{aligned}
u(t) &= a(\sin \omega_r t)\,(\sin \Delta\omega_r t) \quad \text{for} \quad T_{r-1} < t \leq T_r' \\
&= 0 \quad \text{for} \quad T_r' < t \leq T_r
\end{aligned}
\tag{5.10}
$$

This situation is shown in figure 5-8(e). When a sine-beat time history with pauses is specified, we must give the frequencies, the corresponding time intervals, and the corresponding pause times. Typically, the pause time is also reduced with each frequency increment.

The single-frequency time-history relations described in this section are summarized in table 5-1.

Multifrequency Time Histories. In contrast to single-frequency time histories, multifrequency time histories usually appear irregular and can have more than one predominant frequency component at a given time. Two common examples of multifrequency time histories are actual earthquake records and simulated earthquake time histories.

Actual Earthquake Records. Past earthquake records are sample functions of random processes. By analyzing these deterministic records, however, characteristics of the original stochastic processes can be established, provided that the records are sufficiently long. This is possible because of the

Table 5-1
Typical Single-Frequency Time Histories Used in Dynamic Testing

Single-Frequency Acceleration Time Histories	*Mathematical Expression*
Sine sweep	$u(t) = a \sin[\omega(t)t + \phi]$
	$\omega(t) = \omega_{\min} + (\omega_{\max} - \omega_{\min})^{t}T_d$ (linear)
	$\omega(t) = \omega_{\min}\left(\dfrac{\omega_{\max}}{\omega_{\min}}\right)^{t/T_d}$ (exponential)
Sine dwell	$u(t) = a \sin(\omega_r t + \phi_r) \quad T_{r-1} \leq t \leq T_r \quad r = 1, 2, \ldots, n$
Decaying sine	$u(t) = a \exp(-\lambda_r t) \sin(\omega_r t + \phi_r) \quad T_{r-1} \leq t \leq T_r$ $r = 1, 2, \ldots, n$
Sine beat	$u(t) = a(\sin \omega_r t)(\sin \Delta\omega_r t) \quad T_{r-1} \leq t \leq T_r$ $r = 1, 2, \ldots, n, \quad \omega_r/\Delta\omega_r = \text{constant}$
Sine beat with pauses	$u(t) = a(\sin \omega_r t)(\sin \Delta\omega_r t) \quad \text{for} \quad T_{r-1} < t \leq T'_r$ $= 0 \quad \text{for} \quad T'_r < t \leq T_r$

ergodic hypothesis. Results thus obtained are not quite accurate, because the actual earthquakes are nonstationary random processes and hence are not quite ergodic. Nevertheless, the information obtained by a Fourier analysis (see chapter 3) is useful in estimating the amplitude, phase, and frequency-content characteristics of the original earthquake. In this manner, we can pick a past earthquake record that can conservatively represent the design-base earthquake (DBE) for the equipment that needs to be qualified. Such records are usually available as analog recordings on magnetic tapes, which can be directly employed as input signals in seismic-qualification procedures.

Past earthquake time histories can be modified to make them acceptably close to a DBE by using spectral-raising and spectral-suppressing methods. In spectral-raising procedures, a sine wave of required frequency is added to the original time history to improve its capability of excitation at that frequency. The sine wave should be properly phased such that the time of maximum vibratory motion in the original time history is unchanged by the modification. Spectral suppressing is achieved essentially by using a narrow band-reject filter for the frequency band that needs to be removed. Physically, this is realized by passing the time history through a linearly damped oscillator tuned to the frequency to be rejected and connected in series with a second damper. Damping of this damper is chosen to obtain the required attentuation at the rejected frequency.

Simulated Earthquake Time Histories. Random-time-history-generating algorithms can be incorporated easily into digital computers, as is well known to computer users. Also, physical experiments can be developed that have some form of random mechanism as an integral part. A time history from any such random simulation, once generated, is a sample function. If the random phenomenon is accurately programmed or physically developed so as to conservatively represent the DBE, a time history from such a simulation may be employed directly to qualify the equipment analytically or by testing. Such time histories are usually available as analog recordings on magnetic tapes. Spectral-raising and spectral-suppressing techniques, mentioned earlier, also may be considered as methods of simulating earthquake time histories.

Before we conclude this section, it is worthwhile to point out that all time histories considered in this section are oscillatory. Even though the single-frequency time histories considered had little resemblance to actual earthquake ground-motion records, they can be chosen to possess the required decay, magnitude, phase, and frequency-content characteristics. Particularly during qualification by testing, these time histories, if used as input signals, will impose reversible stresses and strains to the equipment whose magnitudes, decay rates, and frequencies are representative of those that would be experienced during an actual earthquake, which is sufficiently strong, that might occur at the equipment site during the design life of the equipment.

Stochastic Time History Representation

To generate a truly stochastic signal, a random phenomenon must be incorporatd into the signal-generating process. The signal is generated only in real time, and its numerical value at a given time is unknown until that time instant is reached. A stochastic signal cannot be completely specified in advance, but its statistical properties may be prespecified.

There are many ways of obtaining random processes, including physical experimentation (for example, by tossing a coin at equal time steps and assigning a value to the magnitude over a given time step depending on the outcome of the toss), observation of processes in nature (such as outdoor temperature), and digital-computer simulation. This last procedure is the one commonly used in signal generation associated with dynamic testing or simulation.

A procedure used for the computer simulation of earthquakes is shown schematically in figure 5-10. The process is initiated by the generation of a series of random numbers according to some probability distribution. This is the stage at which the randomness is introduced into the process. At the next stage, a time signal is constructed using the sequence of random numbers. The random number sequence may be taken as the ordinates of the time history at equal time intervals, for example, and some kind of interpolation may be used over each time interval. Next, the signal is multiplied by an intensity function, which is a deterministic function of time. This function introduces the initial intensity buildup and the final decay behavior of actual earthquake time histories. In this manner, the duration of the strong motion can be controlled. Finally, spectral shaping of the signal is carried out by passing it through a suitable filter. This will remove undesirable frequency components. Typically, the frequency components outside the range 1 to 33 Hz is removed or attenuated, because this is the frequency range in which earthquake ground motions are most intense. Every time the digital simulation is carried out using the scheme shown in figure 5-10, a sample function of the particular random process is generated. This sample function may be used as a simulated earthquake record, as described in the previous section.

Ergodic Random Signals. Random process is a signal that is generated by some random (stochastic) mechanism. Each time the mechanism is operated, a different time history (sample function) generally is generated. The likelihood of any two sample functions becoming identical is governed by some probabilistic law. The random process is denoted by $X(t)$, and any sample function of the process is denoted by $x(t)$. It should be remembered that no numerical computations can be made on $X(t)$ because it is not known for certain. Its Fourier transform, for instance, can be written as an analytical expression but cannot be computed. Once a sample function $x(t)$ is gene-

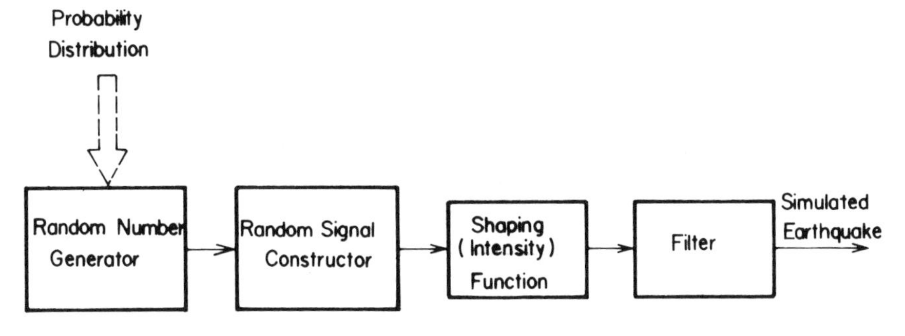

Figure 5-10. Computer Simulation of Earthquakes

rated, however, any numerical computation can be performed on it, because it is a completely known function of time (a time history). This important difference is often confused by users.

At any given time t_1, $X(t_1)$ is a random variable that has a certain probability distribution. Consider a well-behaved function $f[X(t_1)]$ of this random variable (which is also a random variable). Its expected value (statistical mean) is $E[fX(t_1)]$. This is also known as the ensemble average, because it is equivalent to the average value at t_1 of a collection (ensemble) of a large number of sample functions of $x(t)$.

Now consider the function $f\{x(t)\}$ of one sample function $x(t)$ of the random process. Its temporal (time) mean is expressed by

$$\lim_{T \to \infty} \frac{1}{2T} \int_{-T}^{T} f\{x(t)\}\, dt$$

If

$$E[f\{x(t_1)\}] = \lim_{T \to \infty} \frac{1}{2T} \int_{-T}^{T} f\{x(t)\}\, dt \tag{5.11}$$

then the random signal is said to be ergodic. Note that the right-hand side of equation (5.11) does not depend on time. Hence, the left-hand side also should be independent of the time point t_1.

As a result of this relation (known as the ergodic hypothesis), we can obtain the properties of a random process merely by performing computations using one of its sample functions. Ergodic hypothesis is the bridge linking the stochastic domain of imaginations and the deterministic domain of actual numerical computations. Digital Fourier computations, such as correlaton functions and spectral densities, would not be possible for random signals if not for this hypothesis.

Stationary Random Signals. If the statistical properties of a random signal $x(t)$ are independent of the time point considered, it is stationary. In particular, $X(t_1)$ will have a probability density that is independent of t_1, and the joint probability of $X(t_1)$ and $X(t_2)$ will depend only on the time difference $t_2 - t_1$. Consequently, the mean value $E[x(t)]$ of a stationary random signal is independent of t, and the autocorrelation function defined by

$$E[X(t)X(t + \tau)] = \phi_{xx}(\tau) \tag{5.12}$$

depends on τ and not on t. Note that the ergodic signals are always stationary, but the converse is not always true.

Consider Parseval's theorem, given by equation (3.46):

$$\int_{-\infty}^{\infty} x^2(t)\, dt = \int_{-\infty}^{\infty} |X(f)|^2\, df \tag{5.13}$$

This can be interpreted as an energy integral and is usually infinite for random signals. An appropriate measure for random signals is its power. This is given by its root mean square (RMS) value $E[X(t)^2]$. Power spectral density (psd) $\Phi(f)$ is the Fourier transform of the autocorrelation function. Hence,

$$\phi_{xx}(\tau) = \int_{-\infty}^{\infty} \Phi_{xx}(f) \exp(j2\pi f\tau)\, df \tag{5.14}$$

Now, from equations (5.12) and (5.14) we obtain

$$\text{RMS value} = E[X(t)^2] = \phi_{xx}(0) = \int_{-\infty}^{\infty} \Phi_{xx}(f)\, df \tag{5.15}$$

It follows that the RMS value of a stationary random signal is equal to the area under its psd curve.

Independent and Uncorrelated Signals. Two random signals $X(t)$ and $Y(t)$ are independent if their joint distribution is given by the product of the individual distributions. A special case is the uncorrelated signals that satisfy

$$E[X(t_1)Y(t_2)] = E[X(t_1)]E[Y(t_2)] \tag{5.16}$$

Consider the stationary case, with

$$\mu_x = E[X(t)] \tag{5.17}$$

$$\mu_y = E[Y(t)] \tag{5.18}$$

The autocovariance functions are given by

$$\psi_{xx}(\tau) = E[\{X(t) - \mu_x\}\{X(t+\tau) - \mu_x\}] = \phi_{xx}(\tau) - \mu_x^2 \quad (5.19)$$

$$\psi_{yy}(\tau) = E[\{Y(t) - \mu_y\}\{Y(t-\tau) - \mu_y\}] = \phi_{yy}(\tau) - \mu_y^2 \quad (5.20)$$

and the cross-covariance function is given by

$$\psi_{xy}(\tau) = E[\{X(t) - \mu_x\}\{Y(t-\tau) - \mu_y\}] = \Phi_{xy}(\tau) - \mu_x\mu_y \quad (5.21)$$

For uncorrelated signals [equation (5.16)],

$$\phi_{xy}(\tau) = \mu_x\mu_y \quad (5.22)$$

and, from equation (5.21),

$$\psi_{xy}(\tau) = 0 \quad (5.23)$$

The correlation-function coefficient is defined by

$$\rho_{xy}(\tau) = \frac{\psi_{xy}(\tau)}{\sqrt{\psi_{xx}(0)\,\psi_{yy}(0)}} \quad (5.24)$$

which satisfies

$$-1 \le \rho_{xy}(\tau) \le 1 \quad (5.25)$$

For uncorrelated signals, $\rho_{xy}(\tau) = 0$. This function measures the degree of correlation of the two signals.

The correlation of two random signals $X(t)$ and $Y(t)$ is measured in the frequency domain by its ordinary coherence function,

$$\gamma_{xy}^2(f) = \frac{|\Phi_{xy}(f)|^2}{\Phi_{xx}(f)\,\Phi_{yy}(f)} \quad (5.26)$$

which satisfies the condition

$$0 \le \gamma_{xy}^2(f) \le 1 \quad (5.27)$$

Transmission of Random Excitations. When the excitation input to a system is a random signal, the corresponding system response will also be random. Consider the system shown by the block diagram in figure 5-11 (a). The response is given by the convolution integral:

$$Y(t) = \int_{-\infty}^{\infty} h(t_1)U(t - t_1)\, dt_1 \qquad (5.28)$$

in which the response psd is given by the Fourier transform:

$$\Phi_{yy}(f) = \mathscr{F}\{E[Y(t)Y(t + \tau)]\} \qquad (5.29)$$

Now, by using equation (5.28) in equation (5.29), in conjunction with the definition of Fourier transform (see chapter 3):

$$\Phi_{yy}(f) = \int_{-\infty}^{\infty} d\tau \exp(-j2\pi f\tau)\, E\left[\int_{-\infty}^{\infty} dt_1\, h(t_1)U(t - t_1) \times \int_{-\infty}^{\infty} dt_2\, h(t_2)U(t + \tau - t_2)\right]$$

which can be written as

$$\Phi_{yy}(f) = \int_{-\infty}^{\infty} dt_1\, h(t_1) \int_{-\infty}^{\infty} dt_2\, h(t_2) \int_{-\infty}^{\infty} d\tau \exp(-j2\pi f\tau) \times \phi_{uu}(\tau + t_1 - t_2)$$

Now, if we let $\tau' = \tau + t_1 - t_2$,

$$\Phi_{yy}(f) = \left[\int_{-\infty}^{\infty} h(t_1) \exp(j2\pi f t_1) dt_1\right]\left[\int_{-\infty}^{\infty} h(t_2) \exp(-j2\pi f t_2) dt_2\right] \times \left[\int_{-\infty}^{\infty} \phi_{uu}(\tau') \exp(-j2\pi f\tau') d\tau'\right]$$

Note that $U(t)$ is assumed to be *stationary.*

Now, by referring to the definition of frequency-response function (chapter 2), we see that

$$\Phi_{yy}(f) = H^*(f)H(f)\Phi_{uu}(f) \qquad (5.30)$$

in which $H^*(f)$ is the complex conjugate of $H(f)$. Alternatively, if $|H(f)|$ denotes the magnitude of the complex quantity,

$$\Phi_{yy}(f) = |H(f)|^2\, \Phi_{uu}(f) \qquad (5.31)$$

(a)

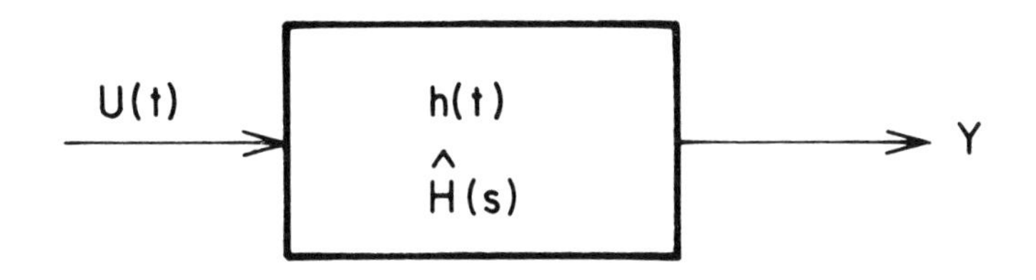

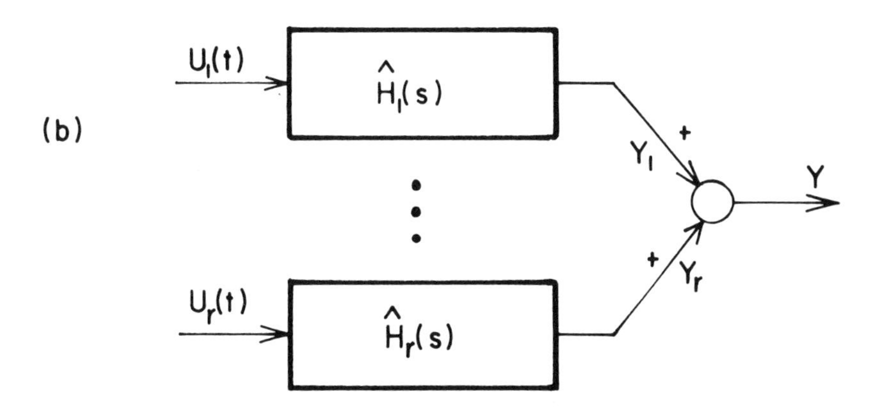

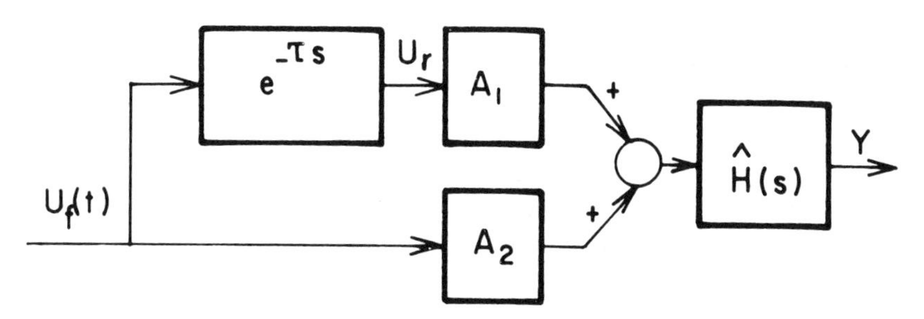

Figure 5-11. Combined Response to Various Random Excitations: (a) System Excited by a Single Input, (b) Response to Several Random Excitations, (c) Response to a Delayed Excitation

By using equation (5.30) or equation (5.31), we can determine the psd of the system response from the psd of the excitation if the system-frequency response function is known.

In a similar manner, it can be shown that the cross-spectral density function

$$\Phi_{uy}(f) = H(f)\Phi_{uu}(f) \tag{5.32}$$

Now consider r stationary, independent, random excitations $U_1, U_2, \ldots, U_r$ (which are assumed to have zero-mean values, without loss of generality) applied to r subsystems, having transfer functions $\hat{H}_1(s), \hat{H}_2(s), \ldots, \hat{H}_r(s)$, as shown in figure 5-11(b). The total response Y consists of the sum of individual responses $Y_1, Y_2, \ldots, Y_r$. It can be shown that $Y_1, Y_2, \ldots, Y_r$ are also stationary, independent, zero-mean, random processes. By definition,

$$\phi_{yy}(\tau) = E[\{Y_1(t) + \cdots + Y_r(t)\}\{Y_1(t+\tau) + \cdots + Y_r(t+\tau)\}] \tag{5.33}$$

Now, for independent, zero-mean Y_i, equation (5.33) becomes

$$\phi_{yy}(\tau) = E[Y_1(t)Y_1(t+\tau)] + \cdots + E[Y_r(t)Y_r(t+\tau)] \tag{5.34}$$

Since Y_i are stationary,

$$\phi_{yy}(\tau) = \phi_{y_1y_1}(\tau) + \cdots + \phi_{y_ry_r}(\tau) \tag{5.35}$$

On Fourier transformation,

$$\Phi_{yy}(f) = \Phi_{y_1y_1}(f) + \cdots + \Phi_{y_ry_r}(f) \tag{5.36}$$

In view of equation (5.31),

$$\Phi_{yy}(f) = \sum_{i=1}^{r} |H_i(f)|^2 \Phi_{u_iu_i}(f) \tag{5.37}$$

from which the response psd can be determined if the input psd's are known.

If all inputs $U_i(t)$ have identical probability distributions (for example, when they are generated by the same mechanism), the corresponding psd's will be identical. Note that this does not imply that the inputs are equal. They could be dependent, independent, correlated, or uncorrelated. Equation (5.37) becomes

$$\Phi_{yy}(f) = \left[\sum_{i=1}^{r} |H_i(f)|^2\right] \Phi_{uu}(f) \tag{5.38}$$

in which $\Phi_{uu}(f)$ is the common input psd.

Finally, consider the linear combination of two excitations $U_f(t)$ and $U_r(t)$, with the latter excitation delayed in time by τ but otherwise identical to the former. This situation is shown in figure 5-11(c). From Laplace transform tables (chapter 2), it is seen that the Laplace transforms of the two signals are related by

$$\hat{U}_r(s) = \exp(-\tau s)\ \hat{U}_f(s) \tag{5.39}$$

From equation (5.39) it follows that [see figure 5-11(c)]

$$\hat{Y}(s) = (A_1 \exp(-\tau s) + A_2)\hat{H}(s)\hat{U}_f(s) \tag{5.40}$$

Consequently,

$$\Phi_{yy}(f) = |(A_1 \exp(-j2\pi f\tau) + A_2)H(f)|^2\ \Phi_{uu}(f) \tag{5.41}$$

From this result, the net response can be determined when the phasing between the two excitations is known.

Response Spectrum

Response spectrum is commonly used to represent signals associated with dynamic testing. A given signal has a certain fixed response spectrum, but many different signals could have the same response spectrum. For this reason, as will be clear shortly, the original signal cannot be reconstructed from its response spectrum. This is a major disadvantage.

If the given signal is passed through a single-degree-of-freedom oscillator (of certain natural frequency), and the response of the oscillator (mass) is recorded, we can determine the maximum (peak) value of the response. Suppose we repeat the process for a number of different oscillators (having different natural frequencies) and plot the peak response values thus obtained against the corresponding oscillator natural frequency. This procedure is shown schematically in figure 5-12. For an infinite number of oscillators (or the same oscillator with continuously variable natural frequency), we get a continuous spectrum of the given signal. It is obvious, however, that the original signal cannot be completely determined from the knowledge of its response spectrum alone. In Figure 5.12, for instance, another signal, when passed through a given oscillator, might produce the same peak response.

Note that we have assumed the oscillators to be undamped; that is, the response spectrum obtained using undamped oscillators, corresponds to $\zeta = 0$. If all the oscillators are damped, however, and have the same damping ratio ζ, the resulting response spectrum corresponds to ζ. It is therefore clear

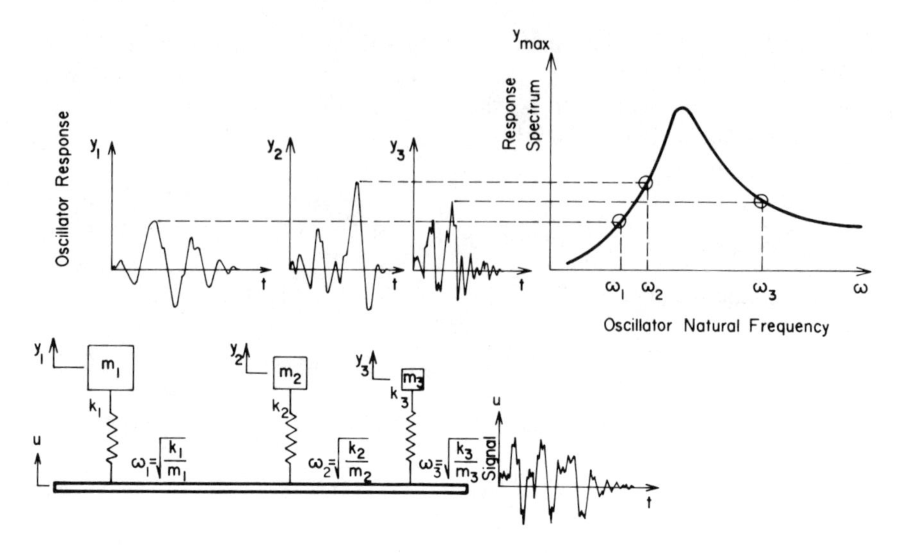

Figure 5-12. Definition of Response Spectrum of a Signal

that ζ is also a parameter in the response-spectrum representation. We also should specify the damping value when we represent a signal by a response spectrum.

Displacement, Velocity, and Acceleration Spectra

It is clear that a motion signal can be represented by the corresponding displacement, velocity, or acceleration time history. First consider a displacement time history $u(t)$. The corresponding velocity time history is $\dot{u}(t)$, and the acceleration time history is $\ddot{u}(t)$. The impulse-response function for an undamped, single-degree-of-freedom oscillator (see chapter 2) having natural frequency ω_n is given by

$$h(t) = \omega_n \sin\omega_n t \tag{5.42}$$

The response $y(t)$ of this oscillator, when excited by the displacement signal $u(t)$, is given by the convolution integral

$$y_d(t) = \omega_n \int_0^{\infty} u(\tau) \sin w_n(t - \tau) d\tau \tag{5.43}$$

The response of the same oscillator, when excited by the velocity signal $\dot{u}(t)$, is given by

$$y_v(t) = \omega_n \int_0^\infty \dot{u}(\tau) \sin \omega_n(t - \tau) d\tau \qquad (5.44)$$

and the response when excited by the acceleration signal $\ddot{u}(t)$ is

$$y_a(t) = \omega_n \int_0^\infty \ddot{u}(\tau) \sin \omega_n(t - \tau) d\tau \qquad (5.45)$$

If the peak value of $y_d(t)$ is plotted against ω_n, we get the displacement-spectrum curve of the displacement signal $u(t)$. If the peak value of $y_v(t)$ is plotted against ω_n, we get the velocity-spectrum curve of the displacement signal $u(t)$ or velocity signal $\dot{u}(t)$. If the peak value of $y_a(t)$ is plotted against ω_n, we get the acceleration-spectrum curve of the displacement signal $u(t)$ or acceleration signal $\ddot{u}(t)$.

Now consider equation (5.44). Integration by parts gives

$$y_v(t) = [\omega_n u(\tau) \sin \omega_n(t - \tau)]_0^\infty + \omega_n^2 \int_0^\infty u(\tau) \cos \omega_n(t - \tau) d\tau \qquad (5.46)$$

From the initial and final conditions for $u(t)$, it follows that the first term in equation (5.46) vanishes. The second term is $\omega_n[y_d(t + \pi/2\omega - \tau)]$, which is clear by noting that $\sin \omega_n(t + \pi/2\omega_n - \tau)$ is equal to $\cos \omega_n (t - \tau)$:

$$y_v(t) = \omega_n y_d\left(t + \frac{\pi}{2\omega_n}\right) \qquad (5.47)$$

If we integrate equation (5.45) by parts twice, and apply end conditions as before, we obtain

$$y_a(t) = -\omega_n^2 y_d(t) \qquad (5.48)$$

By taking the peak values of response time histories, we see from equations (5.47) and (5.48) that

$$v(\omega_n) = \omega_n d(\omega_n) \qquad (5.49)$$

$$a(\omega_n) = \omega_n^2 d(\omega_n) \qquad (5.50)$$

in which $d(\omega_n)$, $v(\omega_n)$ and $a(\omega_n)$ represent the displacement spectrum, the velocity spectrum, and the acceleration spectrum, respectively, of the displacement time history $u(t)$. It follows from equations (5.49) and (5.50) that

$$a(\omega_n) = \omega_n v(\omega_n) \tag{5.51}$$

Response-Spectra Plotting Paper

Response spectra are usually plotted on frequency-velocity coordinate axes or on frequency-acceleration coordinate axes. Values are normally plotted in logarithmic scale, as shown in figure 5-13. First, consider the axes shown in figure 5.13(a). Obviously, constant velocity lines are horizontal for this coordinate system. From equation (5.49), constant-displacement lines correspond to

$$v(\omega_n) = c\omega_n$$

By taking logarithms of both sides,

$$\log v(\omega_n) = \log\omega_n + \log c$$

It follows that the constant-displacement lines have +1 slope on the logarithmic frequency-velocity plane. Similarly, from equations (5.51), constant-acceleration lines correspond to

$$\omega_n v(\omega_n) = c$$

Hence,

$$\log v(\omega_n) = -\log \omega_n + \log c$$

It follows that the constant-acceleration lines have −1 slope on the logarithmic frequency-velocity plane.

Similarly, it can be shown from equations (5.50) and (5.51) that, on the logarithmic frequency-acceleration plane [Figure 5-13(b)], constant-displacement lines have +2 slope, and constant-velocity lines have +1 slope.

On the frequency-velocity plane, a point corresponds to a specific frequency and a specific velocity. The corresponding displacement at the point is obtained [equation (5.49)] by dividing the velocity value by the frequency value at that point. The corresponding acceleration at that point is obtained [equation (5.51)] by multiplying the particular velocity value by the frequency value. Any types of units may be used for displacement, velocity, and acceleration quantities. A typical logarithmic frequency-velocity plotting sheet is shown in figure 5-14. Note that the sheet is already graduated on constant deplacement, velocity, and acceleration lines. Also, a period axis (period = 1/cyclic frequency) is given for convenience in plotting.

Zero Period Acceleration

Frequently, response spectra are specified in terms of accelerations rather than velocities. This is particularly true in dynamic testing associated with seismic qualification, because earthquake ground-motion records are usually available as acceleration time histories. Of course, no information is lost because the logarithmic frequency-acceleration plotting paper can be graduated for velocities and displacements also. It is therefore clear that an acceleration quantity (peak) on a response spectrum has a corresponding velocity quantity (peak), and a displacement quantity (peak). In dynamic testing, however, the motion variable in common usage is the acceleration. Zero-period acceleration (ZPA) is an important parameter that characterizes a response spectrum. It should be remembered, however, that zero-period velocity or zero-period displacement could be similarly defined.

Zero-period acceleration is defined as the acceleration value (peak) at zero period (or infinite frequency) on a response spectrum. Specifically,

$$\mathrm{ZPA} = \lim_{\omega_n \to \infty} a(\omega_n) \tag{5.52}$$

Consider the simple oscillator equation:

$$\ddot{y} + 2\zeta\omega_n\dot{y} + \omega_n^2 y = \omega_n^2 u(t) \tag{5.53}$$

By differentiating equation (5.53) throughout, once or twice, it is seen that if u and y initially refer to displacements, then the same equation is valid when they refer to either velocities or accelerations. Let us consider the case in which u and y refer to input and response acceleration variables. For a sinusoidal signal $u(t)$ given by

$$u(t) = A \sin \omega t \tag{5.54}$$

the response $y(t)$, neglecting the transient components (that is, the steady-state value) is given by

$$y(t) = A \frac{\omega_n^2}{\sqrt{(\omega_n^2 - \omega^2)^2 + 4\zeta^2\omega_n^2\omega^2}} \sin(\omega t + \phi) \tag{5.55}$$

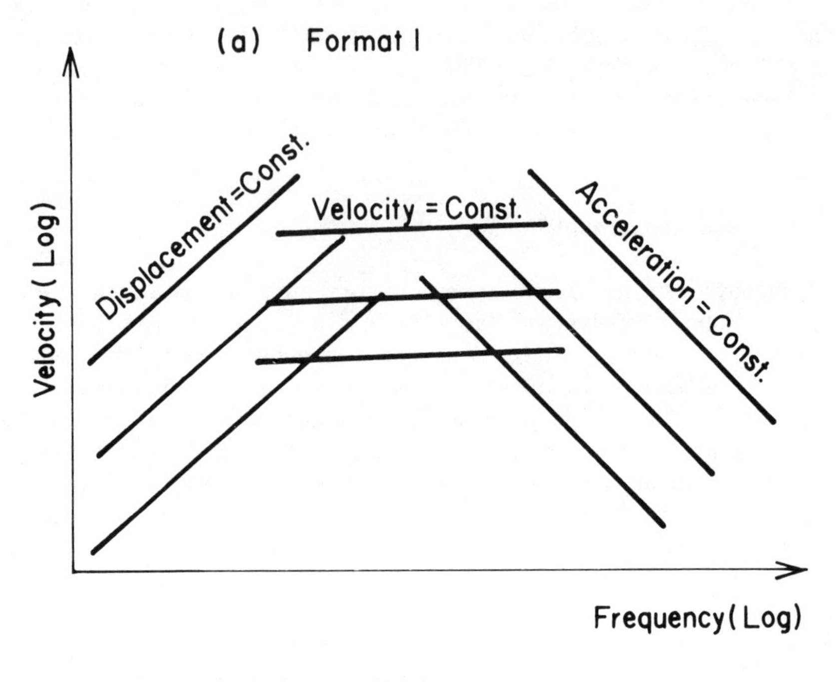

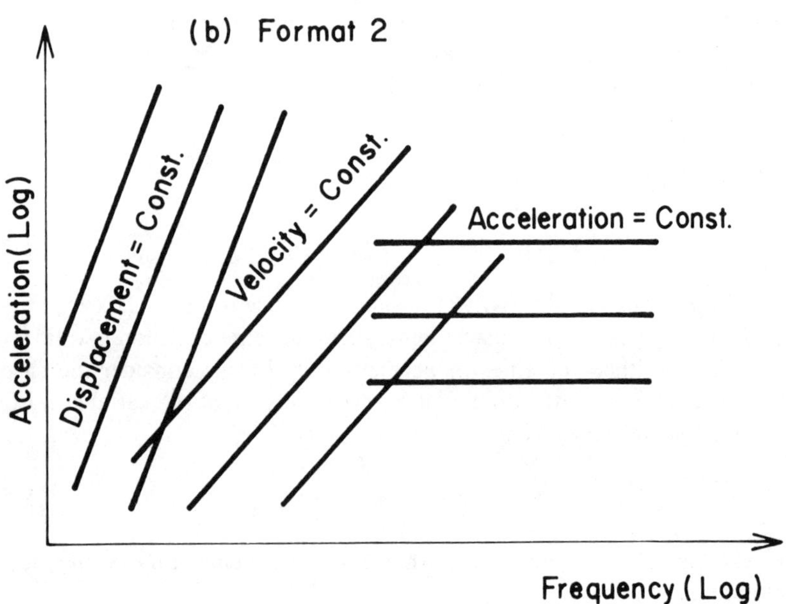

Figure 5-13. Response-Spectra Plotting Formats: (a) Frequency-Velocity Plane, (b) Frequency-Acceleration Plane

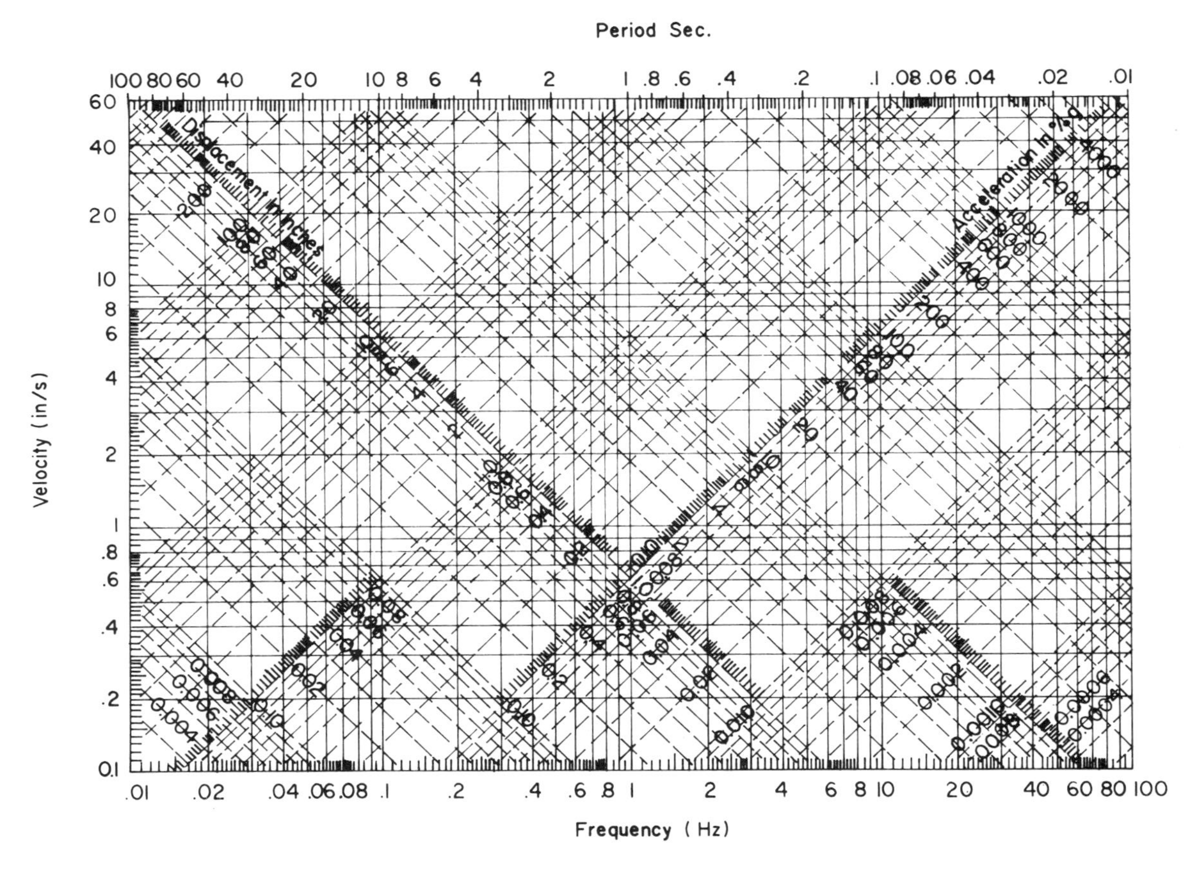

Figure 5-14. Response-Spectra Plotting Sheet (Frequency-Velocity Plane)

Hence, the response spectrum, given by $a(\omega_n) = [y(t)]_{max}$, for a sinusoidal signal of frequency ω and amplitude A is

$$a(\omega_n) = A \frac{\omega_n^2}{\sqrt{(\omega_n^2 - \omega^2)^2 + 4\zeta^2\omega_n^2\omega^2}} \tag{5.56}$$

A plot of this response spectrum is shown in figure 5-15. Note that $a(0) = 0$. Also,

$$ZPA = \lim_{\omega_n \to \infty} a(\omega_n) = A \tag{5.57}$$

It is worth observing that at the point $\omega_n = \omega$ we have $a(\omega_n) = A/(2\zeta)$, which corresponds to an amplification by a factor of $1/(2\zeta)$ over the ZPA value.

Uses of Response Spectra

In dynamic testing, response-spectra curves are employed to specify the dynamic environment to which the test object is required to be subjected. This specified response spectrum is known as the required response spectrum (RRS). In order to conservatively satisfy the test specification, the response spectrum of the test input excitation, known as the test response spectrum (TRS), should envelop the RRS. Note that, when response spectra are used to represent excitation input signals in dynamic testing, the damping value of the hypothetical oscillators used in computing the response spectrum has no bearing on the actual damping that is present in the test object. In this application, the response spectrum is merely a representation of the shaker-input signal, and therefore does not depend on system damping.

Another use of response spectra is in estimating the peak value of the response of a multi-degree-of-freedom or distributed-parameter system when it is excited by a signal whose response spectrum is known. To understand this concept, we should recall the fact that, for a multi-degree-of-freedom or truncated-distributed-parameter system having distinct natural frequencies, the total response can be expressed as a linear combination of the individual modal responses. Specifically, the response $y(t)$ can be written

$$y(t) = \sum_{i=1}^{r} \alpha_i \, a(\omega_i) \exp(-\zeta_i \omega_i t) \sin(\omega_i t + \phi_i) \tag{5.58}$$

in which $a(\omega_i)$ are the amplitude contributions from each mode (simple oscillator equation). Hence, $a(\omega_i)$ corresponds to the value of the response spectrum at frequency ω_i. The linear combination parameters α_i depend on

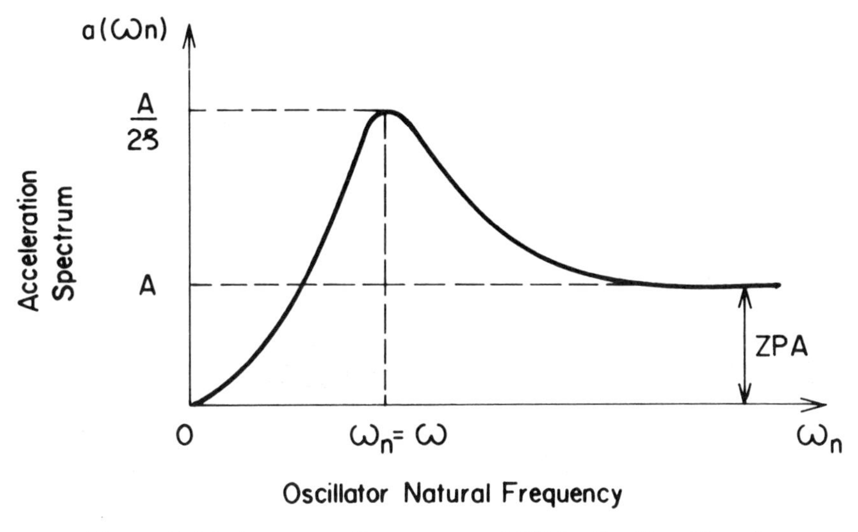

Figure 5-15. Response Spectrum and ZPA of a Sine Signal

modal-participation factors and can be determined from system parameters. Since the peak values of all terms in the summation on the right-hand side of equation (5.58) do not occur at the same time, we observe that

$$[y(t)]_{\text{peak}} < \sum_{i=1}^{r} \alpha_i a(\omega_i) \tag{5.59}$$

It follows that the right-hand side of the inequality (5.59) is a conservative upperbound estimate for the peak response of the multi-degree-of-freedom system. Some prefer to make the estimate less conservative by taking the square root of sum of the squares (SRSS):

$$\left[\sum_{i=1}^{r} \alpha_i^2 a^2(\omega_i)\right]^{1/2}$$

This method, however, has the risk of giving an estimate that is less than the true value. Note that, in this application, the damping value associated with the response spectrum is directly related to modal damping of the system. Hence, the response spectrum $a(\omega_i)$ should correspond to the same damping ratio as that of the mode considered in the summation of the inequality (5.59).

If all modal damping ratios ζ_i are identical or nearly so, the same response spectrum could be used to compute all terms in the inequality (5.59). Otherwise, different response-spectra curves should be used to determine each quantity $a(\omega_i)$, depending on the modal damping ratio ζ_i.

Other Representations

Earlier in this chapter, we listed four types of representations that can be employed for specifying a dynamic environment, particularly in dynamic testing. In previous sections we have discussed the time-history method and the response-spectrum method. In this section, we shall discuss the Fourier spectrum method and the power spectral density method.

Fourier Spectrum Method

Fourier analysis of signals was discussed in detail in chapter 3. Since time domain and frequency domain are related through Fourier transformation, a time history can be represented by its Fourier spectrum. In dynamic testing, a required Fourier spectrum can be specified as the test specification. Then, the actual input signal used to excite the test object should have a Fourier spectrum that envelops the required Fourier spectrum. Generation of a signal to satisfy this requirement might be difficult. Usually, digital Fourier analysis of the control sensor signal (see figure 3-1) is necessary to compare the actual (test) Fourier spectrum with the required Fourier spectrum. If the two spectra do not match in a certain frequency bend, the error is fed back to correct the situation. This process is known as frequency-band equalization. Also, the sample step in DFT analysis should be adequately small to cover the required frequency range of interest in that particular dynamic-testing application. Advantages of using DFT in dynamic testing include flexibility and convenience with respect to the type of the signal that can be analyzed, availability of complex processing capabilities, increased speed of processing, accuracy and reliability, reduction in test cost, practically unlimited repeatability of processing, and reduction in overall size and weight of the analyzer.

Power Spectral Density Method

The operational vibration environment of equipment is usually random. Consequently, a stochastic representation of the test excitation input appears to be suitable for a majority of dynamic-testing situations. One way of representing a stationary random signal is by its power spectral density (psd), defined earlier in this chapter. The numerical computation of psd is not possible, however, unless the ergodic hypothesis is assumed. Using the ergodic hypothesis, we can compute the psd of a random signal simply by using one sample function (one record) of the signal.

Three methods of determining the psd of a random signal are shown in figure 5-16. Frm Parseval's therorem [equation (5.13)], we notice that the mean square value of a random signal may be obtained from the area under the psd curve. This suggests the method shown in figure 5-16(a) for estimating the psd of a signal. The mean square value of a sample of the signal in the frequency band Δf having a certain center frequency is obtained by first extracting the signal components in the band and then squaring them. This is done for several samples and averaged to get a better accuracy. It is then divided by Δf. By repeating this for a range of center frequencies, an estimate for the psd is obtained.

In the second scheme, shown in figure 5-16(b), correlation function is first computed digitally using equation (3.40). Its Fourier transform (by FFT) gives an estimate of the psd.

In the third scheme, shown in figure 5-16(c), the psd is computed using equation (3.41), as described in chapter 3. The Fourier spectrum of the sample record is computed and the psd is estimated directly, without first computing the autocorrelation function.

In these numerical techniques of computing psd, a single sample function would not give the required accuracy, and averaging of results for a number of sample records is usually needed. In real-time digital analysis, the running average and the current estimate are normally computed. In the running average, it is desirable to give a higher weighting to the more recent estimates. The fluctuations in the psd estimate about the local average could be reduced by selecting a larger filter bandwidth Δf (see figure 5-17) and a large record length T. A measure of this fluctation is given by

$$\varepsilon = \frac{1}{\sqrt{\Delta f \; T}} \tag{5.60}$$

It should be noted that large Δf results in reduction of the precision of the estimates while improving the appearance. To offset this, T has to be increased further, or averaging should be done for a several sample records.

Generating a test-input signal with a psd that satisfactorily compares with the required psd could be a tedious task if attempted manually by mixing various signal components. A convenient method is to use an automatic multiband equalizer.[2] By this means, the mean amplitude of the signal in each small frequency band of interest could be made to approach the spectrum of the specified vibration environment (see figure 5-18). Unfortunately, this type of random-signal dynamic testing could be more costly than testing with deterministic signals.

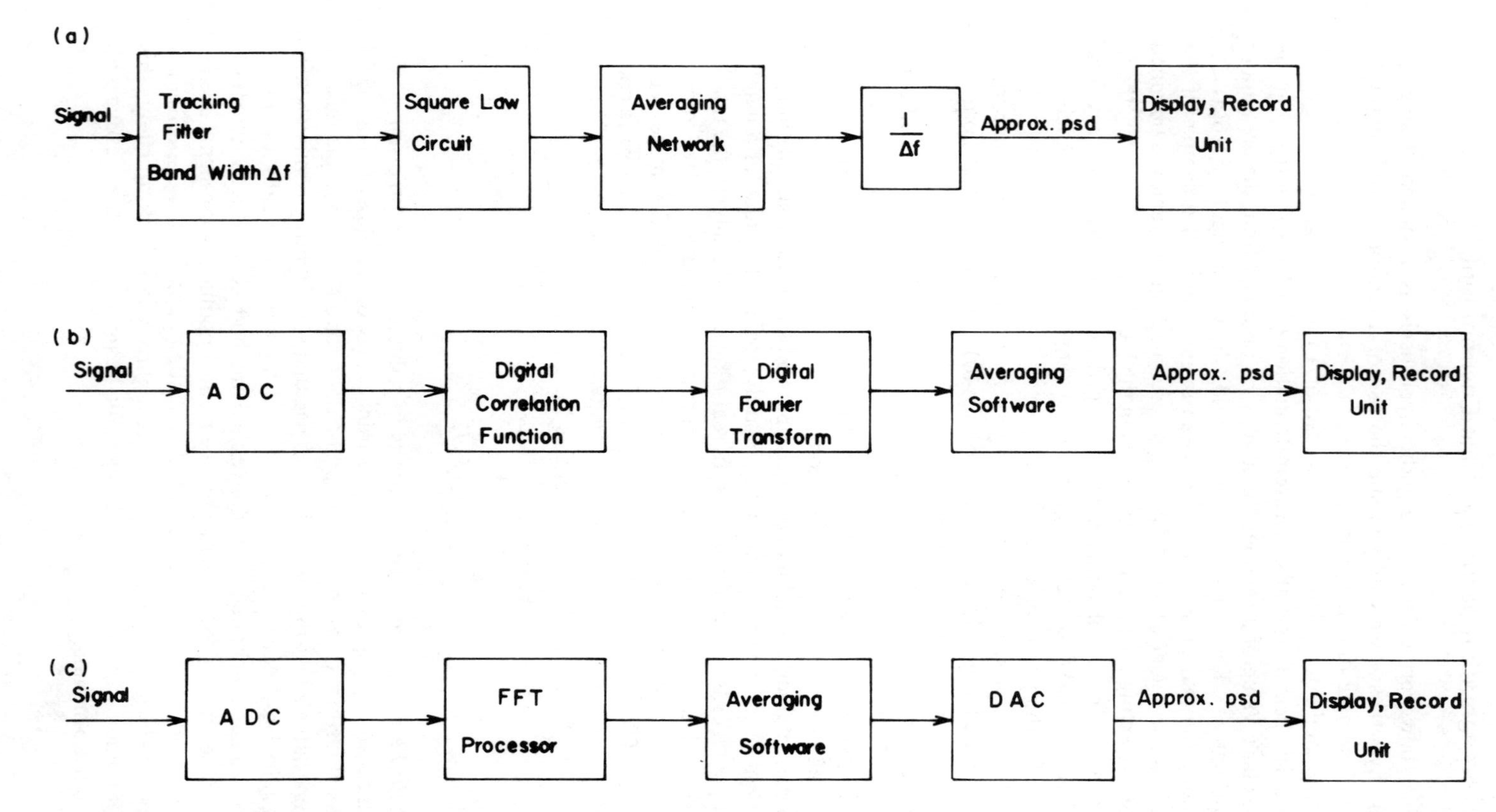

Figure 5-16. Some Methods of psd Determination: (a) Filtering, Squaring, and Averaging Method; (b) Using Autocorrelation Function; (c) Using Direct FFT

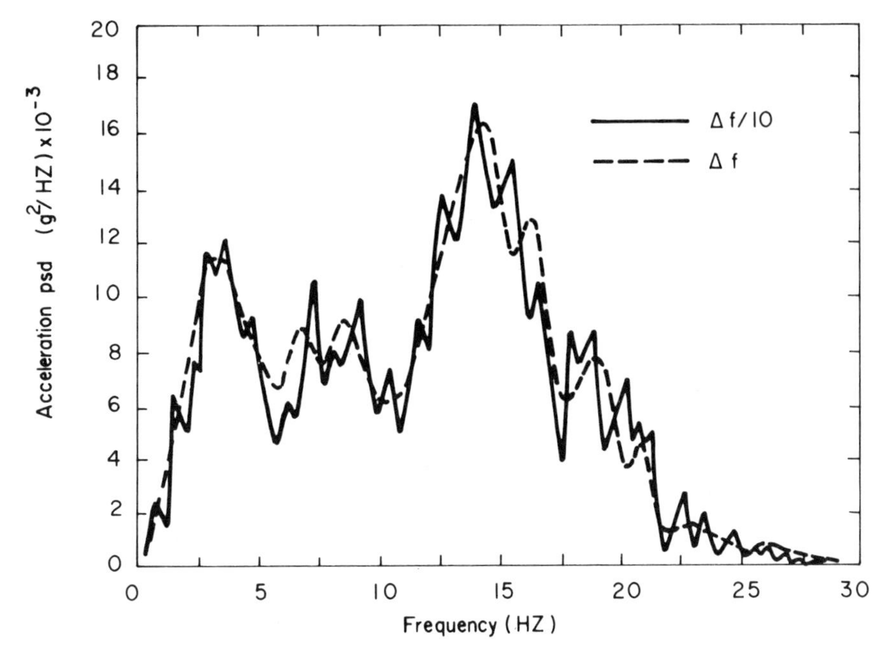

Figure 5-17. Effect of Filter Bandwidth on psd Results

Comparison of Various Representations

In this section, we shall state some major advantages and disadvantages of the four representations of the dynamic environment that we have discussed.

Time-history representation has several advantages. It can be employed to represent either deterministic or random dynamic environment. It is an exact repesentation of a single dynamic event. Also, when performing multiexcitation (multiple shaker) dynamic testing, phasing between the various inputs can be conveniently incorporated simply by delaying each excitation with respect to the others. There are also disadvantages to time-history representation. Since each time history represents just one sample function (single event) of a random environment, it may not be truly representative of the actual random process. This can be overcome by using longer time histories, which, however, will increase the duration of the test, which is limited by test specifications. If the random process is truly ergodic (or at least stationary), this problem will not be serious. Furthermore, the problem does not arise when testing with deterministic signals. An extensive knowledge of the true dynamic environment for which the test object is qualified is necessary, however, in order to conclude that it is stationary or that it could be represented by a deterministic signal. In this sense, time-history representation is difficult to implement.

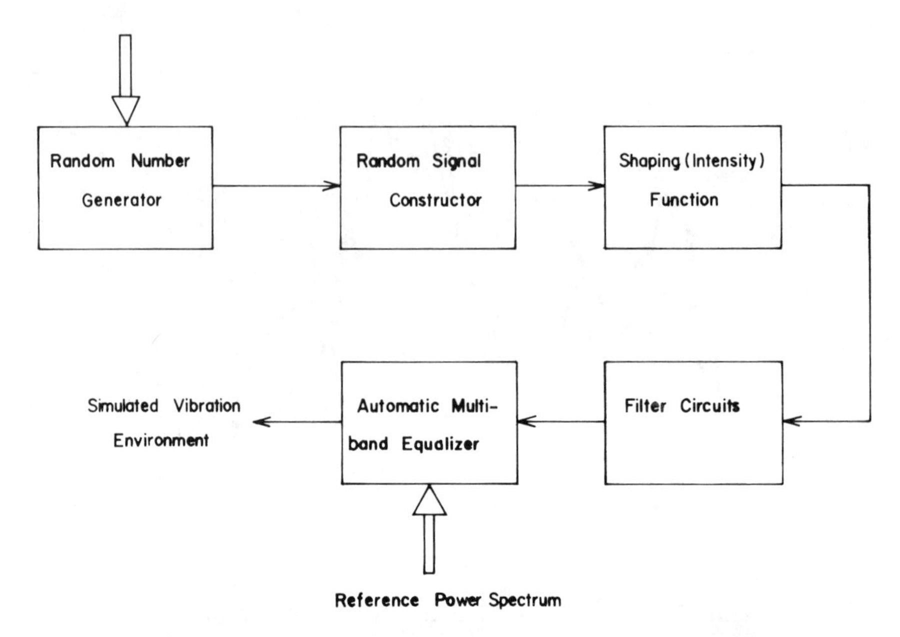

Figure 5-18. Accurate Generation of a Specified Random Dynamic Environment

The response-spectrum method of representing a dynamic environment also has several advantages. It is relatively easier to implement. Since peak response of a simple oscillator is used in its definition, it is representative of peak response or structural stress of simple dynamic systems. An upper bound for the peak response of a multi-degree-of-freedom system can be conveniently obtained by the method outlined in the earlier section on uses of response spectra. Also, by considering the envelope of a set of response spectra at the same damping value, it is possible to use a single response spectrum to conservatively represent more than one dynamic event. The method also has disadvantages. It employs deterministic signals in its definition. Sample functions (single events) of random processes can be used, however. It is not possible to determine the original time history from the knowledge of its response spectrum, because it considers peak values of response of a simple oscillator, and hence more than one time history can have the same response spectrum. Thus, response spectrum cannot be considered a complete representation of a dynamic environment. Also, characteristics such as the transient nature and the duration of the dynamic event cannot be deduced from the response spectrum. For the same reason, it is not possible to incorporate information on excitation signal phasing into the response-spectrum representation. This is a disadvantage in multiple excitation testing.

Fourier spectrum representation also has advantages. Since the actual dynamic environment time history can be obtained by inverse transformation, it has the advantages of time-history representation. In particular, since Fourier spectrum is generally complex test excitation input-phasing information can be incorporated into the Fourier spectra in multiple excitation testing. Furthermore, by considering an envelope Fourier spectrum (like an envelope response spectrum), it can be employed to conservatively represent more than one dynamic environment. Also, it gives frequency-domain information (such as resonances), which is very useful in dynamic-testing situations. The disadvantages of the Fourier spectrum representation include the following. It is a deterministic representation, but, as in response-spectrum method, a sample function (single event) of a random process can be represented by its Fourier spectrum. Transient effects and event duration are hidden in this representation. Also, it is rather difficult to implement, because complex procedures of multiband equalization might be necessary in the signal synthesis associated with this representation.

Power spectral density representation has the following advantages. It takes into account the random nature of a dynamic environment. As in response-spectrum and Fourier spectrum representation, by taking an envelope psd, it can be used to conservatively represent more than one dynamic environment. It can display important frequency-domain characteristics, such as resonances. Its disadvantages include the following. It is an exact representation of truly stationary or ergodic random environments only. In nonstationery situations, as in seismic ground motions, a significant error could result. Also, it is not possible to obtain the original sample function (dynamic event) from its psd. Hence, transient characteristics and duration of the event are not known from its psd. Since mean square values, not peak values, are considered, psd representation is not structural-stress-related. Furthermore, since psd functions are real, we cannot incorporate phasing information into them. This is a disadvantage in multiple excitation testing situations, but this problem can be overcome by considering either the cross spectrum (which is complex) or the cross correlation in each pair of excitations.

Notes

1. N. M. Newmark and E. Rosenblueth, *Fundamentals of Earthquake Engineering,* (Englewood Cliffs, N.J.: Prentice-Hall, 1971).
2. A.G. Ratz and F.R. Barlett, "Vibration Simulation Using Electrodynamic Exciters," *Vibration Testing—Instrumentation and Data Analysis,* AMD - Vol 12, American Society of Mechanical Engineers, New York, 1975.

6 Pretest Procedures

The selection of a test procedure for dynamic qualification of an item should be based on technical information regarding the item and its intended use. Vendors usually prefer to use more established, conventional qualification methods and generally are reluctant to incorporate modifications and improvements. This is primarily due to economic reasons, convenience, testing-time limitations, availability of the equipment and facilities (test-lab limitations), and similar factors. Regulatory agencies, however, usually modify their guidelines from time to time, and some of these requirements are mandatory.

Before conducting a dynamic test on a test object, it is necessary to follow several pretest procedures. Such procedures are necessary in order to conduct a meaningful test. Some important pretest procedures include the following:

1. Understanding the purpose of the test
2. Studying the service functions of the test object
3. Information acquisition on the test object
4. Test-program planning
5. Pretest inspection of the test object
6. Resonance-search pretesting to gather dynamic information about the test object
7. Mechanical aging of the test object

In the following sections, we shall discuss each of these procedures to emphasize how they can contribute to a meaningful test.

Purpose of Testing

Dynamic testing is useful in various stages of development, production and utilization of a product. In the initial design stage, design weaknesses and possible improvements can be determined through dynamic testing of a preliminary design of the unit or a partial product. In the production stage, the quality of workmanship and material of the final product may be evaluated using nondestructive dynamic testing. In the product-utilization stage, the adequacy of a finished product of good quality for a special-purpose application or a range of applications can be determined by qualification testing.

Depending on the outcome of a dynamic test, design modifications or corrective actions can be recommended for a preliminary design or a partial product. To determine the most desirable location (in terms of minimal noise and vibration) for the compressor in a refrigerator unit, for example, a resonance-search test could be employed. As another example, dynamic testing could be employed to determine vibration-isolation material requirements in structures for adequate damping. Such tests fall into the first category of system development tests. They are beneficial for the designer and the manufacturer in improving the quality of performance of the product. Government regulatory agencies do not usually stipulate the requirements for this category of tests, but they might stipulate minimal requirements for safety and performance levels of the final product, which could indirectly affect the test requirements. Custom-made items are exceptions for which the customers could stipulate the design-test requirements.

For special-purpose products, it might also be necessary to conduct a dynamic test on the final product before its installation for service operation. For mass-produced items, it is customary to select representative samples from each batch for these tests. The purpose of such tests is to detect any inferiorities in the workmanship or in the materials used. These tests fall into the second category—quality-assurance tests. These usually consist of a standard series of routine tests that are well established for a given product.

Seismic qualification of devices and components is a good example of the use of the third category—qualification tests. A high-quality product such as a valve actuator, for instance, which is thoroughly tested in the design-development stage and at the final production stage, will need further dynamic tests or analysis if it is to be installed in a nuclear power plant. The purpose in this instance is to determine whether the unit could perform its intended functions when it was subjected to specified levels of seismic excitations. The nature of the tests required for seismic qualification depends on the characteristics of the valve actuator and its intended use—for example, whether it would be crucial for system-safety-related functions. Government regulatory agencies usually stipulate the basic requirements for qualification tests. These tests are specifically application-oriented. The vendor or the customer might employ more elaborate test programs than those stipulated by the regulatory agency, but at least the minimum requirements set by the agency should be met before commissioning the plant.

The purpose of any dynamic test should be clearly understood before incorporating it in a test program. A particular test might be meaningless under some circumstances. If it is known, for instance, that no resonances below 35 Hz exist in certain equipment that requires seismic qualification, then it is not necessary to conduct a resonance search, because the predominant frequency content in seismic excitations occurs below 35 Hz. If, however, the test serves a dual purpose, such as mechanical aging in addition to resonance detection, then it may still be conducted even without resonances in the predominant frequency range of excitation.

If testing is performed on one test item selected from a batch of equipment to assure the quality of the entire batch or to qualify the entire batch, it is necessary to establish that all items in the batch are of identical design. Otherwise, testing of all items in the batch might be necessary unless some form-design similarity can be identified. Qualification by similarity is done in this manner.

The nature of dynamic testing employed is usually governed by the test purpose. Single-frequency tests, using deterministic test excitation inputs, for example, are well suited to design-development and quality-assurance applications. The main reason for this choice is that the test-input excitations can be completely defined; consequently, a complete analysis can be performed with relative ease, based on existing theories and dynamic models. Random or multifrequency tests are more realistic in qualification tests, however, because under typical service conditions, the dynamic environments to which equipment is subjected are random and multifrequency in nature (for example, seismic disturbances, ground-transit road disturbances, aerodynamic disturbances). Since random excitation tests are relatively more expensive and complex in terms of signal generation and data processing, single-frequency tests might also be employed in qualification tests. Under some circumstances, single-frequency tests could add excessive conservatism to the test excitation. It is known, for instance, that single-frequency tests are justified in the qualification of line-mounted equipment, which can encounter in-service disturbances that are amplified because of resonances in the mounting structure.

Service Functions

For dynamic qualification by testing, it is required that the test object remain functional and maintain its structural integrity when subjected to a certain prespecified dynamic environment. In seismic qualification of equipment, for instance, the dynamic environment is an excitation that adequately represents the amplitude, phasing, frequency content, and transient characteristics (decay rate and signal duration) of the motion at the equipment-support locations caused by the most severe seismic disturbance that is expected, with a reasonable probability, during the design life of the equipment. Monitoring the proper performance of in-service functions (functional-operability monitoring) of a test object during dynamic testing could be crucial in the qualification decision.

The intended service functions of the test object should be clearly defined prior to testing. For active equipment, functional operability is necessary during dynamic testing. For passive equipment, however, only the structural integrity need be maintained during testing.

Active Equipment

Equipment that should perform a mechanical motion (for example, valve closure, relay contact) or that produces a measurable signal (for example, electrical signal, pressure, temperature, flow) during the course of performing its intended functions is termed active equipment. Some examples of active equipment are valve actuators, relays, motors, pumps, transducers, control switches, and recorders.

Passive Equipment

Passive equipment typically performs containment functions and consequently should maintain a certain minimum structural strength or pressure boundary. Such equipment usually does not perform mechanical motions or produce measurable output signals, but it may have to maintain displacement tolerances. Some examples of passive equipment are piping, tanks, cables, supporting structures and heat exchangers.

Functional Testing

When defining intended functions of equipment for test purposes, the following information should be gathered for each active component of the unit that will be tested:

1. The maximum number of times a given function should be performed during the design life of the equipment
2. The best achievable precision (or monitoring tolerance) for each functional-operability parameter and the time duration for which a given precision is required
3. Mechanisms and states of malfunction or failure
4. Limits of the functional-operability parameters (electrical signals, pressures, temperatures, flow rates, mechanical displacements and tolerances, relay chatter, and so forth) that correspond to a state of malfunction or failure.

It should be noted that, under a state of malfunction, the equipment would not perform the intended function properly. Under a failure state, however, the equipment would not perform its intended function at all.

For equipment consisting of an assembly of several crucial components, it should be determined how a malfunction or failure of each component could result in malfunction or failure of the entire unit. In such cases, any hardware redundancy (that is, when component failure does not necessarily cause unit failure) and possible interactive and chain effects (such as failure in one component overloading another, which could result in subsequent failure of the second component, and so on) should be identified. In considering functional precision, it should be noted that high precision usually means increased complexity of the test procedure. This is further complicated if a certain precision is required at a prescribed instant.

It is common practice for the equipment supplier (customer) to define the functional test, including acceptance criteria and tolerances for each function, for the benefit of the test engineer. This information eventually is used in determining acceptance criteria for the qualification tests of active equipment. Complexity of the required tests also depends on the precision requirements for the intended functions of the test object.

Examples of functional failure are sensor and transducer (measuring instrumentation) failure, actuator (motors, and so forth) failure, chatter in relays, gyroscopic drift, and discontinuity of electrical signals because of short-circuiting. It should be noted that functional failures caused by mechannical excitations are often linked with structural integrity of the test object. Such functional failures are primarily caused in two ways: (1) when displacement amplitude exceeds a certain critical value once or several times, or (2) when vibrations (or dynamic motions) of moderate amplitudes occur for an extended period of time. Functional failures in the first category include, for example, short-circuiting, contact errors, and instabilities and nonlinearities in amplifier outputs. Such failures are usually reversible, so that, when the excitation intensity drops, the system would function normally. Under the second category, slow degradation of components would occur because of wear and fatigue, which could cause subsequent malfunction or failure. This kind of failure is usually irreversible. We must emphasize that the first category of functional failure can be better simulated using high-intensity single-frequency testing and shock testing, and the second category by multifrequency or broadband random testing and low-intensity single-frequency testing.

For passive devices, a damage criterion should be specified. This could be expressed in terms of parameters such as cumulative fatigue, deflection tolerances, wearout limits, pressure drops, and leakage rates. Often, damage or failure in passive devices can be determined by visual inspection.

Information Acquisition

In additon to information concerning service functions, as discussed in the previous section, and dynamic characteristics determined from a resonance search, as discussed later in this chapter, there are other characteristics of the test object that need to be studied in the development of a dynamic-testing program. In particular, there are characteristics that cannot be described in exact quantitative terms. In determining the value of equipment, for instance, the monetary value (or cost) might be relatively easy to estimate, whereas it could be very difficult to assign a dollar value to its significance under service conditions. One reason for this could be that the particular piece of equipment alone might not determine the proper operation of a complex system. Interaction of a particular unit with other subsystems in a complex operation would determine the importance attached to it and, hence, its value. In this sense, the true value of a test object is a relatively complex consideration. Service function of the test object is also an important consideration in determining its value. The value of a test object is important in planning a test program, because the cost of a test program and the effort expended therein are governed mainly by this factor.

Many features of test object that are significant in planning a test program can be deduced from manufacturer's data. The following information is representative:

1. Drawings (schematic or to scale when appropriate) of principal components and the whole assembly, with manufacturer's name, identification numbers, and dimensions clearly indicated
2. Materials used, design strengths, fatigue life, and so on, of various components, and factors determining the structural integrity of the unit
3. Component weight and total weight of the unit
4. Design ratings, capacities, and tolerances for in-service operation of each crucial component
5. Description of intended functions of each component and of the entire unit, clearly indicating the parameters that determine functional operability of the unit
6. Interface details (intercomponent as well as for the entire assembly), including in-service mounting configurations and mounting details
7. Details of the probable operating site (particularly with respect to seismic activities, if seismic qualification is intended)
8. Details of any previous testing or analysis performed on that unit or a similar one

Scale drawings and component-weight information describe the size and geometry of the test object. This information is useful in determining the following:

1. The locations of sensors (accelometers, strain gauges, and the like) for monitoring dynamic response of the test object during tests
2. The ratings of dynamic-test (shaker) apparatus (power, force, stroke, and so on)
3. The degree of dynamic interaction between the test object and the test apparatus
4. The level of coupling between various degrees of freedom and modal interactions in the test object
5. The assembly level of the test object (for example, whether it can be treated as a single component, as a subsystem consisting of several components, or as an independent, stand-alone system).

In general, as the size and the assembly level increase, the tests become increasingly complex and difficult to perform. To test heavy, complex test objects, we would need a large test apparatus with high power ratings and multiple-excitation location capability. In this case, the number of operability parameters monitored and the number of dynamic observation locations also will increase.

Reliability Considerations for Multicomponent Units

Equipment that has several components that are crucial to its operation can have more than one mode of failure. Each failure mode will depend on some combination of failure of the components. Component failure is governed by the laws of probability. We shall first consider some fundamentals of probability theory that are useful in the reliability or failure analysis of multicomponent units.

Reliability. The probability that the component will perform satisfactorily over a specified time period t (component age) under given operating conditions is called reliability. It is denoted by R. Hence,

$$R(t) = \mathscr{P}\ (\text{survival}) \tag{6.1}$$

in which $\mathscr{P}\ (\cdot)$ denotes "the probability of."

Unreliability. The probability that the component will malfunction or fail during the time period t is called its unreliability, or its probability of failure. It is denoted by F. Hence,

$$F(t) = \mathscr{P}\ (\text{failure}) \tag{6.2}$$

Since we know as a certainty that the component will either survive or fail during the specified time period t,

$$R(t) + F(t) = 1 \tag{6.3}$$

The probability of survival of a component usually decreases with age. Consequently, typical $R(t)$ is a monotonically decreasing function of t, as shown in figure 6-1. If it is known as a certainty that the component is good in the beginning, then $R(0) = 1$. Because of manufacturing defects, damage during shipping, and the like, however, we usually have $R(0) \leq 1$. For a satisfactory component, $R(t)$ should not drop appreciably during its design life T_d. The drop is faster initially, however, because of infant mortality (again due to manufacturing defects and the like), and, as it exceeds its design life, because of old age (wear, fatigue, and so on).

It is clear from equation (6.3) that the unreliability curve is completely defined by the reliability curve. As shown in figure 6-1, transforming one to the other is a simple matter of reversing the axis.

The Inclusion-Exclusion Formula. Consider two events, A and B, that are schematically represented by areas (as in figure 6-2). Each event consists of a set of outcomes. The total area covered by the two sets denoted by A and B

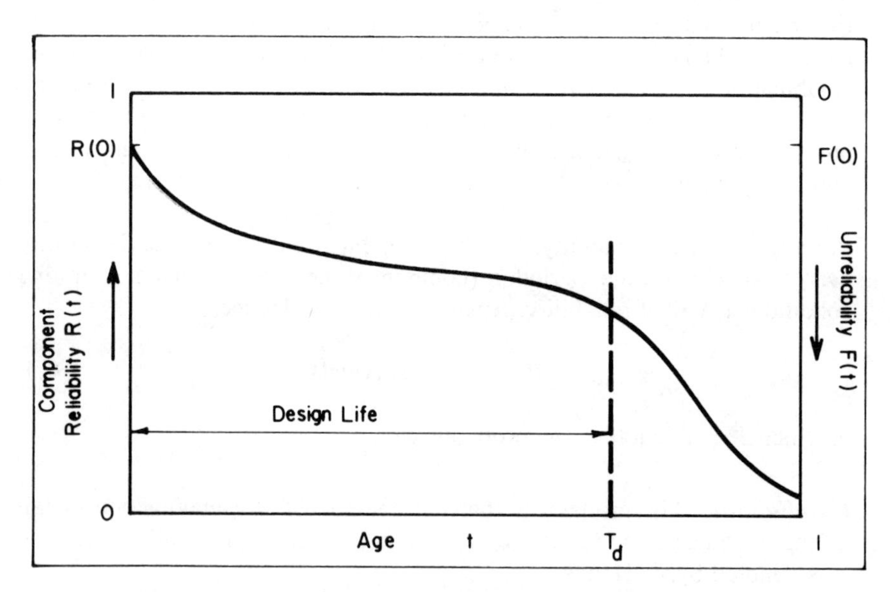

Figure 6-1. A Typical Reliability (Unreliability) Curve

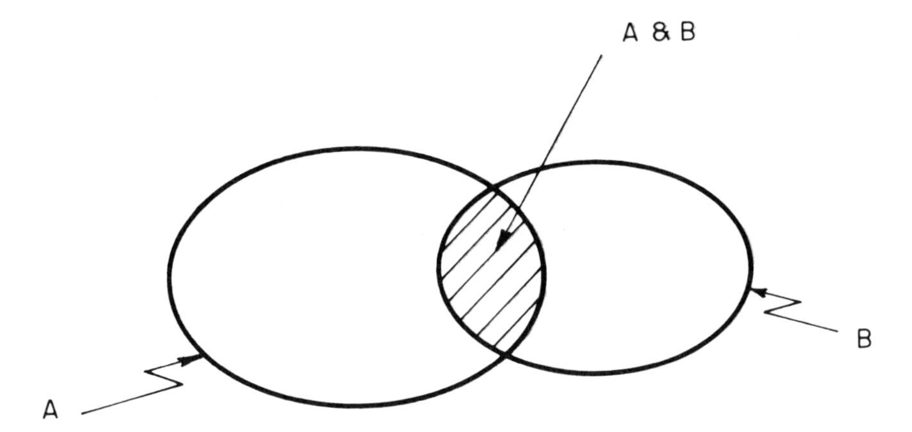

Figure 6-2. Venn Diagram Illustrating the Inclusion-Exclusion Formula

is given by adding the area of A to the area of B and subtracting the common area. This procedure can be expressed as

$$\mathscr{P}\ (A \text{ or } B) = \mathscr{P}\ (A) + \mathscr{P}\ (B) - \mathscr{P}\ A \text{ and } B) \qquad (6.4)$$

Consider the rolling of a fair die, for example. The set of total outcomes consists of six elements forming the space

$$S = \{1, 2, 3, 4, 5, 6\}$$

Each outcome has a probability of 1/6. Now consider the two events

$$A = \{\text{outcome is odd}\}$$
$$B = \{\text{outcome is divisible by 3}\}$$

Then,

$$A = \{1, 3, 5\}$$
$$B = \{3, 6\}$$

Consequently,

$$A \text{ or } B = \{1, 3, 5, 6\}$$
$$A \text{ and } B = \{3\}$$

It follows that

$$\mathscr{P}(A) = 3/6, \mathscr{P}(B) = 2/6, \mathscr{P}(A \text{ or } B) = 4/6, \mathscr{P}(A \text{ and } B) = 1/6.$$

These values satisfy equation (6.4).

If the events A and B do not have common outcomes, they are said to be mutually exclusive. Then the common area of intersection of sets A and B in figure 6-2 would be zero. Hence,

$$\mathscr{P}(A \text{ and } B) = 0 \tag{6.5}$$

for mutually exclusive events.

Bayes's Theorem. A simplified version of Bayes's theorem can be expressed as

$$\begin{aligned} \mathscr{P}(A \text{ and } B) &= \mathscr{P}(A/B)\mathscr{P}(B) \\ &= \mathscr{P}(B/A)\mathscr{P}(A) \end{aligned} \tag{6.6}$$

in which $\mathscr{P}(A/B)$ denotes the conditional probability that event A occurs, given the condition that event B has ocurred.

In the previous example of rolling a fair die, if it is known that event B has occurred, the outcome must be either 3 or 6. Then, the probability that event A would occur is simply the probability of picking 3 from the set {3, 6}. Hence, $\mathscr{P}(A/B) = 1/2$. Similarly, $\mathscr{P}(B/A) = 1/3$. It should be noted that equation (6.6) holds for this example.

Product Rule for Independent Events. If the two events A and B are independent of each other, then the occurrence of event B has no effect whatsoever on determining whether event A occurs. Consequently,

$$\mathscr{P}(A/B) = \mathscr{P}(A) \tag{6.7}$$

for independent events. Then, it follows from equation (6.6) that

$$\mathscr{P}(A \text{ and } B) = \mathscr{P}(A)\mathscr{P}(B) \tag{6.8}$$

for independent events. Equation (6.8) is the product rule, which is applicable to independent events.

It should be emphasized that, even though independence implies that the product rule holds, the converse is not necessarily true. In the example on rolling of a fair die, $\mathscr{P}(A/B) = \mathscr{P}(A) = 1/2$. Suppose, however, that it is not a fair die and that the probabilities of the outcomes {1, 2, 3, 4, 5, 6} are {1/3, 1/6, 1/6, 0, 1/6, 1/6}. Then,

$$\mathscr{P}(A) = 1/3 + 1/6 + 1/6 = 2/3$$

whereas

$$\mathscr{P}(A/B) = \frac{1/6}{1/6 + 1/6} = 1/2$$

This shows that A and B are not independent events in this sample.

Failure Rate. The function $F(t)$ defined by equation (6.2) is the probability-distribution function of the random variable denoting the time to failure. We shall define the rate functions:

$$r(t) = \frac{dR(t)}{dt} \tag{6.9}$$

$$f(t) = \frac{dF(t)}{dt} \tag{6.10}$$

In equation (6.10), $f(t)$ is the probability-density function corresponding to the time to failure. It follows that

$$\mathscr{P}(\text{component survived up to } t\text{, failed within next duration } dt) = \mathscr{P}(\text{failed within } t,\ t + dt) = dF(t) = f(t)dt \tag{6.11}$$

Also,

$$\mathscr{P}(\text{component survived up to } t) = R(t) \tag{6.12}$$

Let us define the function $\beta(t)$ such that

$$\mathscr{P}(\text{failed within next duration } dt/\text{survived up to } t) = \beta(t)dt \tag{6.13}$$

By substituting equations (6.11) through (6.13) into equation (6.6) we obtain

$$f(t)dt = \beta(t)\, dt\ R(t)$$

or

$$\beta(t) = \frac{f(t)}{R(t)} = \frac{f(t)}{1 - F(t)} \tag{6.14}$$

Let us suppose that there are N components. If they all have survived up to t, then, on the average, $N\beta(t)\,dt$ components will fail during the next dt. Consequently, $N\beta(t)$ corresponds to the rate of failure for the collection of components at time t. For a single component ($N = 1$), the rate of failure is $\beta(t)$. For obvious reasons, $\beta(t)$ is sometimes termed conditional failure. Other names for this function include intensity function and hazard function, but *failure rate* is the most common name.

In view of equation (6.10), we can write equation (6.14) as a first-order linear, ordinary differential equation with variable parameters:

$$\frac{dF(t)}{dt} + \beta(t)F(t) = \beta(t) \tag{6.15}$$

Assuming a good component initially,

$$F(0) = 0 \tag{6.16}$$

The solution of equation (6.15) subject to equation (6.16) is

$$F(t) = 1 - \exp\left(- \int_0^t \beta(\tau)\, d\tau \right) \tag{6.17}$$

in which τ is a dummy variable. Then, from equation (6.3),

$$R(t) = \exp\left(- \int_0^t \beta(\tau)\, d\tau \right) \tag{6.18}$$

It is observed from equation (6.18) that the reliability curve can be determined from the failure-rate curve, and the reverse.

A typical failure-rate curve for an engineering component is shown in figure 6-3. It has a characteristic bathtub shape, which can be divided into three phases, as in the figure. These phases might not be so distinct in a real situation. Initial burn-in period is characterized by a sharp drop in the failure rate. Because of such reasons as poor workmanship, material defects, and poor handling during transportation, a high degree of failure can occur during a short initial period of the design life. Following that, the failures typically will be due to random causes. The failure rate is approximately constant in this region. Once the design life is exceeded (third phase), rapid failure can occur because of wearout, fatigue, and other types of cumulative damage, and eventual collapse will result.

It is frequently assumed that the failure rate is constant during the design life of a component. In this case, equation (6.18) gives the exponential reliability function:

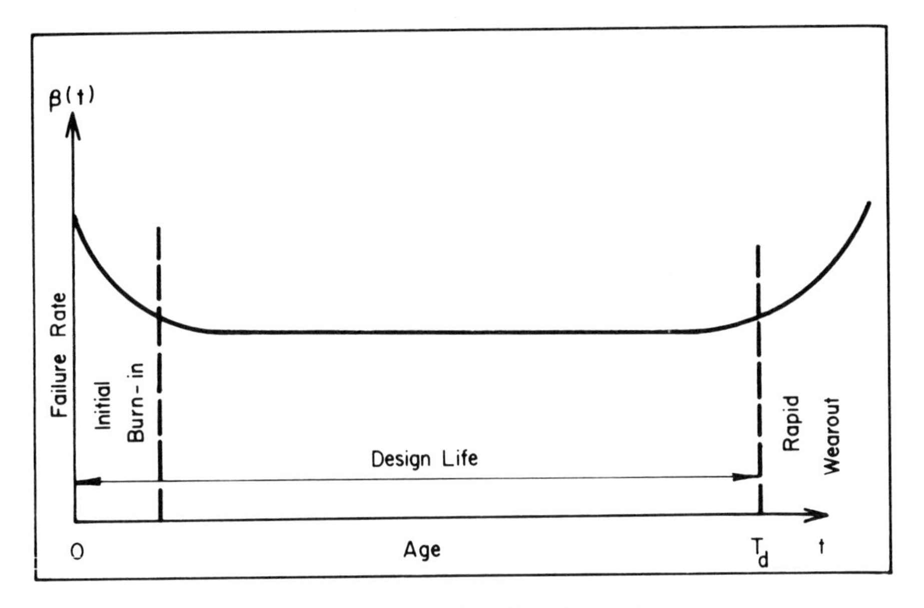

Figure 6-3. A Typical Failure-Rate Curve

$$R(t) = \exp(-\beta t) \tag{6.19}$$

This situation is represented in figure 6.4. This curve is not comparable to the general reliability curve shown in figure 6-1. As a result, constant failure rate should not be used for relatively large durations of time (that is, for a large portion of the design life), unless it has been verified by tests. For short durations, however, this approximation is normally used and results in considerable analytical simplicity.

Product Rule for Reliability. For multicomponent equipment, if we assume that the failure of one component is independent of the failure of any other, the product rule given by equation (6.8) can be used to determine the overall reliability of the equipment. The reliability of N-component equipment, with independently failing components, is given by

$$R(t) = R_1(t)R_2(t) \ldots R_N(t) \tag{6.20}$$

in which $R_i(t)$ is the reliability of the ith component. If there is no component redundancy, which is assumed in equation (6.20), none of the components should fail (that is, $R_i(t) \neq 0$ for $i = 1, 2, \ldots, N$) for the equipment to operate properly (that is, $R(t) \neq 0$). This follows from equation (6.20).

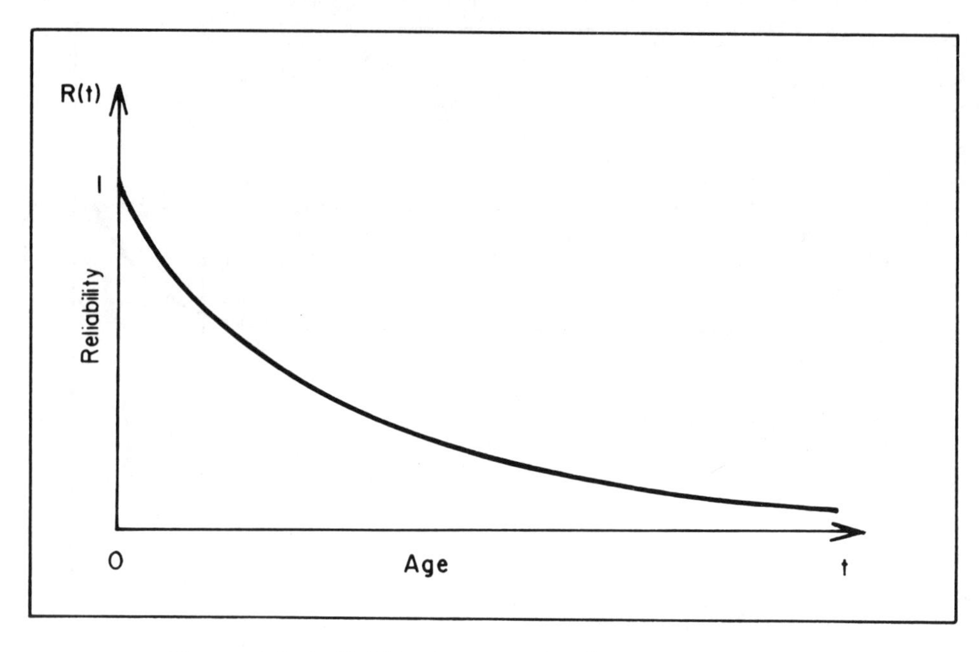

Figure 6-4. Reliability Curve under Constant Failure Rate

In dynamic testing, a primary objective is to maximize the risk of component failure when subjected to the test environment (so that the probability of failure is less in the actual in-service environment). One way of achieving this is by maximizing the test-strength-measure function (see chapter 10 for details), given by

$$TS = \sum_{i=1}^{r} F_i(T)\,\Phi_i \tag{6.21}$$

in which $F_i(T)$ is the probability of failure (unreliability) of the ith component for the test duration T, and Φ_i is a dynamic-response measure at the location of the ith component. The parameters of optimization could be the input direction and the frequency of excitation for a given input intensity.

Regarding component redundancy, consider the simple situation of r_i identical subcomponents connected in parallel (r_ith-order redundancy) to form the ith component. The component failure requires the failure of all r_i subcomponents. The failure of one subcomponent is assumed to be independent of the failure state of other subcomponents. Then, the unreliability of the ith component can be expressed as

$$F_i = (F_{0i})^{r_i} \qquad (6.22)$$

in which F_{0i} is the unreliability of each subcomponent in the ith component. This simple model for redundancy may not be valid in some situations.

There are two basic types of redundancy: active redundancy and standby redundancy. In active redundancy, all redundant elements are permanently connected and active during the operation of the equipment. In standby redundancy, only one of the components in a redundant group is active during the equipment operation. If that component fails, an identical second component will be automatically connected.

For standby redundancy, some form of switching mechanism is needed, which means that the reliability of the switching mechanism itself must be accounted for. Component aging is relatively less, however, and the failure of components within the redundant group is mutually independent. In active redundancy, however, there is no need for a switching mechanism. The failure of one component in the redundant group can overload the rest, however, increasing their probability of failure (unreliability). Consequently, component failure within the redundant group is not mutually independent in this case. Also, component aging is relatively high because the components are continuously active.

Interface Details

The dynamics of a piece of equipment depends on the way the equipment is attached to its support structure. In addition to mounting details, equipment dynamic response also is affected by other interfacing linkages, such as wires, cables, conduits, pipes, and auxiliary instrumentation. In dynamic testing of equipment, such interface characteristics should be simulated appropriately. Dynamics of the test fixture and the details of the test object–fixture interface are very important considerations that affect the overall dynamics of the test object. If interface characteristics are not properly represented during testing, a nonuniform test could result, in which case some parts of the test object would be overtested and other parts undertested. This situation can bring about failures that are not representative of the failures that could take place in actual service. In effect, the testing could become meaningless if interface details are not simulated properly.

The test fixture is a structure attached to the shaker table and used to mount the test object (see figure 6-5). Test-fixture dynamics can significantly modify shaker-table motion before reaching the test object. Such modifications include filtering of the shaker motion and introduction of auxiliary (cross-axis) motions. In the test setup shown in figure 6-5, for example, the direct motion will be modified to some extent by fixture dynamics. In addition, some transverse and rotational motion components will be transmitted to the test object by the test fixture because of its overhang.

To minimize interface-dynamic effects in dynamic-testing situations, an attempt should be made (1) to make the test fixture as light and as rigid as is feasible; (2) to simulate in-service mounting conditions at the test object–fixture interface; and (3) to simulate other interface linkages, such as cables, conduits, and instrumentation, to represent in-service conditions. Very often, design of a proper test fixture can be a costly and time-consuming process. A trade-off is possible by locating the control sensors (accelerometers) at the mounting locations of the test object and, if the error between the actual and the desired excitations is used to control the mounting-location excitations during testing, by using feedback.

Effect of Neglecting Interface Dynamics. We shall consider a simplified model to study some important effects of neglecting interface dynamics. In the model shown in figure 6-6, the equipment and the mounting interface are modeled separately as single-degree-of-freedom systems. Capital letters are used to denote the equipment parameters (mass M, stiffness K, and damping coefficient C). When mounting-interface dynamics are included, the model appears as in figure 6-6(a). When the mounting-interface dynamics are neglected, we obtain the single-degree-of-freedom model shown in figure 6-6(b). Note that, in the later case, shaker motion $u(t)$ is directly applied to the equipment mounts, whereas, in the former case, it is applied through the interface. If the equipment response in the two cases is denoted by y and $\widetilde{y}$, respectively, it can be shown by considering system-frequency transfer functions $Y(\omega)/U(\omega)$ and $\widetilde{Y}(\omega)/U(\omega)$ that

$$\frac{\widetilde{Y}(w)}{\widetilde{Y}(\omega)} = \frac{Ms^2}{(Ms^2 + Cs + K)} \frac{(Cs + K)}{(cs + k)} + \frac{(ms^2 + cs + k)}{(cs + k)} \tag{6.23}$$

with $s = j\omega$

The following nondimensional parameters are defined:

$$\text{Mass ratio } \alpha = \frac{m}{M} \tag{6.24}$$

$$\text{Natural frequency ratio } \beta = \frac{\omega_n}{\Omega_n} \tag{6.25}$$

$$\text{Normalized excitation frequency } \bar{\omega} = \frac{\omega}{\Omega_n} \tag{6.26}$$

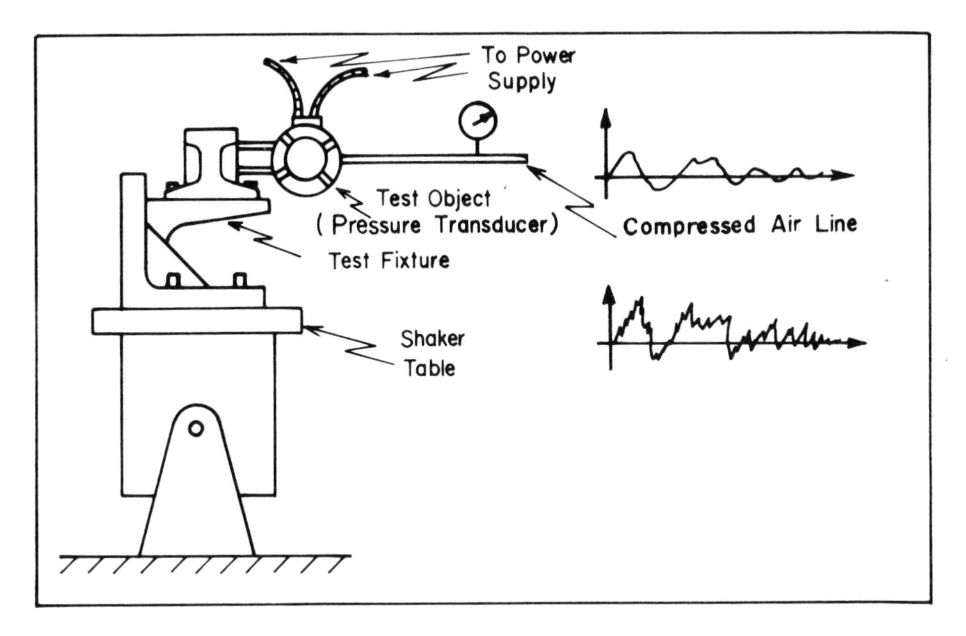

Figure 6-5. Influence of Test Fixture on Test Input Excitation

in which the natural frequency (undamped) of the equipment is

$$\Omega_n = \sqrt{\frac{K}{M}} \tag{6.27}$$

and the mounting-interface natural frequency is

$$\omega_n = \sqrt{\frac{k}{m}} \tag{6.28}$$

Then, equation (6.23) can be written

$$\frac{\widetilde{Y}(\omega)}{\widetilde{Y}(\omega)} = \frac{(\beta^2 + 2j\zeta\beta\bar{\omega} - \bar{\omega}^2)}{(\beta^2 + 2j\zeta\beta\bar{\omega})} - \frac{\bar{\omega}^2(1 + 2jZ\bar{\omega})}{\alpha(\beta^2 + 2j\zeta\beta\bar{\omega})(1 + 2jZ\bar{\omega} - \bar{\omega}^2)} \tag{6.29}$$

in which ζ and Z denote the damping ratios of the interface and the equipment, respectively.

The ratio $\widetilde{Y}(\omega)/Y(\omega)$ is representative of the equipment-response amplification when interface-dynamic effects are neglected for a harmonic excitation.

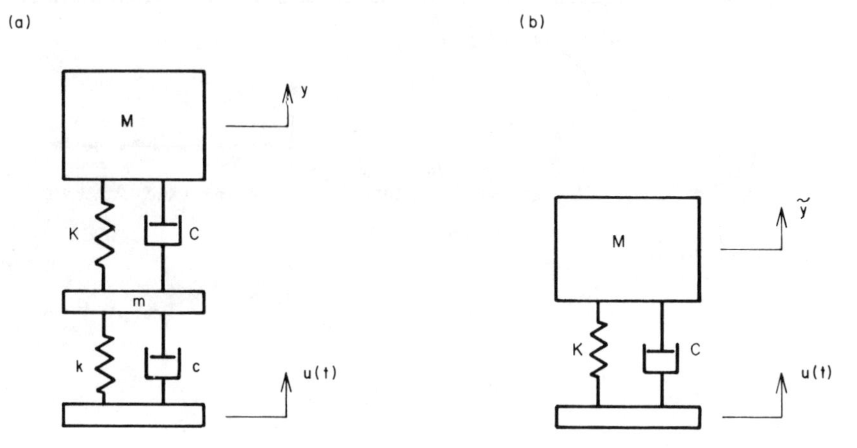

Figure 6-6. A Simplified Model to Study the Effect of Interface Dynamics: (a) With Interface Dynamics, (b) Without Interface Dynamics

Figure 6-7 shows eight curves, corresponding to equation (6.29), for the parameter combinations given in table 6-1. Interpretation of the results becomes easier when peak values of the response ratios are compared for various parameter combinations. Sample results are given in table 6-1.

Effect of Damping. By comparing cases 1, 2, 3, and 4 in table 6-1 with cases 5, 6, 7, and 8, respectively, we see that increasing interface damping has reduced the peak response (a favorable effect), irrespective of the values of interface mass and natural frequency (α and β values).

Effect of Inertia. By comparing cases 2, 3, 6, and 7 with 4, 1, 8, and 5, respectively, we see that interface inertia has a favorable effect in decreasing

Table 6-1
Response Amplification Caused by Neglecting Interface Dynamics

Case (Curve No.)	*Parameter Combination*				*Peak Value of Response Ratio*
	ζ	α	β	Z	
1	0.1	2.0	2.0	0.1	1.11
2	0.1	0.5	0.5	0.1	38.80
3	0.1	0.5	2.0	0.1	2.77
4	0.1	2.0	0.5	0.1	10.80
5	0.2	2.0	2.0	0.2	0.89
6	0.2	0.5	0.5	0.2	18.40
7	0.2	0.5	2.0	0.2	1.71
8	0.2	2.0	0.5	0.2	5.98

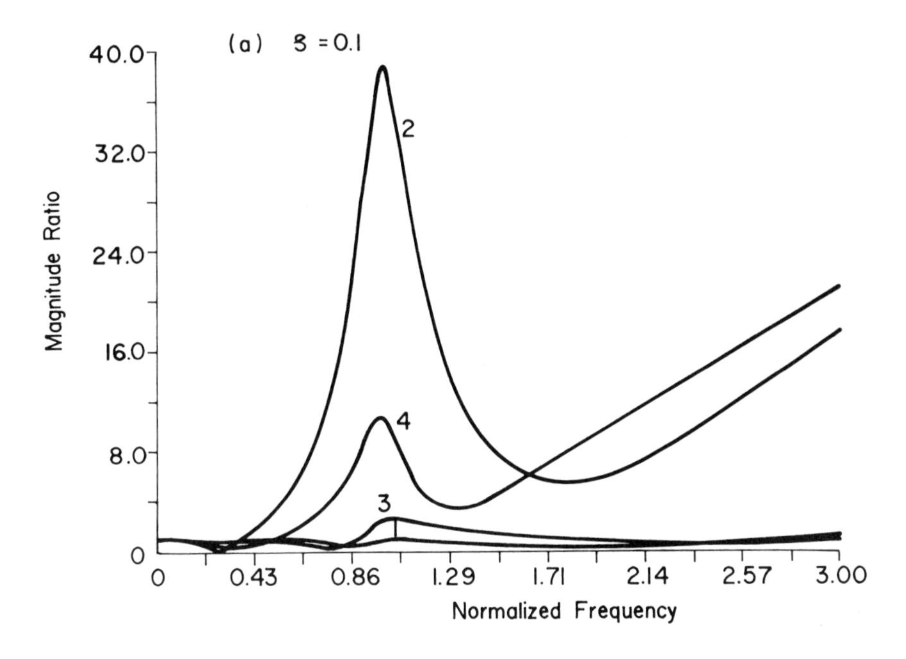

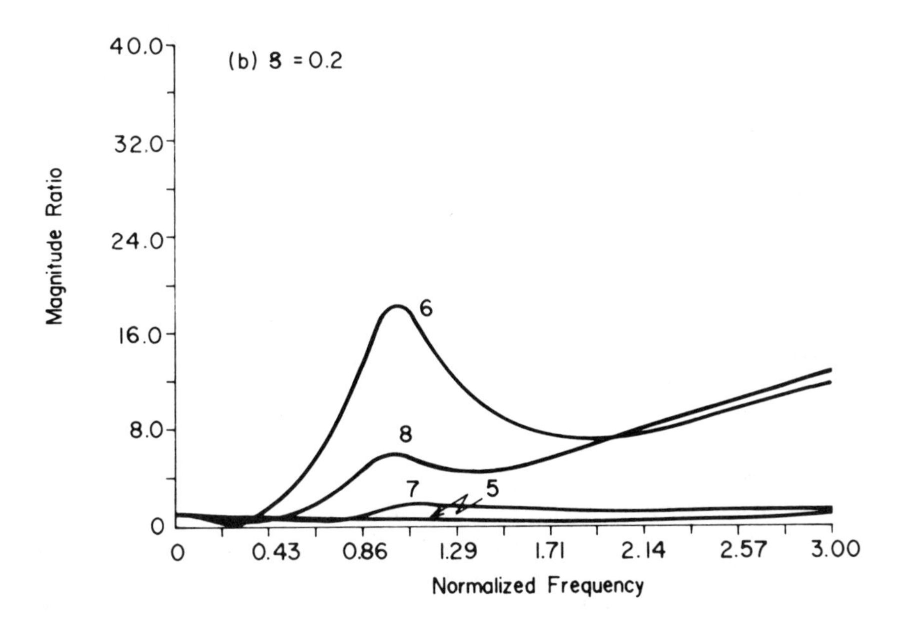

Figure 6-7. Response Amplification When Interface Dynamic Interactions Are Neglected

dynamic interactions, irrespective of interface damping and natural frequency.

Effect of Natural Frequency. By comparing cases 2, 4, 6, and 8 with 3, 1, 7, and 5, respectively, we see that increasing interface natural frequency has a favorable effect in decreasing dynamic interactions, irrespective of interface damping and inertia.

Effect of Excitation Frequency. All response plots (see figure 6-7) diverge to ∞ and $\bar{\omega}$ increases. This indicates that, at very high excitation-input frequencies, dynamic-testing results could become meaningless because of the interactions with interface dynamics.

It can be concluded that, to reduce dynamic interactions caused by interface we should (1) increase interface damping as much as is feasible, (2) increase interface mass as much as is feasible, (3) increase interface natural frequency as much as is feasible, and (4) avoid testing at relatively high frequencies of excitation input. It should be noted that, in the foregoing analysis and discussion, interface also includes test fixtures and the shaker table.

Other Effects of Interface. The type of dynamic test used sometimes depends on interface characteristics. An example is the testing of line-mounted equipment. Single-frequency testing is preferred for such equipment to add a certain degree of conservatism, because, as a result of interface resonances, line-mounted equipment could be subjected to higher levels of narrow-band excitations through the support structure.

In dynamic testing of multicomponent equipment, it is customary to test the cabinet first, with the components replaced by dummy weights, and then to test the individual components separately, using different test-excitation inputs depending on the component locations and their mounting characteristics. Interface details of individual components are important in such situations. As a result, interface information is an important constituent of the pretest information collected for a test object.

Most of the interface data, particularly information related to size and geometry (for example, mass, dimensions, configurations, and locations), can be gathered simply by observing the test object and using scale drawings supplied by the manufacturer. Size and number of anchor bolts used or weld thickness, for example, can be obtained in this manner. When analysis is also used to augment dynamic testing, however, it is often necessary to know the loads transmitted (forces, moments, and so on), relative displacements, and stiffness values at interface under in-service conditions. These must be determined by tests, by analysis (static or dynamic) of a suitable model, or from manufacturer's data.

Test-Program Planning

The test program to which a test object is subjected depends on several factors, including the following:

1. The objectives and specific requirements of the test;
2. In-service conditions, including equipment-mounting features, dynamic environment, and specifications of the test environment;
3. The nature of the test object, including complexity, assembly level, and functional-operability parameters to be monitored;
4. Test-laboratory capabilities, testing apparatus available, past experience, conventions, and established practices of testing.

Some of these factors are based on solid technical reasons, whereas others depend on economics, convenience, and personal likes and dislikes.

Initially, it is not necessary to develop a detailed test procedure; this is required only at the stage of actual testing. In the initial stage, it is only necessary to select the appropriate test method, based on factors such as those listed at the beginning of this section. Before conducting the tests, however, a test procedure should be prepared in sufficient detail. In essence, this is a pretest requirement.

Objectives and specific requirements of a test depend on such considerations as whether the test is conducted at the design stage, the quality-control stage, or the utilization stage. The objective of a particular test could be to verify the outcome of a previously conducted test. In that case, it is necessary to assess the adequacy of one or a series of tests conducted at an earlier time (for instance, when the specifications and government regulations were less stringent). Often, this could be done by analysis alone. Some testing might be necessary at times, but it usually is not necessary to repeat the entire test program. If the previous tests were conducted for the frequency range 1-25 Hz, for example, and the present specifications require a wider range of 1-35 Hz, it might be adequate merely to demonstrate (by analysis or testing) that there are no significant resonances in the test object in the 25-35 Hz range.

If it is necessary to qualify the test object for several different dynamic environments, a generic test that represents (conservatively, but without the risk of overtesting) all these dynamic environments could be used. For this purpose, special test-excitation inputs must be generated, taking into account the variability of the dynamic excitation characteristics under the given set of environments. Alternatively, several tests might be conducted if the dynamic environments for which the test object is to be qualified are significantly different. Operating-basis earthquake (OBE) tests and safe-shutdown earthquake (SSE) tests in seismic qualification of nuclear power plant equipment, for example, represent two significantly different test conditions. Consequently, they cannot be represented by a single test. When qualifying an equipment for several geographic regions or locations, however, we might be able to combine all OBE tests into a single test and all SSE tests into another single test.

Another important consideration in planning a test program is the required accuracy for the test, including the accuracy for the excitation inputs, response and operability measurements, and analysis. This is related to the value of the test object and the objectives of the test.

When it is required to evaluate or qualify a group of equipment by testing a sample, it is first necessary to establish that the selected sample unit is truly representative of the entire group. When the items in the batch are not identical in all respects, some conservatism could be added to the tests to minimize the possibility of incorrect qualification decision. It might be necessary to test more than one sample unit in such situations.

When planning a test procedure, we should clearly identify the standards, government regulations, and specifications that are applicable to a particular test. The pertinent sections of the applicable documents should be noted, and proper justification should be given if the tests deviate from regulatory-agency requirements.

Excitation input that is employed in a dynamic test depends on the in-service dynamic environment of the test object (see chapter 5). The number of tests needed also will depend on this to some extent. Test orientation depends mainly on mounting features and interface details of the test object under in-service conditions. Mounting features might govern the nature of the test-excitation inputs used for a particular test.

Two distinct mounting types can be identified for most equipment: line-mounted equipment and floor-mounted equipment. Line-mounted equipment is equipment that is mounted upright or hanging from pipelines, cables, or similar line structures that are not rigid. Generally, devices such as valves, nozzles, valve actuators, and transducers are considered line-mounted equipment. Any equipment that is not line-mounted is considered floor-mounted. The supporting structure is considered relatively rigid in this case. Examples of such mounting structures are floors, walls, and rigid frames. Typical examples of floor-mounted equipment include motors, compressors, and cabinets of relays and switchgear.

Wide-band ground disturbances are filtered by line structures. Consequently, the environmental disturbances to which line-mounted equipment is subjected would generally be narrow-band disturbances. Accordingly, dynamic testing of line-mounted equipment is best performed using narrow-band random test excitations or single-frequency deterministic test excitations. Relatively higher test intensities might be necessary for line-mounted equipment, because any low-frequency resonances that might be present in the mounting structure (which is relatively flexible in this case) could amplify the excitations before reaching the equipment.

Floor-mounted equipment often requires relatively wide-band random test excitations. As an example, consider a pressure transducer mounted on (1) a rigid wall, (2) a rigid I-section frame, (3) a pressurized gasline, or (4) a cabinet. In cases (1) and (2), wide-band random excitation inputs with response spectra approximately equal to the floor-response spectra could be employed for dynamic testing of the pressure transducer. For cases (3) and (4), however, flexibility of the support structure should be taken into consideration in developing the required response spectra (RRS) specifications for dynamic testing. In case (3), a single-frequency deterministic test, such as a sine-beat test or a sine-dwell test, could be employed, giving sufficient attention to testing at the equipment-resonant frequencies. In case (4), single-frequency tests could also be employed if the cabinet is considerably flexible and not rigidly attached to a rigid structure (a floor or a wall). Alternatively, a wide-band test on the cabinet itself, with the pressure transducer mounted on it, could be used.

Size, complexity, assembly level, and related features of a test object can significantly complicate and extend the test procedure. In such cases, testing the entire assembly might not be practical and testing of individual components or subassemblies might not be adequate because, in the in-service dynamic environment, the motion of a particular component could be significantly affected by the dynamics of other components in the assembly, the mounting structure, and other interface subsystems.

Functional-operability parameters to be monitored during testing should be predetermined. They depend on the purpose of the test, the nature of the test object, and the availability and characteristics of the sensors that are required to monitor these parameters. Malfunction or failure criteria should be related in some way to the monitored operability parameters; that is, each operability parameter should be associated with one or several components in the test object that are crucial to its operation.

The decision of whether to perform an active test (for example, whether a valve should be cycled during the test) and the determination of actuation time requirements (for example, the number of times the valve is cycled and at what instants during the test) should be made at this stage. The loading conditions for the test (that is, in-service loading simulation) also should be defined.

An important nontechnical factor that determines the nature of a dynamic test is the availability of hardware (test apparatus) in the test laboratory. This is especially true when nonconventional dynamic tests are required. Some specifications require three-degree-of-freedom test inputs (see chapter 7), for example, but most test laboratories have only one-degree-of-freedom or two-degree-of-freedom test machines. When two-degree-of-freedom or one-degree-of-freedom tests are used in place of three-degree-of-freedom tests, it is first required to determine what additional orientations of the test object should be tested in order to add the required conservatism. Also, it should be verified by analysis or testing that the modified series of tests does not cause significant undertesting or overtesting of certain parts of the test object. Otherwise, some other form of justification should be provided for replacing the test.

It may be necessary in some cases to modify the test procedure, for economic reasons, at the customer's request. If the tests are merely for the personal information of the customer, this would not cause any major problems. If regulatory standards and specifications are involved, however, some justification should be given for the modifications incorporated.

Test plans prepared in the pretest stage should include an adequate description of the following important items:

1. Test purpose
2. Test-object details
3. Test environment, specifications, and standards
4. Functional-operability parameters and failure or malfunction criteria
5. Pretest inspection
6. Aging requirements
7. Test outline
8. Instrumentation requirements
9. Data-processing requirements
10. Methods of evaluation of test results

Testing of Cabinet-Mounted Equipment

In dynamic testing of cabinet- or panel-mounted equipment, the following is a standard procedure:

Step 1 Test the cabinet or panel with the equipment replaced by a dummy weight.

Step 2 Obtain cabinet response at equipment-mounting locations and, based on these observations, develop the required dynamic environment for testing (the RRS) of the equipment.

Step 3 Test the equipment separately, using the excitation inputs developed in step 2.

This procedure may not be satisfactory if there is a considerable degree of dynamic interaction between the equipment and the mounting cabinet. This could be illustrated by using a simplified model to represent cabinet-mounted equipment. The cabinet and the equipment are represented separately by single-degree-of-freedom systems, as shown in figure 6-8. Cabinet para-

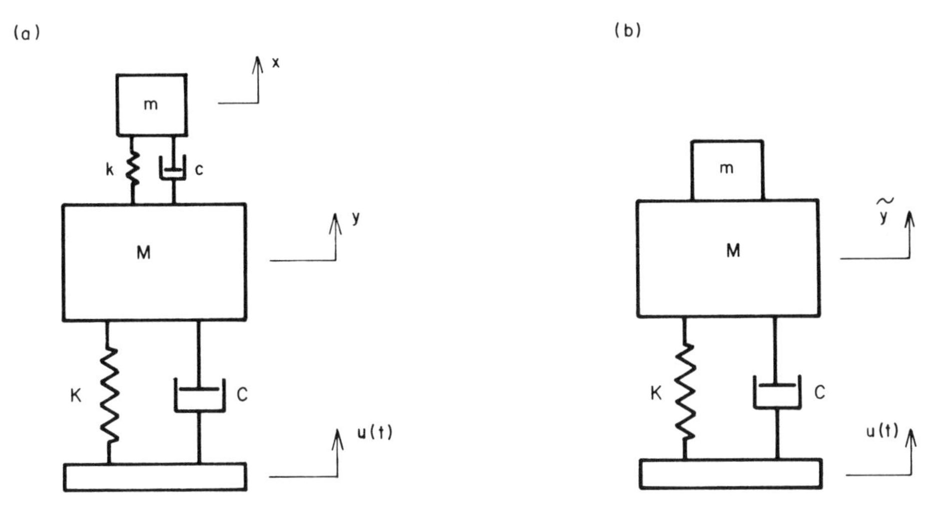

Figure 6-8. A Simplified Model to Study Limitations in Dummy Weight Tests: (a) Equipment Test Model, (b) Dummy-Weight Test Model

meters are represented by capital letters and equipment parameters by lower case letters. The cabinet response when the equipment is mounted on it is denoted by $y(t)$. The cabinet response when the equipment is replaced by a dummy weight of equal mass is denoted by $\tilde{y}(t)$. The excitation input applied to the cabinet base is denoted by $u(t)$. It can be shown that the frequency-response ratio in the two cases is given by

$$\frac{\tilde{Y}(\omega)}{Y(\omega)} = \frac{[ms^2(cs + k)]/(ms^2 + cs + k) + MS^2 + Cs + K}{(M + m)s^2 + Cs + K} \qquad (6.30)$$

with $s = j\omega$. Using the nondimensional parameters defined by equations (6.24) through (6.28), we obtain

$$\frac{\tilde{Y}(\omega)}{Y(\omega)} = \frac{(j\bar{\omega})^2 \alpha(2\zeta\beta j\bar{\omega} + \beta^2) + (j\bar{\omega})^2 + 2Zj\bar{\omega} + 1}{(1 + \alpha)(j\bar{\omega})^2 + 2Zj\bar{\omega} + 1} \qquad (6.31)$$

in which ζ denotes the equipment damping ratio and Z denotes the cabinet damping ratio.

The ratio $\tilde{Y}(\omega)/Y(\omega)$ represents the amplification in the cabinet response when the equipment is replaced by a dummy weight for a harmonic excitation. Figure 6-9 shows eight curves obtained for the ζ, α, β, and Z

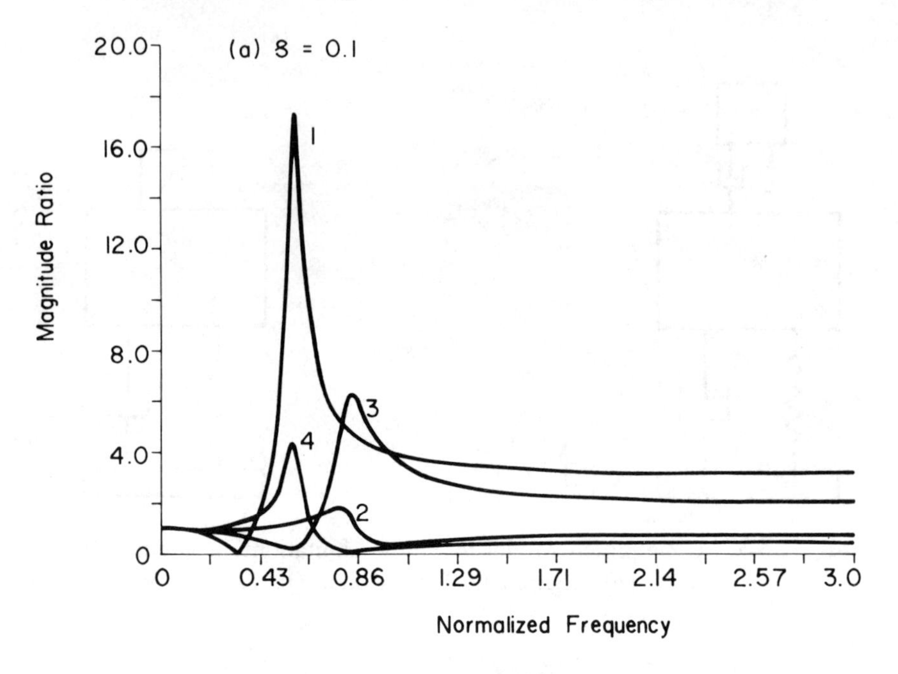

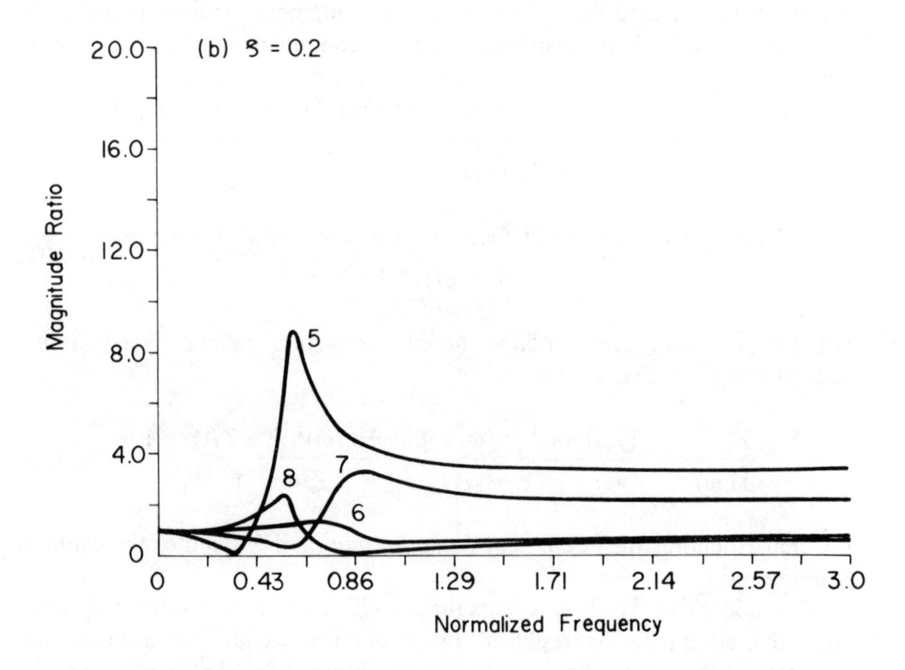

Figure 6-9. Cabinet Response Amplification in Dummy-Weight Tests

combinations, as given in table 6-1. It is seen that the best response ratio is obtained in curve 6. It can be concluded that dummy test procedure for cabinet-mounted equipment is satisfactory when the equipment inertia and natural frequency are small in comparison to the values for the cabinet. Also, increasing damping has a favorable effect on test results.

Pretest Inspection

Pretest inspection of a test object is important at least for two major reasons. First, if the equipment supplied for testing is different from the piece of equipment or the group of equipment that is required to be qualified, then these differences must be carefully observed and recorded in sufficient detail. In particular, deviations in the model number, mounting features and other interface details, geometry, size, and significant dynamic features should be recorded. Second, before testing, the test object should be inspected for any damage, deficiencies, or malfunctions. Structural integrity usually can be determined by visual inspection alone. To determine malfunctions by operability monitoring, however, the test object must be actuated and the operating environment should be simulated.

If the equipment supplied for testing is not identical to that required to be tested, adequate justification must be provided for the differences to guarantee that the objectives of the test can be achieved by testing the equipment that is supplied. Otherwise, the test should be abandoned pending the arrival of the correct test object.

If any structural failure or operational malfunction is noted during pretest inspection, no corrective action should be taken by the test laboratory personnel unless those actions are notified to and fully authorized by the supplier of the test object. Otherwise, the test should be abandoned and the customer should be promptly notified of the anomalies.

It is important that the functional-operability pretest inspection be performed in the same functional environment as that experienced under normal in-service conditions. When monitoring functional-operability parameters, it is necessary to guarantee that the monitoring instrumentation meets the required accuracy. Instrumentation data should be provided to the customer for review. This assures that the observed malfunction is real and not a false alarm caused by a malfunction in the monitoring instrumentation and channels. The monitoring-equipment accuracy should be higher than that required for the operability parameter itself.

Justification is needed if some components in the test object were not actuated and monitored during pretest inspection. Also, the warm-up period and the total time of actuation should be justified. In particular, if the proper operation of the equipment is governed by the continuity of a parameter (such as an electrical signal), the time duration of monitoring should be noted. If, however, the proper function is governed by a change of state (such as opening or closing a valve, a switch, or a relay), the number of cycles of actuation is important.

Resonance Search

Vibration test programs usually require a resonance-search pretest. This is usually carried out at a much lower excitation intensity than that used for the main test, in order to minimize the damage potential (overtesting). The primary objective of a resonance-search test is to determine resonant frequencies of the test object. More elaborate tests are employed, however, to determine mode shapes and modal damping ratios in addition to resonant frequencies. Such frequency-response data on the test object are useful in planning and conducting the main test.

Frequency-response data usually are available as a set of complex frequency-response functions (see chapter 2). There are tests that determine the frequency-response functions of a test object, and simpler tests are available to determine resonant frequencies alone. Some of the uses of frequency-response data are as follows:

1. A knowledge of the resonant frequencies of the test object is important in conducting the main test. More attention should be given, for example, when performing a main test in the vicinity of resonant frequencies. In resonance neighborhoods, lower sweep rates should be used if sine sweep is used in the main test, and larger dwell periods should be used if a sine dwell is part of the main test. Frequency-response data give the most desirable frequency range for conducting main tests.

2. From frequency-response data, it is possible to determine the most desirable test excitation input directions and the corresponding input intensity (see chapter 10).

3. The degree of nonlinearity and the time variance in system parameters of the test object can be estimated by conducting more than one frequency-response test at different excitation levels. If the deviation in the frequency-response functions thus obtained is sufficiently small, then a linear, time-invariant dynamic model is considered satisfactory in the analysis of the test object.

4. If no resonances are observed in the test object over the frequency range of interest, as determined by the operating environment for a given application (for example, 0-33 Hz for seismic qualification tests), then a static analysis would be adequate to qualify the test object.

5. A set of frequency-response functions can be considered a dynamic

model for the test specimen. This model can be employed in further study of the test specimen by analytical means.

Methods of Determining Frequency-Response Functions

Three methods of determining frequency response functions are outlined here. The theoretical bases of these methods have been covered in previous chapters (chapters 2, 3, and 5).

Fourier Transform Method. If $y(t)$ is the response at location B of the test object, when a transient input $u(t)$ is applied at location A, then the frequency-response function $H(f)$ between locations A and B is given by the ratio of the Fourier integral transforms of the output $y(t)$ and the input $u(t)$:

$$H(f) = \frac{Y(f)}{U(f)} \tag{6.32}$$

In particular, if $u(t)$ is a unit impulse, $H(f) = Y(f)$.

Spectral-Density Method. If the input excitation is a random signal, the frequency-response function between the input point and the output point can be determined as the ratio of the cross-spectral density $\Phi_{uy}(f)$ of the input $u(t)$ and the output $y(t)$, and the power spectral density $\Phi_{uu}(f)$ of the input,

$$H(f) = \frac{\Phi_{uy}(f)}{\Phi_{uu}(f)} \tag{6.33}$$

Harmonic Excitation Method. If the input signal is sinusoidal (harmonic) with frequency f, the output also will be sinusoidal with frequency f at steady state, but with a change in the phase angle. Then, the frequency-response function is obtained as a magnitude function and a phase-angle function. The magnitude $|H(f)|$ = steady-state amplification of the output signal, and the phase angle $\angle H(f)$ = steady-state phase lead of the output signal.

Resonance-Search Test Methods

There are three basic types of resonance-search test methods. They are categorized according to the nature of the excitation used in the test, specifically, (1) impulsive excitation, (2) initial displacement, or (3) forced

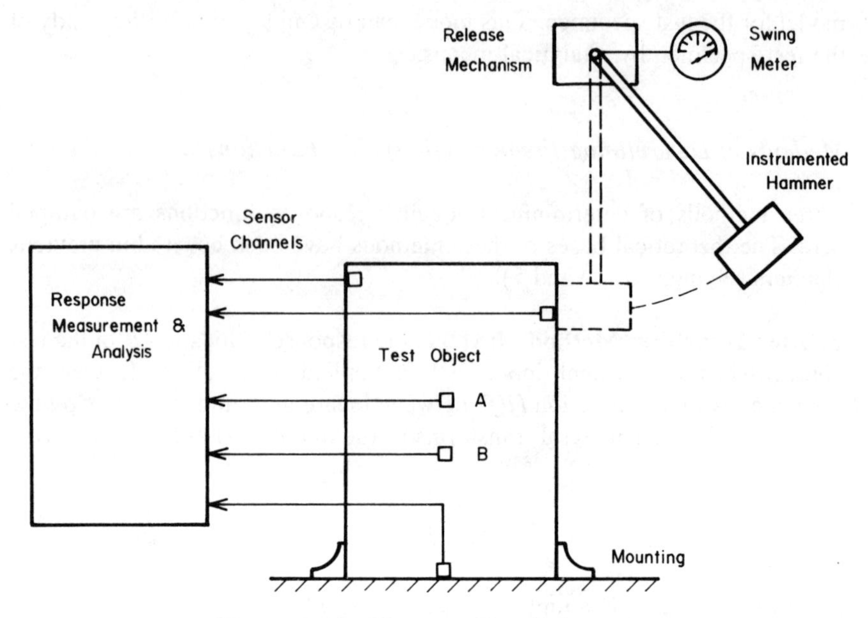

Figure 6-10. Hammer Test Schematic.

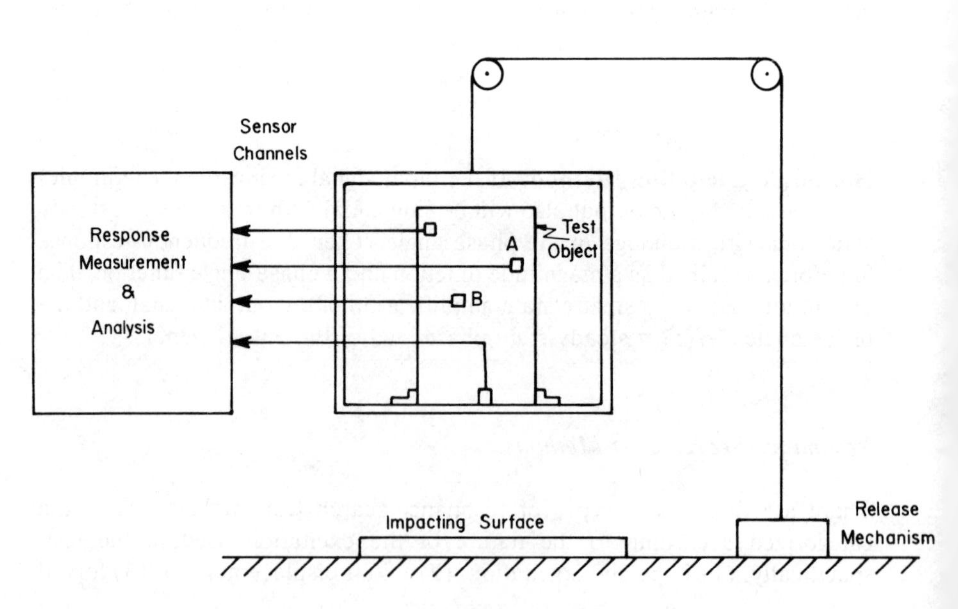

Figure 6-11. Drop Test Schematic

vibration. The first two categories are free-vibration tests; that is, response measurements are made on free decay of the test object following a momentary excitation.

Typical tests belonging to each of these categories are described in the following sections.

Hammer (Bump) Test and Drop Test. In resonance search by the impulsive-excitation method, an impulsive force (a large magnitude of force acting over a short duration) is applied to a suitable location of the test object, and its transient response is observed, preferably at several locations. This is equivalent to applying an initial velocity to the test object and letting it vibrate freely. By Fourier analysis of the response data, it is possible to obtain the resonant frequencies, corresponding mode shapes, and modal damping.

Hammer tests and drop tests are in the impulsive-excitation category. A schematic of the hammer-test arrangement is shown in figure 6-10. A schematic of the drop-test arrangement is shown in figure 6-11. The angle of swing of the hammer or the drop height of the object determine the intensity of the applied impulse. Alternatively, the impulse could be generated by explosive cartridges (for relatively large structures) located suitably in the test object, or by firing small projectiles at the test object. The response is monitored at several locations of the test object. The response at the point of application of the impulse is always monitored. Response analysis can be done in real time, or the response can be recorded for subsequent analysis. A major concern in these tests is making sure that all significant resonances in the required frequency range are excited under the given excitation. Several tests for different configurations of the test object might be necessary to achieve this.

Proper selection of the response-monitoring locations is also important in obtaining meaningful test results. By changing the impulsive-force intensity and repeating the test, any significant nonlinear (or time-variant-parameter) behavior of the test object can be determined. A common practice is to monitor the impulsive-force signal during impact. In this way, poor impacts (for example, low-intensity impacts, multiple impacts caused by bouncing back of the hammer) can be detected and the corresponding test results can be rejected.

Pluck Test. Resonance search on a test object can be performed by applying a displacement initial condition (rather than a velocity initial condition, as in impulsive tests) to a suitably mounted test object and measuring its response at various locations as it executes free vibrations. By properly selecting the locations and the magnitudes of the initial displacements, it is sometimes possible to excite various modes of vibration, provided that these modes are reasonably uncoupled.

The pluck test is the most common test using the initial-displacement method. A schematic of the test setup is shown in figure 6-12. The test object is initially deformed by pulling it with a cable. When the cable is suddenly released, the test object will undergo free vibrations about its static-equilibrium position. The response is observed for several locations of the test object and analyzed to obtain the required parameters.

In figures 6-10 to 6-12, the frequency-response function between two locations (A and B, for example) is obtained by analyzing the corresponding two signals, using either the Fourier transform method, equation (6.32), or the spectral-density method, equation (6.33). These frequency-domain techniques will automatically provide natural-frequency and modal-damping information. Alternatively, modal damping could be determined using time-domain methods—for example, by evaluating the logarithmic decrement (see chapter 4) of the response after passing it through a filter having a center frequency adjusted to the predetermined natural frequency of the test object for that mode. The effect of filtering is shown schematically in figure 6-13. Accuracy of the estimated modal-damping value can be improved significantly by such filtering methods.

Often, the most difficult task in a natural-frequency search is exciting a single mode. If two natural frequencies are close together, modal interactions of the two invariably will be present in most response measurements. Because of the closeness of the frequencies, the response curve will display a

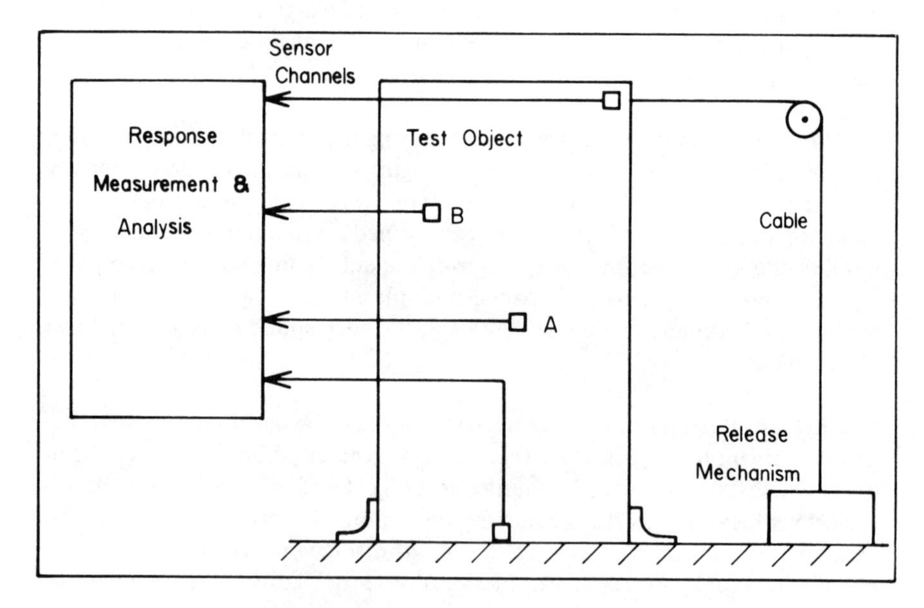

Figure 6-12. Pluck Test Schematic

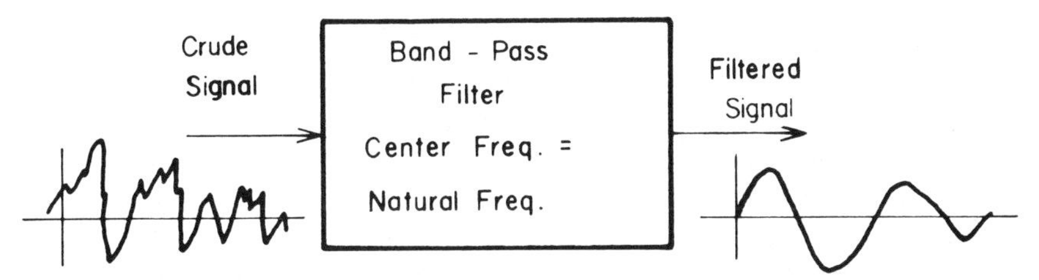

Figure 6-13. Use of Filtering to Determine Modal Damping

beat phenomenon, as shown in figure 6-14, which makes it difficult to dertermine damping by the logarithmic-decrement method. It is difficult to distinguish between decay caused by damping and rapid drop-off caused by beating. In this case, one of the frequency components must be filtered out, using a very narrow band-pass filter, before computing damping.

The required testing time for the impulsive-excitation and initial-displacement test methods is relatively small in comparison to forced-vibration test durations. For this reason, these tests are often preferred in preliminary (exploratory) testing before main tests. Directions and locations of impact or initial displacements should be properly chosen, however, so that as many significant modes as possible are excited in the desired frequency range. If the impact is applied at a node point for a particular mode, for instance, it will be virtually impossible to detect that mode from the response data. Sometimes a large number of monitoring locations might be necessary to accurate-

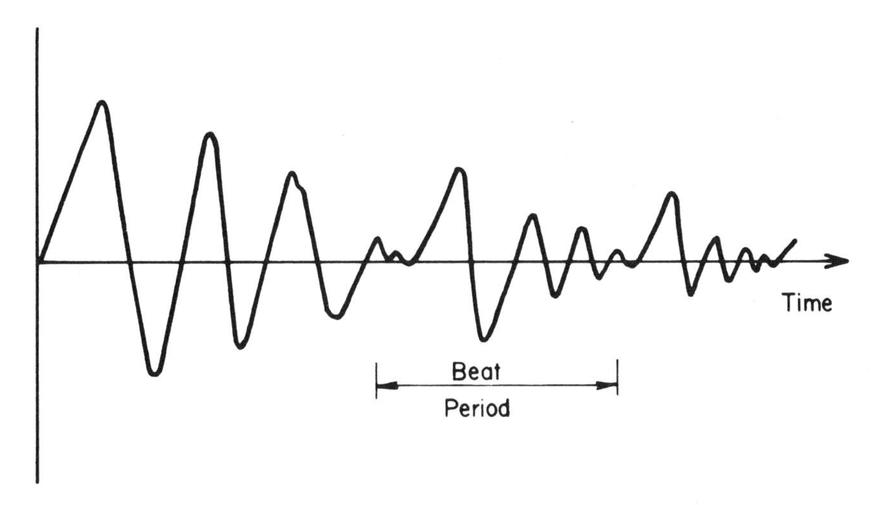

Figure 6-14. Beat Phenomenon Resulting from Interaction of Closely Spaced Modes

ly determine mode shapes of the test object. This depends primarily on the size and dynamic complexity of the test object and the particular mode number. This, in turn, necessitates the use of more sensors (accelerometers and the like) and recorder channels. If a sufficient number of monitoring channels is not available, the test will have to be repeated each time, using a different set of monitoring locations. Under such circumstances, it is advisable to keep one channel (monitoring location) unchanged and to use it as the reference channel. In this manner, any deviations in the test-excitation input could be detected for different tests and properly adjusted or taken into account in subsequent analysis (for example, by normalizing the response data).

Shaker Tests. A convenient method of resonance search is using a continuous excitation. The forced excitation, which typically is a sinusoidal signal or a random signal, is applied to the test object by means of a shaker, and the response is continuously monitored. The test setup is shown schematically in figure 6-15. For sinusoidal excitations, signal amplification and phase shift over a range of excitations will determine the frequency-response function. For random excitations, equation (6.33) may be used to determine the frequency-response function.

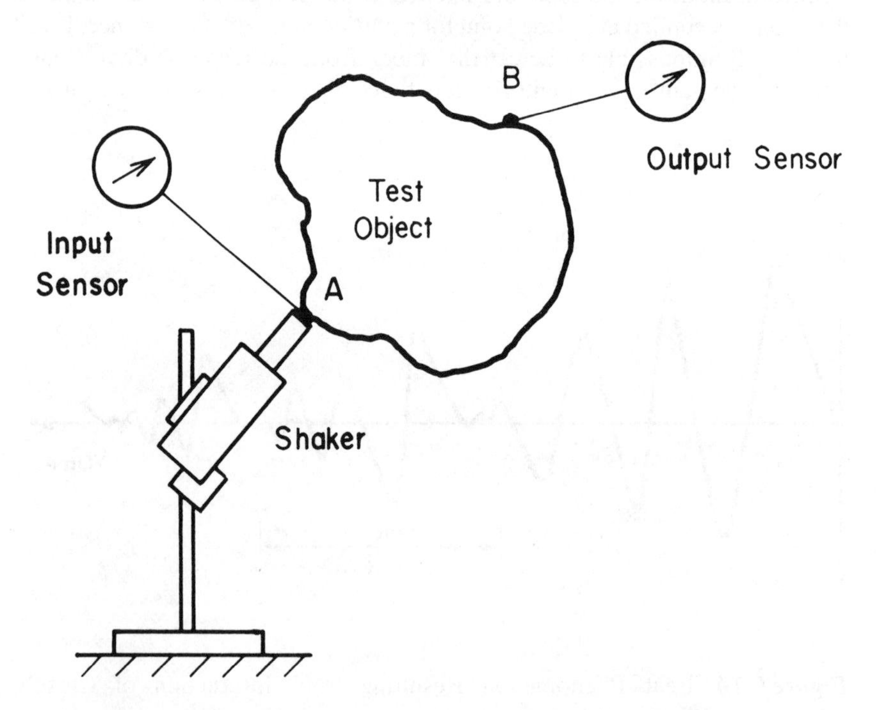

Figure 6-15. Shaker Test Schematic for Resonance Search

One or several portable shakers, or a large shaker table similar to that used in the main dynamic test, can be employed to excite the test object. The number and the orientations of the shakers, and the mounting configurations and monitoring locations of the test object, should be chosen depending on the size and complexity of the test object, the required accuracy of the resonance-search results, and the modes of vibration that need to be excited. The shaker-test method has the advantage of being able to control the nature of the test-excitation input (for example, frequency content, intensity, sweep rate), although it might be more complex and costly. The results from shaker tests are relatively more accurate and more complete.

Test objects usually display a change in resonant frequencies when the shaker amplitude is increased. This is caused by inherent nonlinearities in complex structural systems. Usually, the change appears as a spring-softening effect, which results in lower resonant frequencies at higher shaker amplitudes. If this nonlinear effect is significant, resonant frequencies for the main test level cannot be accurately determined using a resonance search at low intensity. Some form of extrapolation of the test results, or analysis using an appropriate dynamic model, might be necessary in this case to determine the resonant-frequency information that might be required to perform the main test.

Mechanical Aging

Before performing a qualification dynamic test, it is usually necessary to age the test object to put it into a condition that represents the state following its operation for a predetermined period under in-service conditions. In this manner, it is possible to reduce the probability of burn-in failure (see figure 6-3) during testing. Some tests, such as design-development tests and quality-assurance tests, might not require prior aging.

The nature and degree of aging that is required depends on such factors as the intended function of the test object, the operating environment, and the purpose of the dynamic test. In qualification tests, it may be necessary to demonstrate that the test object still has adequate capability to withstand an extreme dynamic environment toward the end of its design life (that is, the period in which it can be safely operated without requiring corective action). In such situations, it is necessary to age the test object to an extreme deterioration state, representing the end of the design life of the test object.

Test objects are aged by subjecting them to various environmental conditions (for example, high temperatures, radiation, humidity, vibrations). Usually, it is not practical to age the equipment at the same rate as it would age under a normal service environment. Consequently, accelerated aging procedures are used to reduce the test duration and cost. Furthermore, the operating environment may not be fully known at the testing stage. This makes the simulation of the true operating environment virtually impossible. Usually, accelerated aging is done in sequence by subjecting the test equipment to the various environmental cnditions one at a time. Under in-service conditions, however, these effects occur simultaneously, with the possibility of interactions between different effects. Therefore, when sequential aging is employed, some conservatism should be added. The type of aging used should be consistent with the environmental conditions and operating procedures of the specific application of the test object. Often these conditions are not known in advance. In that case, standardized aging procedures should be used.[1]

Our main concern in this section is mechanical aging, although other environmental conditions can significantly affect the dynamic characteristics of the test object. The two primary mechanisms of mechanical aging are material fatigue and mechanical wearout. The former mechanism plays a primary role if in-service operation consists of cyclic loading over relatively long periods of time. Wearout, however, is a long-term effect casued by any type of relative motion between components of the test object. It is very difficult to analyze component wearout, even if only the mechanical aspects are considered (that is, neglecting the effects of corrosion, radiation, and the like). Some mechanical wearout processes resemble fatigue aging, however, depending simultaneously on the number of cycles of load applications and the intensity of the applied load. Consequently, only the cumulative damage phenomenon, which is related to material fatigue, is usually treated in the literature.

Although mechanical aging is often considered a pretest procedure (for example, resonance-search test), it actually is part of the main test. If, in a dynamic qualification program, the test object malfunctions during mechanical aging, this amounts to failure in the qualification test. Furthermore, exploratory tests, such as resonance-search tests, are sometimes conducted at higher intensities than what is required to introduce mechanical aging into the test object.

Equivalence for Mechanical Aging

It is usually not practical to age a test object under its normal operating environment, primarily because of time limitations and the difficulty in simulating the actual operating environment. Therefore, it may be necessary to subject the test object to an accelerated aging process in a dynamic environment of higher intensity than that present under normal operating conditions.

Two aging processes are said to be equivalent if the final aged condition attained by the two processes is identical. This is virtually impossible to realize in practice, particularly when the environment is complex and the interactions of many dynamic causes have to be considered. In this case, a single most severe aging effect is used as the standard for comparison to establish the equivalence. The equivalence should be analyzed in terms of both the intensity and the nature of the dynamic excitations used for aging.

Excitation-Intensity Equivalence. A simplified relationship between the dynamic-excitation intensity U and the duration of aging T that is required to attain a certain level of aging, keeping the other environmental factors constant, may be given as

$$T = \frac{c}{U^r} \tag{6.34}$$

in which c is a proportionality constant and r is an exponent. These parameters depend on such factors as the nature and sequence of loading and characteristics of the test object. It follows from equation (6.34) that, by increasing the excitation intensity by a factor η, the aging duration can be reduced by a factor of η^r. In practice, however, the intensity-time relationship is much more complex, and caution should be exercised when using equation (6.34). This is particularly so if the aging is caused by multiple dynamic factors of varying characteristics that are acting simultaneously. Furthermore, there is usually an acceptable upper limit to η. It is unacceptable, for example, to use a value that would produce local yielding or any such irreversible damage to the equiment that is not present under normal operating conditions.

It is not necessary to monitor functional operability during mechanical aging. Furthermore, it could happen that, during accelerated aging, the equipment would malfunction but, when the excitation is removed, it would operate properly. This type of reversible malfunctions is acceptable in accelerated aging.

The time to attain a given level of aging is usually related to the stress level at a critical location of the test object. Since this critical stress can be related, in turn, to the excitation intensity, the relationship given by equation (6.34) is justified.

Dynamic-Excitation Equivalence. Equivalence of two dynamic excitations that have different time histories can be represented, using methods employed to represent dynamic excitations (for example, response spectrum, Fourier spectrum, and power spectral density, as described in chapter 5). If the maximum (peak) excitation is the factor that primarily determines aging in a given system under a particular dynamic environment, then response-spectrum representation is well suited for establishing the equivalence of two excitations. If, however, the frequency characteristics of the excitation are the major determining factor for mechanical aging, then Fourier spectrum representation is favored for establishing the equivalence of two deterministic excitations, and power spectral density representation is suited for random excitations.

When two excitation environments are represented by their respectives psd's, $\Phi_1(\omega)$ and $\Phi_2(\omega)$, and if the significant frequency range for the two exitations is (ω_1, ω_2), then the degree of aging under the two excitations may be compared using the ratio

$$\frac{A_1}{A_2} = \frac{\int_{\omega_1}^{\omega} \Phi_1(\omega)\, d\omega}{\int_{\omega_1}^{\omega_2} \Phi_2(\omega)\, d\omega} \tag{6.35}$$

in which A denotes a measure of aging. If the two excitations have different frequency ranges of interest, a range consisting of both ranges might be selected for the integrations in equation (6.35).

Cumulative Damage Theory

Miner's linear cumulative damage theory[2] may be used to estimate the combined level of aging resulting from a set of excitation conditions. Consider m excitations acting separately on a system. Suppose that each of these excitations produces a unit level of aging in $N_1, N_2, \ldots, N_m$ loading cycles, respecively, when acting separately. If, in a given dynamic environment, $n_1, n_2, \ldots, n_m$ loading cycles, respectively, from the m excitations actually have been applied to the system (possibly all excitations acting simultaneously), the level of aging attained can be given by

$$A = \sum_{i=1}^{m} \frac{n_i}{N_i} \tag{6.36}$$

The unit level of aging is achieved, theoretically, when $A = 1$. Equation (6.36) corresponds to Miner's linear cumulative damage theory.

Because of various interactive effects produced by different loading conditions, when some or all of the m excitations act simultaneously, it is usually not necessary to have $A = 1$ under the combined excitation to attain the unit level of aging. Furthermore, it is extremely difficult to estimate N_i,

$i = 1, \ldots, m$. For such reasons, the practical value of A in equation (6.36) to attain a unit level of aging could vary widely (typically, from 0.3 to 3.0).

Notes

1. *IEEE Trial-Use Guide for Type Test of Class I Electric Valve Operators for Nuclear Power Generating Stations,* IEEE Standard 382-1972.

2. M.A. Miner, "Cumulative Damage in Fatigue," *Journal of Applied Mechanics* 12:A 159, 1945.

7 Seismic Qualification

It is often necessary to determine whether a given piece of equipment is capable of withstanding a preestablished seismic environment. This process is known as seismic qualification. Electric utility companies, for example, should qualify their equipment for seismic capability before installing it in earthquake-prone geographic localities. Also, safety-related equipment in nuclear power plants requires seismic qualification. Regulatory agencies usually specify the general procedures to follow in seismic qualification.

Seismic qualification by testing is appropriate for complex equipment, but, in such cases, the equipment size is a limiting factor. For large systems that are relatively simple to model, qualification by analysis is suitable. Often, however, both testing and analysis are needed in the qualification of a given piece of equipment. Seismic qualification of equipment by testing is accomplished by applying a dynamic excitation by means of a shaker to the equipment, which is suitably mounted on a test table, and monitoring structural integrity and functional operability of the equipment. Special attention should be given to the development of the dynamic-test environment, mounting features, the operability variables that should be monitored, the method of monitoring functional operability and structural integrity, and the acceptance criteria used to decide qualification.

In monitoring functional operability, the test facility would normally require auxiliary systems to load the test object or to simulate in-service operating conditions. Such systems include actuators, dynamometers, electrical-load and control-signal circuitry, fluid-flow and pressure loads, and thermal loads. In seismic qualification by analysis, a suitable model is first developed for the equipment, and static or dynamic analysis is performed under an analytically defined dynamic environment. The analytical dynamic environment is developed on the basis of the specified dynamic environment for seismic qualification. By analytically determining system response at various locations, and by checking for such crucial parameters as relative deflections, stresses, and strains, qualification can be established.

Stages of Seismic Qualification

Consider the construction of a nuclear power plant. In this context, the plant owner is the customer. Actual construction of the plant is done by the plant builder, who is directly responsible to the customer concerning all equipment purchased from the equipment supplier or vendor. The vendor is often the equipment manufacturer as well. The equipment may be purchased by the customer and handed over to the plant builder or directly purchased by the plant builder. Accordingly, the purchaser could be the plant builder or the plant owner. A regulatory agency might stipulate seismic-excitation capability requirements for the equipment used in the plant, or the regulatory agency might specify the qualification requirements for various categories of equipment. The customer is directly responsible to the regulatory agency for adherence to the stipulations. The vendor, however, is responsible to the plant builder and the customer for the seismic capability of the equipment. The vendor may perform seismic qualification on the equipment according to required specifications. More often, however, the vendor hires the services of a test laboratory, which is the contractor for seismic qualification of equipment in the plant. The qualification procedure and report usually developed by the test laboratory, which adheres to the qualification requirements, may be reviewed by a reviewer, who is hired by the plant builder or the customer. A flowchart for test-object movement and for associated information interactions between various groups in the qualification program, is illustrated in figure 7-1.

A basic step in any qualification program is the preparation of a qualification procedure. This is a document that describes in sufficient detail such particulars as the tests that will be conducted on the test object, pretest procedures, the nature of test-input excitations and the method of generating these signals, inspection and response-monitoring procedures during testing, definitions of equipment malfunction, and qualification criteria. If analysis is also used in the qualification, the analytical methods and computer programs that will be used should be described adequately in the qualification procedure. The qualification procedure is prepared by the test laboratory (contractor); equipment particulars are obtained from the vendor or the purchaser; and the test environment for which the equipment will be qualified is usually supplied by the purchaser.

Before the qualification tests are conducted, the test procedure is submitted to the purchaser for approval. The purchaser normally hires a reviewer to determine whether the qualification procedure satisfies the requirements of both the regulatory agency and the purchaser. There could be several stages of revision of the test procedure until it is finally accepted by the purchaser on the recommendation of the reviewer.

The approved qualification procedure is sent to the test laboratory, and qualification is performed according to it. The test laboratory prepares a qualification report, which also includes the details of static or dynamic analysis when incorporated. The qualification report is sent to the purchaser for evaluation. The purchaser might obtain the services of an authority to review the qualification report. The report might have to be revised, and even

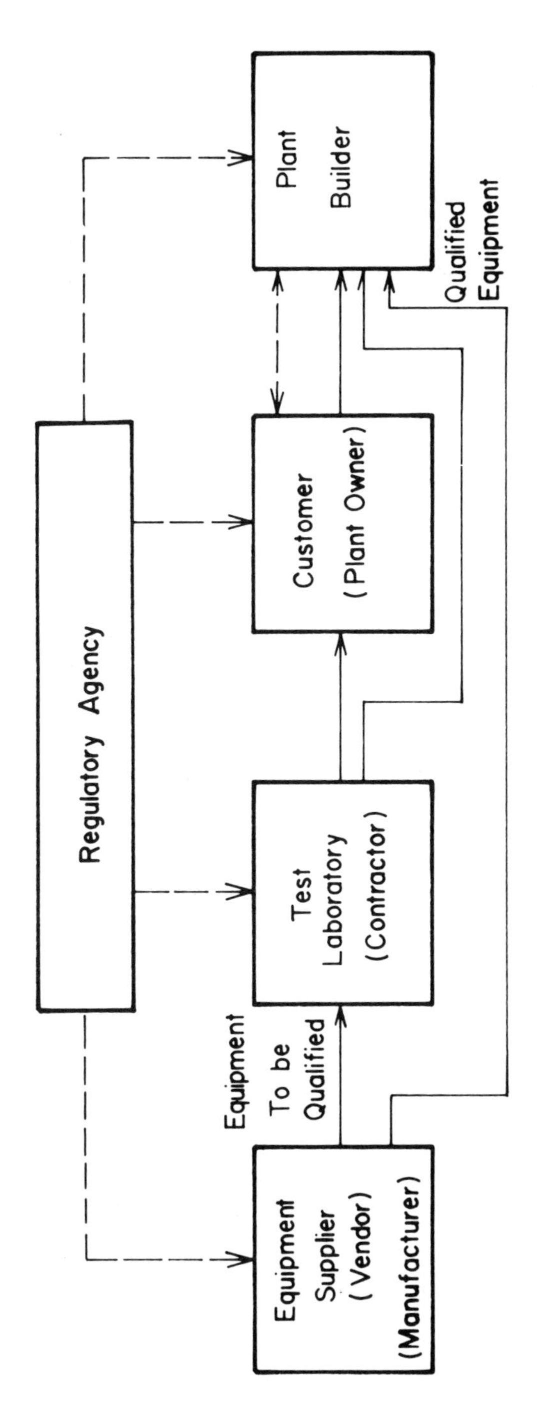

Figure 7-1. Test-Object Movement and Information Interactions in Seismic Qualification

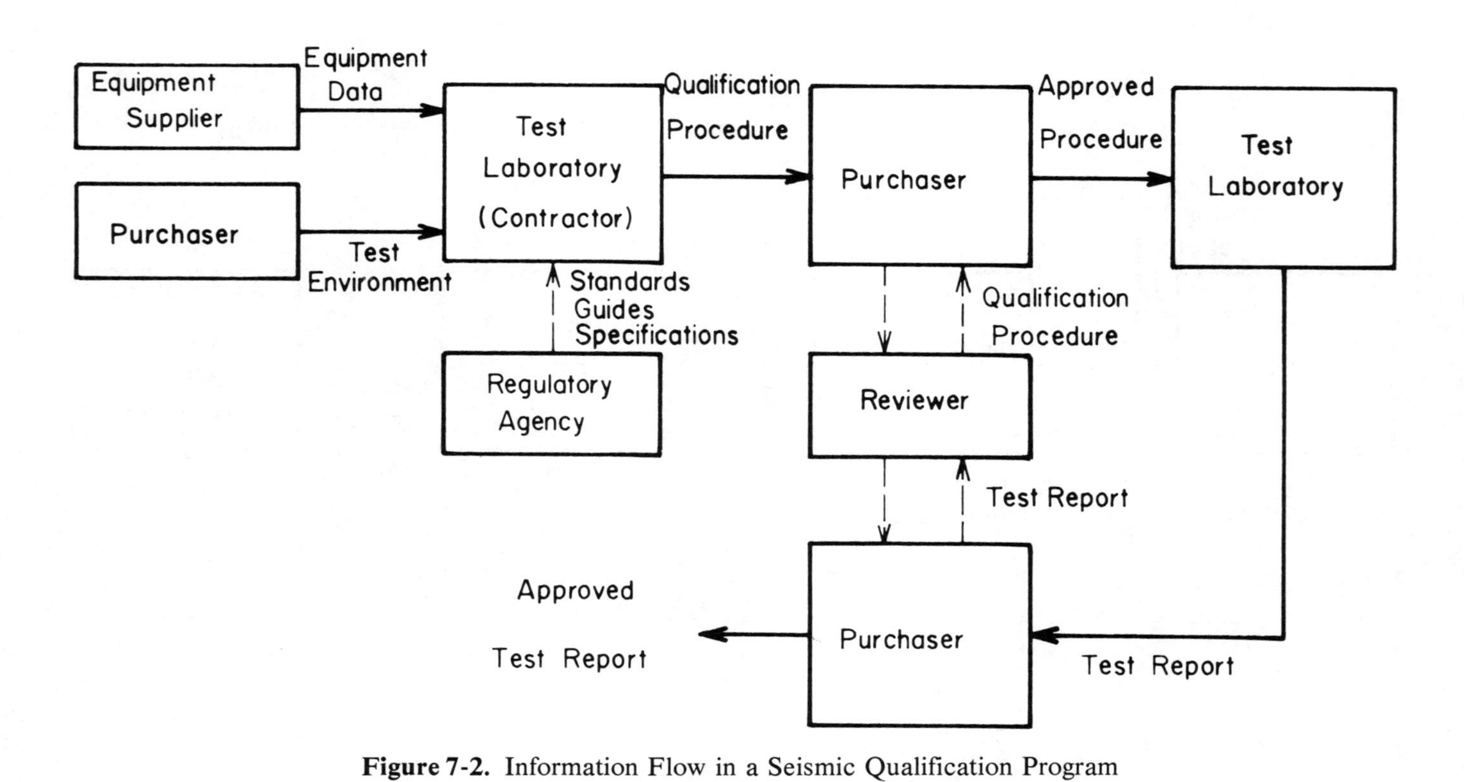

Figure 7-2. Information Flow in a Seismic Qualification Program

analysis and tests might have to be repeated before the final decision is made on the qualification of the equipment. Information flow in a typical qualification program is shown in figure 7-2.

Test Preliminaries

Seismic qualification tests are usually conducted by one of two methods, depending on whether single-frequency or multifrequency excitation inputs (see chapter 5) are employed in the main tests. The two test categories are (1) single-frequency tests and (2) Multifrequency tests. At present, the second test method is more common in seismic qualification by testing, although the first method is used under certain conditions, depending on the nature of the test object and its mounting features (for example, line-mounted versus floor-mounted equipment). Typically, multifrequency excitation inputs are preferred in qualification tests, and single-frequency excitation inputs are favored in design-development and quality-assurance tests.

In single-frequency testing, amplitude of the excitation input is specified by a required input motion (RIM) curve, similar to that shown in figure 7-3. If single-frequency dwells (for example, sine dwell, sine beat) are employed, the excitation input is applied to the test object at a series of selected frequency values in the frequency range of interest for that particular test environment. In such cases, dwell times (and number of beats per cycle, when sine beats are employed) at each frequency point should be specified. If a single-frequency sweep (such as a sine sweep) is employed as the excitation

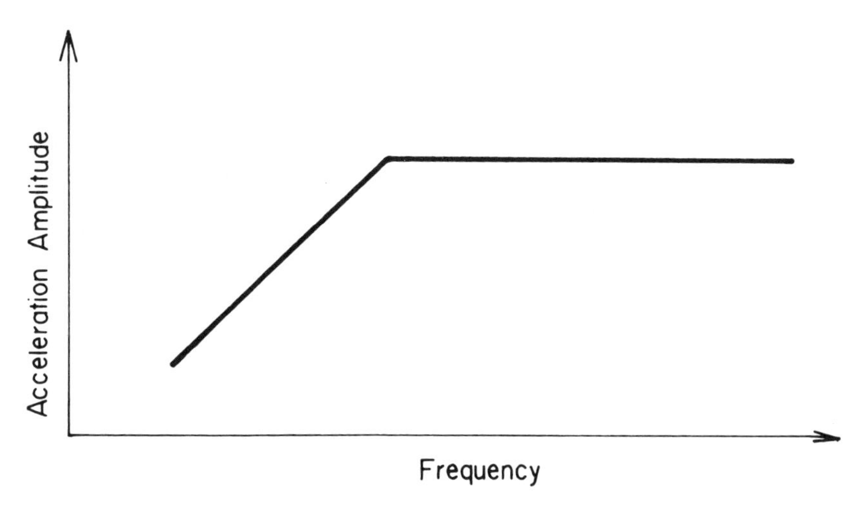

Figure 7-3. A Typical Required Input Motion (RIM) Curve

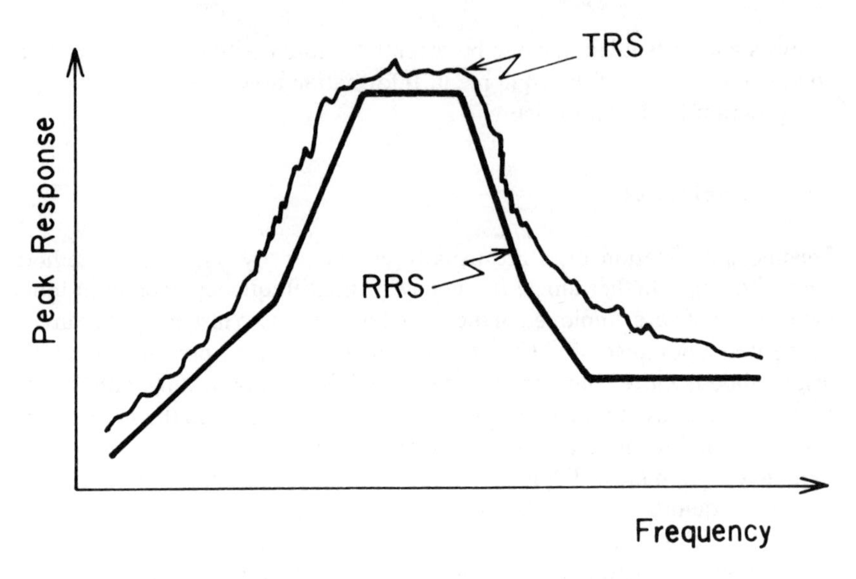

Figure 7-4. The TRS Enveloping the RRS in a Multifrequency Test

input, the sweep rate should be specified. When the single-frequency test-excitation input is specified in this manner, the tests are conducted very much like multifrequency tests.

Multifrequency tests are normally conducted employing the response-spectra method to represent the test-input environment. Basically, the test object is excited using a signal whose response spectrum, known as the test response spectrum (TRS), envelops a specified response spectrum, known as the required response spectrum (RRS). Ideally, the TRS should equal the RRS, but it is practically impossible to achieve this condition. Hence, multifrequency tests are conducted using a TRS that envelops the RRS so that, in significant frequency ranges, the two response spectra are nearly equal (see figure 7-4). Excessive conservatism, however, which would result in overtesting, should be avoided. It is usually acceptable to have TRS values below the RRS at a few frequency points, as illustrated in figure 7-5.

The RRS is part of the data supplied to the test laboratory before the qualification tests are conducted. Two types of RRS are provided, representing (1) the operating-basis earthquake (OBE) and (2) the safe-shutdown earthquake (SSE). The response spectrum of the OBE represents the most severe motions produced by an earthquake under which the equipment being tested would remain functional without undue risk of malfunction or safety hazard. If the equipment is allowed to operate at a disturbance level higher

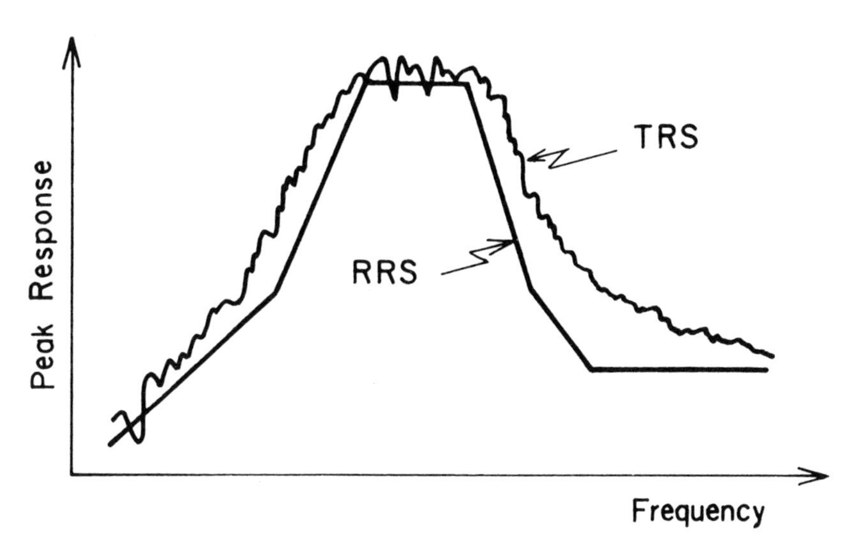

Figure 7-5. The TRS Almost Enveloping the RRS in a Multifrequency Test

than the OBE level for a prolonged period, however, there would be a significant risk of malfunction.

The response spectrum of the SSE represents the most severe motions produced by an earthquake that the equipment being tested could safely withstand while the entire nuclear power plant is being shut down. Prolonged operation (that is, more than the duration of one earthquake), however, could result in equipment malfunction; in other words, equipment is designed to withstand only one SSE in addition to several OBEs.

A typical seismic qualification test would involve subjecting the equipment to several OBE-level excitations, primarily to age the equipment mechanically to its end-of-design-life condition, and then subjecting it to one SSE-level excitation.

When providing RRS test specifications, it is customary to supply only the SSE requirement. The OBE requirement is then taken as a fraction (typically, 0.5 or 0.7) of the SSE requirement.

Test response spectra corresponding to the excitation input signal are generated by the test laboratory during testing. The purchaser usually supplies the test laboratory with an FM tape containing frequency components that should be combined in some ratio to generate the test-input signal.

Qualification tests are conducted according to the test procedure approved and accepted by the purchaser. The main steps of seismic qualification testing are outlined in the following sections.

Pretest Inspection

Test objects might experience harsh conditions while in transit to the test lab. Any damage that occurs during this time could lead to incorrect test results. To avoid erroneous qualification decisions based on component malfunctions or structural failure resulting from poor handling before testing, the test object should be carefully inspected when it is received at the test laboratory. Initially, inspection is done visually to detect any loose or broken parts, including bolts, nuts, and welded joints, and other structural damage to the test object or its housing. Next, the test object should be checked for functional operability before testing, if the test procedure requires functional-operability monitoring during testing. Any structural damage or malfunction should be recorded in sufficient detail (preferably including sketches and photographs), and the purchaser should be notified promptly. If the structural damage is excessive or if component malfunctions are detected, the approval of the purchaser is necessary before any corrective actions and subsequent testing. It may be necessary to obtain a replacement test object if the damages are excessive or the detected malfunction is crucial. In any event, a description of such anomalies and the corrective measures taken should be incorporated in the test report.

TRS Generation

It is customary for the purchaser to provide the test laboratory with a multichannel (for example, 14 channels) FM tape containing the components of the excitation input signal that should be used in the qualification test. Alternatively, the purchaser may request that the test laboratory generate the FM tape containing the required signal components, under the purchaser's supervision. If sine beats are combined to generate test excitation inputs, each FM tape should be supplemented by tabulated data giving the channel number, the beat frequencies (Hz) in that channel, and the amplitude (g) of each sine-beat component. The RRS curve that is enveloped by that particular input should also be specified.

The excitation signal that is applied to the shaker-table actuator is generated by combining the contents of each channel in an appropriate ratio so that the response spectrum of the excitation actually felt at the mounting locations of the test object (the TRS) satisfactorily matches the RRS supplied to the test laboratory. Matching is performed by passing the contents of each channel through a variable-gain amplifier and mixing the resulting components. These operations are performed by a waveform mixer, as shown in figure 7-6. The adjustment of the amplifier gains is done by trial and error (see the later section on the dummy-weight test). The phase of the individual signal components should be maintained during the mixing process.

Each channel may cntain a single-frequency component (such as sine beat) or a multifrequency signal of fixed duration (for example, 20 sec). If the RRS is complex, each channel might have to carry a multifrequency signal to achieve close matching of the TRS and the RRS. Also, a large number of channels might be necessary. The test excitation-input signal is generated continuously by playing the FM tape loop of fixed duration repetitively.

In seismic qualification, response spectra are usually specified in units of acceleration due to gravity (g). Consequently, the contents in each channel of the test-input FM tape represent acceleration motions. For this reason, the signal from the waveform mixer must be integrated twice before using it to actuate the shaker table (see figure 7-6). The actuator itself is driven by a displacement signal, and its control is usually done by means of a displacement-sensor (see chapter 8 for LVDT) signal as well. In typical test facilities, the double integration unit is built into the shaker system. It is then possible to use any type of signal (displacement, velocity, acceleration) as the excitation input and to decide simultaneously on the number of integrations that are necessary. If the input signal is a velocity time history, for example, one integration should be chosen and so on.

The tape speed should be specified (for example, 7.5 in/s, 15 in/s) when signals recorded on tapes are provided to generate input signals for dynamic testing. This is important, so that the frequency content of the signal is not distorted. Speeding up the tape has the effect of scaling up each frequency component in the signal. It has also the effect, however, of filtering out very high frequency components in the signal. If the excitation-input signals are available as digital records, then a digital-to-analog converter (DAC) is needed to convert it into an analog signal.

Instrument Calibration

The test procedure normally contains accuracy requirements and tolerances for various critical instruments used in seismic qualification testing. It is desirable that these instruments have current calibration records that are aggreable to an accepted standard, such as that of the National Bureau of Standards (NBS). Instrument manufacturers usually provide these calibration records. Accelerometers, for example, may have calibration records for several temperatures (for example, −65°F, 75°F, 350°F) and for a range of frequencies (such as 1-1,000 Hz). Calibration records for accelerometers are

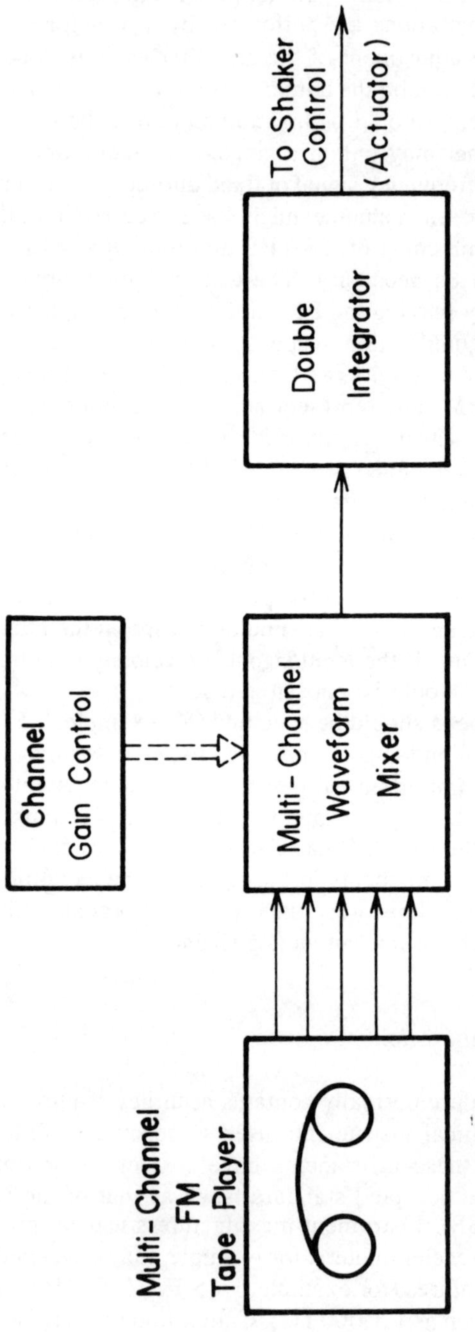

Figure 7-6. Test-Input Generation by Combining Signal Components Stored on an FM Tape

given in voltage sensitivity (mV/g) and charge sensitivity (pC/g), along with percentage-deviation values. These tolerances and peak deviations for various test instruments should be provided for the purchaser's review before they are used in the test apparatus.

From tolerance data for each sensor or transducer, it is possible to estimate peak error percentages in various monitoring channels in the test setup, particularly in the channels used for functional-operability monitoring. The accuracy associated with each channel should be adequate to measure expected deviations in the monitored operability parameter.

It is good practice to calibrate sensor or transducer units, such as accelerometers and associated auxiliary devices, daily or after each test. These calibration data should be recorded for different scales when a particular instrument has multiple scales, and for different instrument settings.

Dummy-Weight Method of TRS Matching

Before subjecting a test object to the test excitation input, we should make sure that the corresponding TRS envelops the RRS. If the appropriate test excitation input signal is generated while the test object is mounted on the shaker table, then, during the process of response-spectrm matching (by adjusting the channel gains in the waveform mixer), the test object would be subjected to prolonged dynamic excitations. This could cause excessive mechanical aging of the test object, which could subsequently produce erroneous test results. The time taken for TRS matching could be comparable to the main test duration, and overtesting would result. This unacceptable situation could result in an incorrect qualification decision. If, however, the TRS is matched to the RRS (so that the former envelops the latter) before the test object is mounted on the shaker table, then, after the test object is mounted, the excitation and, consequently, the TRS would be distorted because of dynamic interaction of the test object and the test table. The resulting TRS might no longer envelop the RRS. This problem situation can be eased by using a dummy-weight test before the main test.

A dummy weight, consisting of a slab or a plate whose weight is approximately equal to the test-object weight, is placed on the shaker table, and the TRS matching is performed. This setup includes the inertia effects of the test object but not its flexibility effects. As a result, the TRS will still be distorted when the dummy weight is replaced by the actual test object. The degree of distortion will be less than that which would result if the dummy weight were not used, however. In the dummy-weight test, the adjustment of the channel gains is done first to get the required shape of the TRS (that is, a shape similar to the RRS). When the shape is satisfactory, the overall gain of the waveform mixer is increased to achieve the required enveloping of the RRS.

The TRS is determined from the acceleration signal sensed by the control accelerometer. This is the accelerometer that measures the shaker-table excitation that is felt at the mounting locations of the test object (or the dummy weight) in the direction of excitation (that is, in the direction of freedom of the test motion). This signal is generally different from the signal generated by the signal generator (for example, a tape player), primarily because of dynamic interactions of the shaker table and the test object.

A response-spectrum analyzer is employed to determine the response spectrum of the control accelerometer signal (TRS). The resulting TRS will automatically appear on the display scope of the analyzer. After making the necessary channel-gain adjustments to realize spectrum matching, a permanent record of the TRS is obtained using a X-Y plotter (or a hard-copy unit). The TRS plots should be superimposed on the corresponding RRS specifications for easy comparison and verification of the validity of the qualification test.

Test-Object Mounting

When a test object is being mounted on a shaker table, care should be taken to simulate all critical interface features under normal installed conditions for the intended operation. This should be done as accurately as is feasible. Critical interface requirements are those that could significantly affect the dynamics of the test object. If the mounting conditions in the test setup significantly deviate from those under installed conditions for normal operation, adequate justification should be provided to show that the test is conservative (that is, the motions produced under the test mounting conditions are more severe). In particular, local mounting, which is not present under normal installation conditions, should be avoided in the test setup.

In simulating in-service interface features, the following details should be considered as a minimum:

1. Test orientation of the test object should be its in-service orientation, particularly with respect to direction of gravity (vertical), available degrees of freedom, and mounting locations.
2. Mounting details at the test object–fixture interface should represent in-service conditions with respect to the number, size, and strength of welds, bolts and nuts, and other hold-down hardware.
3. Additional interface linkages, such as wires, cables, conduits, pipes, instrumentation (dials, meters, gauges, sensors, transducers, and so on), and their supporting brackets should be simulated at least in terms of mass and stiffness, and preferably in terms of size as well.

4. Any dynamic effects of adjacent equipment, cabinets, and supporting structures under in-service conditions should be simulated or taken into account in analysis.
5. Operating loads, such as those resulting from fluid flow, pressure forces, and thermal effects, should be simulated if they appear to significantly affect test object dynamics. In particular, the nozzle loads should be simulated in magnitude, direction, and location.

The required interface details of the test object are obtained by the test laboratory at the information-acquisition stage (see chapter 6). Any critical interface details that are simulated during testing should be included in the test report.

At least three control accelerometers should be attached to the shaker table in the neighborhood of the mounting location of the test object. One control accelerometer measures the excitation-acceleration component applied to the test object in the vertical direction. The other two measure the excitation-acceleration components in two horizontal directions at right angles. The two horizontal (control) directions are chosen to be along the two major freedom-of-motion directions (or dynamic principal axes) of the test object. Engineering judgment is used in deciding these principal directions of high response in the test object. Often, geometric principal axes are used. The control accelerometer signals are passed through a response-spectrum analyzer (or a suitably programmed digital computer) to compute the TRS in vertical and two horizontal directions, which are perpendicular.

Qualification tests generally require monitoring of the dynamic response at several critical locations of the test object. In addition, the tests may call for determining mode shapes and natural frequencies of the test object. For this purpose, a sufficient number of accelerometers should be attached to various key locations in the test object. The qualification test procedure should carry a sketch of the test object, indicating the accelerometer locations. Also, the type of accelerometers employed (for example, crystal type or strain-gauge type), their magnitude and direction of sensitivity (see chapter 8), and the tolerances should be included in the final test report.

Exploratory Tests

The primary purpose of an exploratory test is to determine the dynamic characteristics of the test object that would be useful in the main test. Exploratory tests usually determine the frequency-response functions between the point of excitation and a set of critical locations in the test object. This information will give transmissibility curves (see chapter 2), inpedance functions, natural frequencies, damping ratios, mode shapes, and other useful data regarding the test object. Such frequency-response data are very useful in planning and conducting the main test. When conducting the main test in the neighborhood of a test-object resonance, for example, slower sweep rates (for example, 1-2 octaves per min.) should be used if the excitation consists of a sine sweep, and larger dwell periods should be employed if the excitation is a sine dwell or a sine beat. Frequency-response data also are useful in determining the most desirable test-input directions and intensities for the main test (see chapter 10). The degree of nonlinearity and parameter-variance with time in the test object can be estimated by conducting several frequency-response tests at different input intensities. If the deviation of the frequency-response function with respect to input intensity is sufficiently small, then linear, time-invariant parameter assumption is considered satisfactory. Often, frequency-response tests are conducted on a shaker table at full test intensity. In such cases, this is considered part of the main test rather than an exploratory test.

To determine frequency-response functions, the input location of the test object may be excited using one of the following types of excitation signals: (1) deterministic signals, (2) steady sinusoidal signals, or (3) random signals. Low excitation input levels (typically, 0.2 g acceleration levels) are used normally. The test-object response at an output location is measured and the complex frequency-response function (magnitude and phase angle) is determined from the input-output data. Three methods of computing the frequency-response function are outlined in chapter 6.

In seismic qualification by analysis, previous exploratory-test data are employed to determine the most suitable analysis method for a given situation. If there are no resonances in the frequency range of interest (typically, 1-33 Hz for seismic qualification), for example, then the item to be qualified may be analyzed using static-analysis methods.

Exploratory tests are not always conclusive, because of the following factors:

1. Because of nonlinearity and parameter variance with time in equipment, resonance might not be detected at certain levels of shaking intensity at certain times.
2. Because of the complexity and delicacy of the test object, certain components in it may not be accessible for instrumentation and response monitoring.
3. Mounting structures, mounting instruments, and test orientation can modify dynamic characteristics of the test object, resulting in considerable deviations in major resonances.

The shape of the frequency-response curves, as obtained from an exploratory test, can be employed to determine the degree and the nature of the nonlinearities that are present in the test object. If a high degree of nonlinearity is present in the test object, when the test is repeated at two

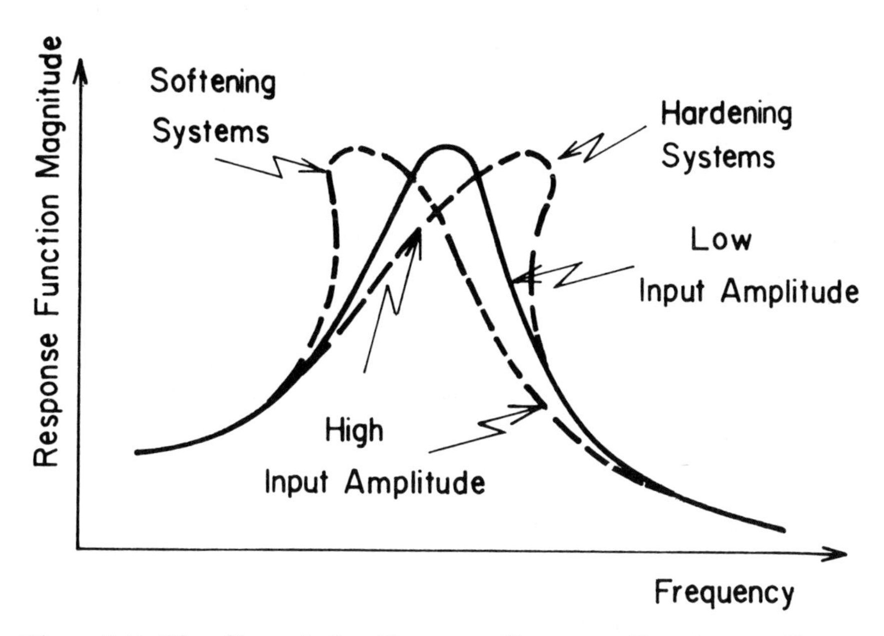

Figure 7-7. The Use of the Frequency-Response Function to Detect System Nonlinearity

different levels of excitation input the resulting response curves would be significantly different. A typical situation is shown in figure 7-7. At low test excitation-input amplitudes, the frequency response would be a typical linear-system curve. As the input level is increased, either a spring-softening or spring-hardening type of deviation would be apparent (see figure 7-7).

Test-Input Considerations

In seismic qualification testing, a significant effort goes into the development of test excitation inputs. Not only the nature but also the number and the directions of the excitation inputs could have a significant effect on the outcomes of a test. This is so because the excitation input characteristics determine the nature of a test.

Test Nomenclature

The accepted practice for seismic qualification by testing is to apply a dynamic excitation to a test object that is appropriately mounted on a shaker table. Customarily, only translatory excitations, as generated by linear actuators, are employed. Nevertheless, the resulting motion of the test object usually consists of rotational components as well. A typical seismic environment consists of three-dimensional motions (see chapter 5). The specification of a three-dimensional test environment, however, is a complex task, even after omitting the rotational motions at the mounting locations of the equipment to be qualified. Furthermore, as discussed in chapter 5, seismic environments are random, and their representation with sufficient accuracy can be done only in a probabilistic sense.

Very often, the type of qualification testing used is governed mainly by the capabilites of the test laboratory to which the qualification contract is granted. Test laboratories conduct qualification tests using their previous experience and engineering judgment. Extensive improvements to existing tests can be very costly and time-consuming, and this is not warranted from the point of view of the customer or the vendor. Regulatory agencies usually allow simpler qualification tests if sufficient justification can be provided indicating that a particular test is conservative with respect to regulatory requirements.

Complexity of a shaker-table apparatus is governed primarily by the number of actuators that are employed and the number of independent directions of simultaneous excitation that it is capable of producing. Terminology for various tests is based on the number of independent directions of excitation used in the test. It is advantageous to standardize this terminology to be able to compare different test procedures. Unfortunately, the terminology used to denote different types of qualification tests usually depends on the particular test laboratory and the specific application. Attempts to standardize various test methods have become tedious, partly because of the lack of a universal nomenclature for dynamic testing. A justifiable grouping of test configurations is presented in this section. Figure 7-8 illustrates the various test types.

In test nomenclature, the degree of freedom refers to the number of directions of independent motions that can be generated simultaneously by means of independent actuators in the shaker table. According to this concept, three basic types of tests can be identified:

1. Single-degree-of-freedom (or rectilinear) testing, in which the shaker table employs only one actuator, producing test-table motions along the axis of that actuator. The actuator may not necessarily be in the vertical direction.
2. Two-degree-of-freedom testing, in which two independent actuators, oriented at right angles to each other, are employed. The most common configuration consists of a vertical actuator and a horizontal actuator. Theoretically, the motion of each actuator can be specified independently.

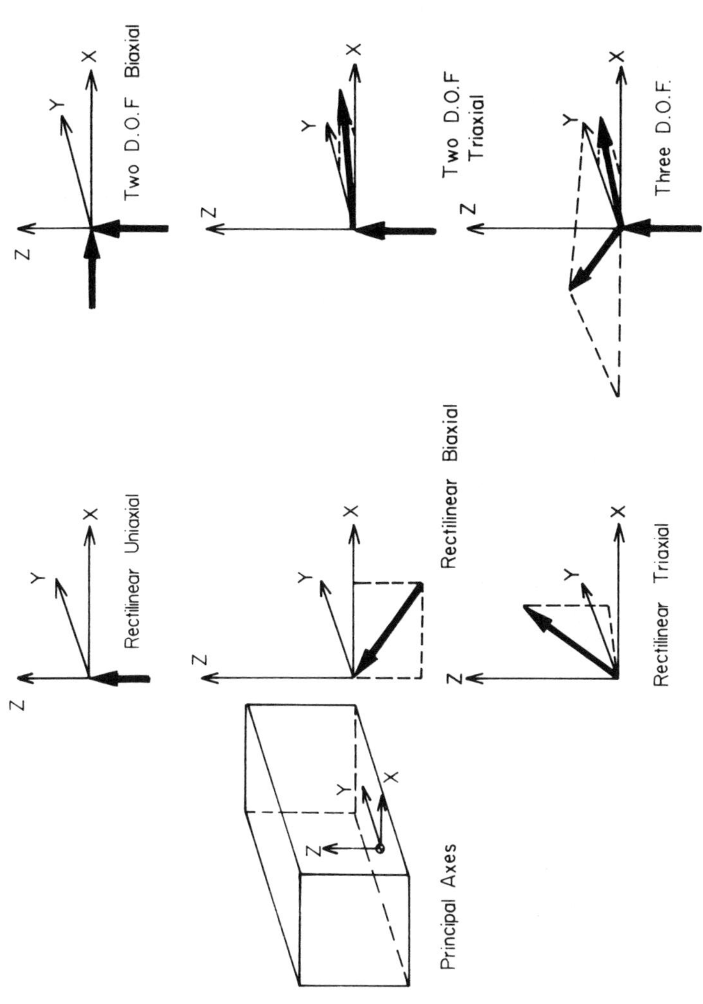

Figure 7-8. Dynamic-Test Configurations

3. Three-degree-of-freedom testing, in which three actuators, oriented at mutually right angles, are employed. A desirable configuration consists of a vertical actuator and two horizontal actuators. Theoretically, at least, the motion of each actuator can be specified independently.

It is common practice to specify the directions of excitation with respect to the geometric principal axes of the test object. This practice is somewhat questionable, primarily because it does not take into account the flexibility of the object. Flexibility elements in the test object have a significant influence on the level of dynamic coupling present in a given pair of directions. In this respect, it is more appropriate to consider dynamic principal axes rather than geometric principal axes of the test object when determining the test configuration. Dynamic principal axes should be defined in terms of dynamic characteristics of the test object. One useful definition is in terms of eigenvectors of an appropriate three-dimensional frequency-response function matrix that takes into account the response at every critical location in the test object. The details of this method are given in chapter 10. The only difficulty in this method is that prior frequency-response testing or analysis is needed to determine the test input direction. For practical purposes, the vertical axis (direction of gravity) is taken as one principal axis. Two horizontal axes at right angles to each other should be selected as the other two principal axes. Often, this is done by using geometric considerations alone. Dynamic characteristics of the test object should be taken into account in determining test input directions if more meaningful testing is desired.

The single-degree-of-freedom (rectilinear) test configuration has three subdivisions, based on the orientation of the actuator with respect to the principal axes of the test object. It is assumed that one principal axis of the test object is the vertical axis and that the three principal axes are mutually perpendicular. The three subdivisions are as follows:

1. Rectilinear uniaxial testing, in which the single actuator is oriented along one of the principal axes of the test object.
2. Rectilinear biaxial testing, in which the single actuator is oriented on the principal plane containing the vertical and one of the two horizontal principal axes. The actuator is inclined to both principal axes in the principal plane.
3. Rectilinear triaxial testing, in which the single actuator is inclined to all three orthogonal principal axes of the test object.

The two-degree-of-freedom test configuration has two subdivisions, based on the orientation of the two actuators with respect to the principal axes of the test object, as follows:

1. Two-degree-of-freedom biaxial testing, in which one actuator is directed along the vertical principal axis and the other along one of the two horizontal principal axes of the test object.
2. Two-degree-of-freedom triaxial testing, in which one actuator is positioned along the vertical principal axis and the other actuator is horizontal but inclined to both horizontal principal axes of the test object.

Testing with Uncorrelated Excitations

Consider the motion at a particular location in a building caused by seismic ground disturbances. For simplicity, we shall neglect rotatory motions and consider only translatory motions. At a given instant, there will be many translatory motion components in different directions (in three dimensions) at the point considered. These components can be added vectorially to get a single vector. If this resultant displacement vector is determined a short time later, it will have a different direction and magnitude; that is, the direction of the resultant motion vector will change with time. This is because of the randomness of the motion amplitudes in different directions in space. It follows that a motion vector with fixed direction (single-degree-of-freedom motion) cannot adequately represent motions caused by earthquake ground excitations.

The amplitude and direction of the resultant motion vector produced in a dynamic system by seismic excitations are random processes. Since a vector oriented in an arbitrary direction can be resolved into three components along three fixed orthogonal coordinate axes, the motion vector with random amplitude and direction can be replaced by three random motion components having fixed directions (along the three coordinate axes) and random amplitudes. It follows that, if rotatory motions are neglected, seismic excitations felt at equipment mounting locations can be simulated by three orthogonal random excitations. These three components generally are not statistically independent. IEEE Standard 344-1975, however, recommends independent random excitations in three orthogonal directions to simulate seismic environments for qualification tests.

When the level of uncertainty in the dynamic environment is high, it is desirable to have the maximum level of freedom in controlling the test environment. This is presumably the primary reason for favoring three-degree-of-freedom testing with statistically independent excitations (IEEE Standard 344-1975) in seismic qualifications by testing. The requirement of statistical independence is replaced, in practice, by a less stringent requirement of uncorrelated excitations. Equivalently, incoherent excitations over the desired band of frequencies (in the frequency domain) could be employed in seismic qualification. For signals that have low coherence, the phase angle of the cross spectrum varies irregularly with frequency. For this reason, uncorrelated (incoherent) signals are sometimes referred to as phase-incoherent signals.

Simultaneous inputs in three orthogonal directions often produce responses (accelerations, stresses, and so forth) that are very different from what is obtained by vectorially summing the responses to individual inputs acting alone, one at a time. This is primarily because of the nonlinear, time-variant nature of test specimens and test apparatus, their dynamic coupling, and the randomness of excitation signals. If these effects are significant, it is theoretically not possible to replace a three-degree-of-freedom test, for example, by a sequence of three single-degree-of-freedom tests. In practice, however, some conservatism can be introduced into two-degree-of-freedom and single-degree-of-freedom tests to account for these effects and these tests with added conservatism may be employed when three-degree-of-freedom testing is not feasible. It should be clear by now that rectlinear triaxial testing is generally not equivalent to three-degree-of-freedom testing, because the former merely applies an identical excitation in all three orthogonal directions, except for a scaling factor (direction cosines). One obvious drawback of rectlinear triaxial testing is that the input excitation in a direction at right angles to the actuator axis is theoretically zero, and the excitation is maximum along the actuator. In three-degree-of-freedom testing using uncorrelated random excitations, however, no single direction has zero excitation at all times, and the probability is zero that the maximum excitation occurs along a fixed direction at all times.

Three-degree-of-freedom testing is mentioned infrequently in the literature on vibration testing.[1] A major reason for this lack of three-degree-of-freedom testing might be the practical difficulty in building test tables that can generate truly uncorrelated input motions in three orthogonal directions. The actuator interactions caused by dynamic coupling through the test table and mechanical constraints at the table supports are primarily responsible for this. Another difficulty arises because it is virtually impossible to synthesize perfectly uncorrelated random signals to drive the actuators. Two-degree-of-freedom testing is more common. In this case, the test must be repeated for a different orientation of the test object (for example, with a 90° rotation about the vertical axis) unless some form of dynamic axis symmetry is present in the test object.

Test programs frequently specify uncorrelated excitations in two-degree-of-freedom testing for the two actuators. This requirement lacks solid justification, because two uncorrelated excitations applied at right angles do not necessarily produce uncorrelated components in a different pair of orthogonal directions, unless the mean square values of the two excitations are equal. To demonstrate this, consider the two uncorrelated excitations u and v shown in figure 7-9. The components u' and v' in a different pair of

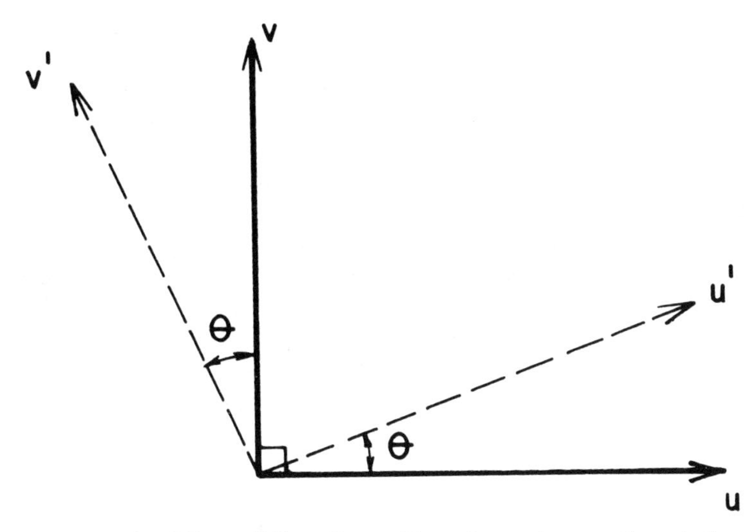

Figure 7-9. Effect of Coordinate Transformation on Correlation

orthogonal directions, obtained by rotating the original coordinates through an angle θ in the counterclockwise direction, is given by

$$u' = u\cos\theta + v\sin\theta \tag{7.1}$$

$$v' = -u\sin\theta + v\cos\theta \tag{7.2}$$

Without loss of generality, we can assume that u and v have zero means. Then, u' and v' will have zero means. Also, since u and v are uncorrelated,

$$E(uv) = E(u)E(v) = 0 \tag{7.3}$$

From equations (7.1) and (7.2),

$$E(u'v') = E[(u\cos\theta + v\sin\theta)(-u\sin\theta + v\cos\theta)]$$

which, when expanded and substituted by equation (7.3), becomes

$$E(u'v') = \sin\theta\cos\theta\,[E(v^2) - E(u^2)] \tag{7.4}$$

Since θ is any general angle, the excitation components u' and v' become uncorrelated if and only if

$$\mathrm{E}(v^2) = E(u^2) \tag{7.5}$$

This is the required result. Nevertheless, a considerable effort of digital Fourier analysis is expended by dynamic-testing laboratories to determine the degree of correlation in test signals employed in two-degree-of-freedom testing. Results from such analysis are presented in figure 7-10. The horizontal and vertical excitation time histories used are shown in figures 7-10(a) and 7-10(b), respectively. The autocorrelation functions of the horizontal excitation and the vertical excitation are shown in figures 7-10(c) and 7-10(d), respectively. The cross-correlation function of the two excitations is shown in figure 7-10(e). The corresponding cross-spectral density magnitude is shown in Figure 7-10(f), and the phase angle is shown in Figure 7-10(g). The ordinary coherence function of the two excitations is given in Figure 7-10(h). This shows that there is a considerable amount of correlation in the two excitations. The maximum magnitude of the cross-correlation function [Figure 7-10(e)] occurs at a time delay of 0.5 sec; that is, if the horizontal

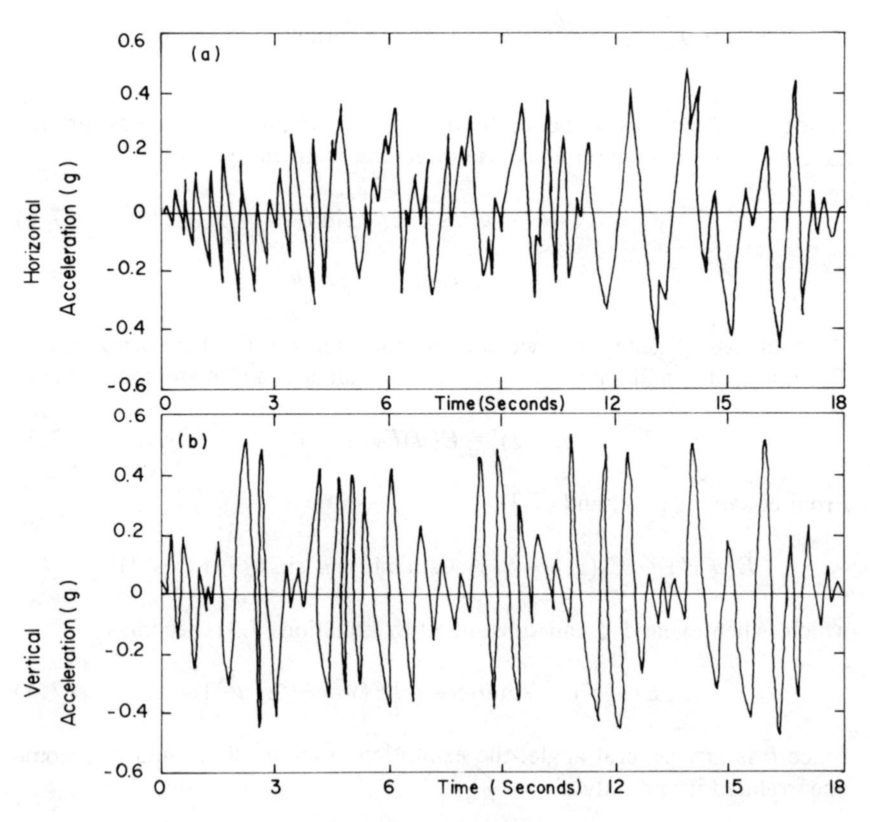

Figure 7-10. Correlation Analysis Example for a Two-Degree-of-Freedom Test

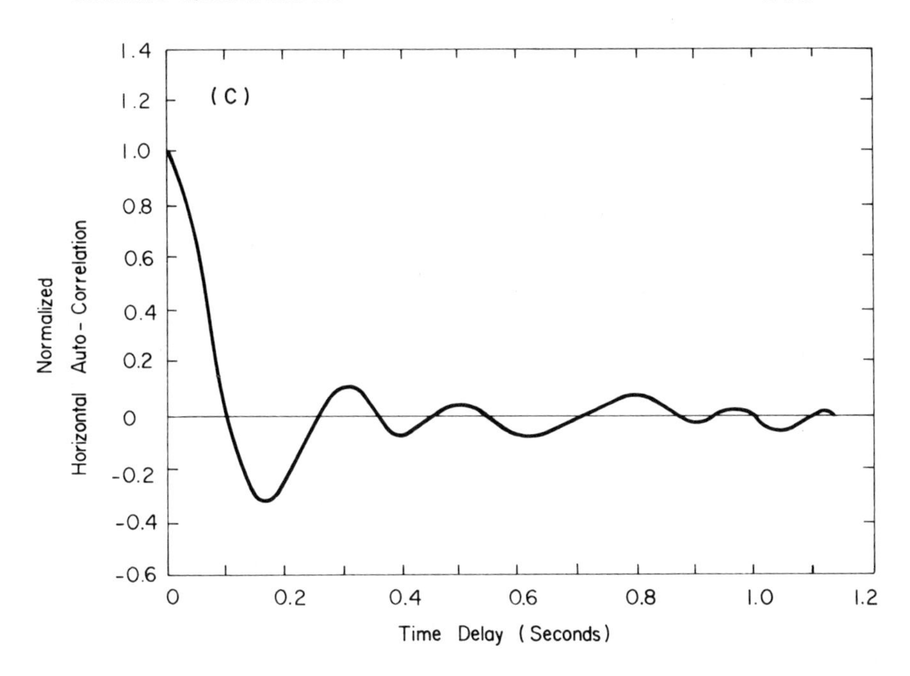

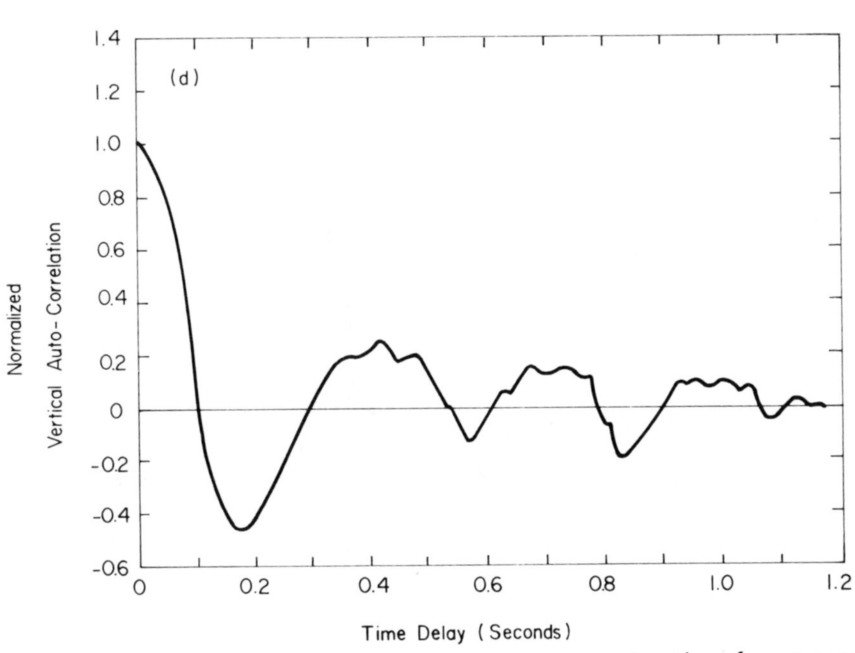

(continued next page)

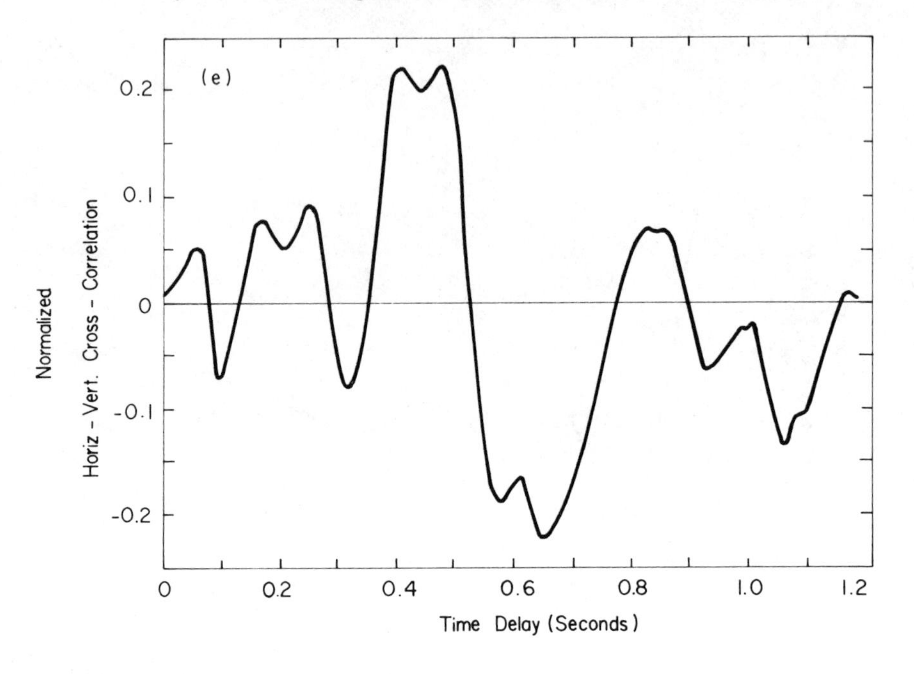

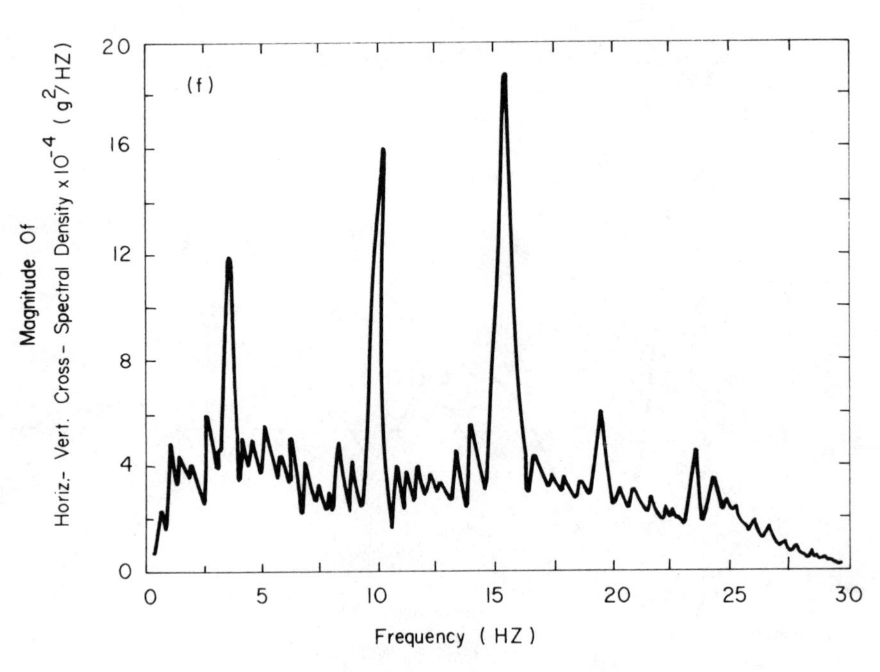

Figure 7-10 (continued)

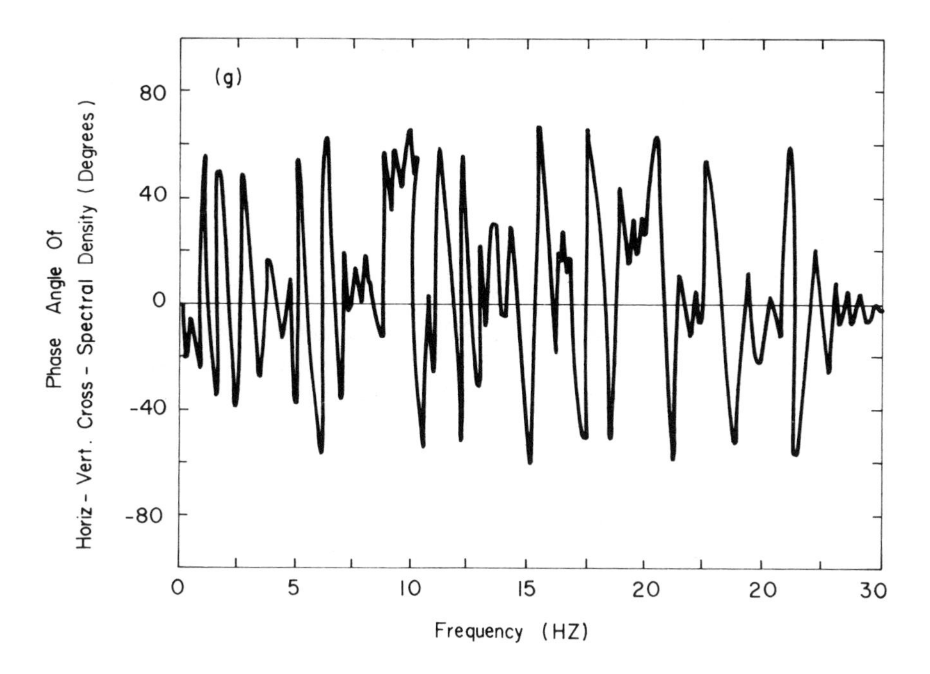
(g)
Phase Angle Of
Horiz - Vert. Cross - Spectral Density (Degrees)
80
40
0
-40
-80
0
5
10
15
20
20
30
Frequency (HZ)

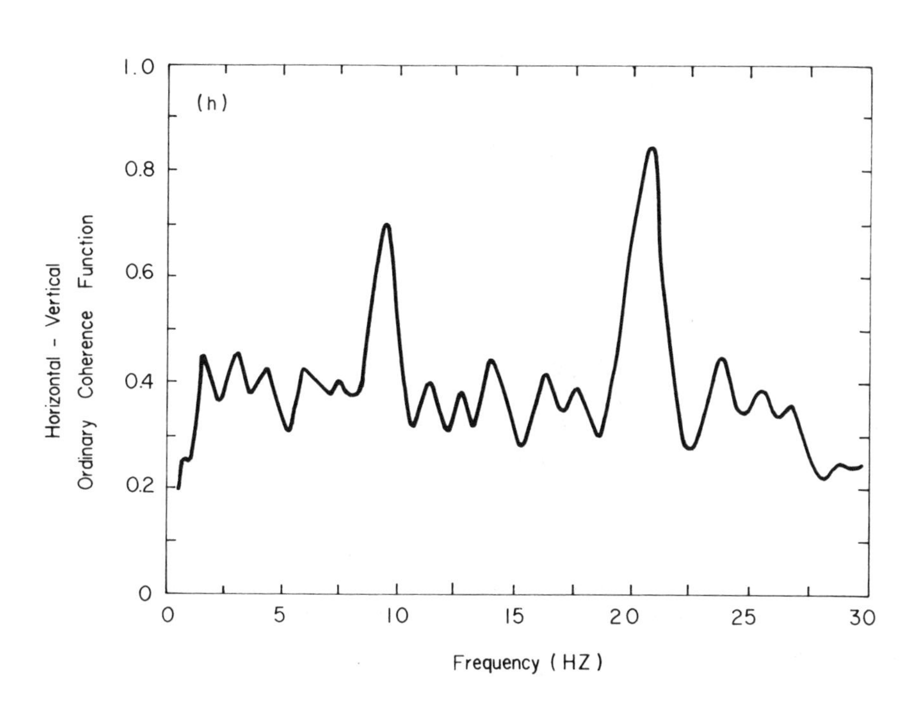
(h)
Horizontal - Vertical
Ordinary Coherence Function
1.0
0.8
0.6
0.4
0.2
0
0
5
10
15
20
25
30
Frequency (HZ)

excitation is delayed with respect to the vertical excitation by this amount, a higher correlation between the two excitations would result. This indicates that the phasing of the two excitations can be adjusted to obtain different levels of correlation.

Symmetrical Rectilinear Testing

Single-degree-of-freedom (rectilinear) testing, which is performed with the test excitation applied along the line of symmetry with respect to an orthogonal system of three principal axes of the test object mainframe, is termed symmetrical rectilinear testing. In seismic qualification literature, this test is often referred to as the 45° test. The direction cosines of the input orientation are $(1/\sqrt{3}, 1/\sqrt{3}, 1/\sqrt{3})$ for this test configuration. The single-actuator input intensity is amplified by a factor of $\sqrt{3}$ in order to obtain the required excitation intensity in the three principal directions. Note that symmetrical rectilinear testing falls into the category of rectilinear triaxial testing, as defined earlier. This is one of the widely used testing configurations in seismic qualification.

Geometry versus Dynamics. In seismic qualification, the emphasis is on the dynamic behavior rather than the geometry of the equipment. For a simple three-dimensional body that has homogeneous and isotropic characteristics, it is not difficult to correlate the geometry to its dynamics. A symmetrical rectilinear test makes sense for such systems. The equipment we come across is often much more complex, however. Furthermore, our interest is not merely in determining the dynamics of the mainframe of the equipment. We are more interested in the dynamic reliability of various critical components located on the mainframe. Unless we have some previous knolwedge of the dynamic characteristics in various directions of the system components, it is not possible to draw a correlation between the geometry and the dynamics of the tested equipment.

Some Limitations. In a typical symmetrical rectilinear test, we deal with black-box equipment whose dynamics are completely unknown. The excitation is applied along the line of symmetry of the principal axes of the mainframe. A single test of this type does not guarantee excitation of all critical components located inside the equipment. Figure 7-11 illustrates this further. Consider the plane perpendicular to the direction of excitation. The dynamic effect caused by the excitation is minimal along any line on this plane. (Any dynamic effect on this plane is caused by dynamic coupling among different body axes.) Accordingly, if there is a component (or several components) inside the equipment whose direction of sensitivity lies on this

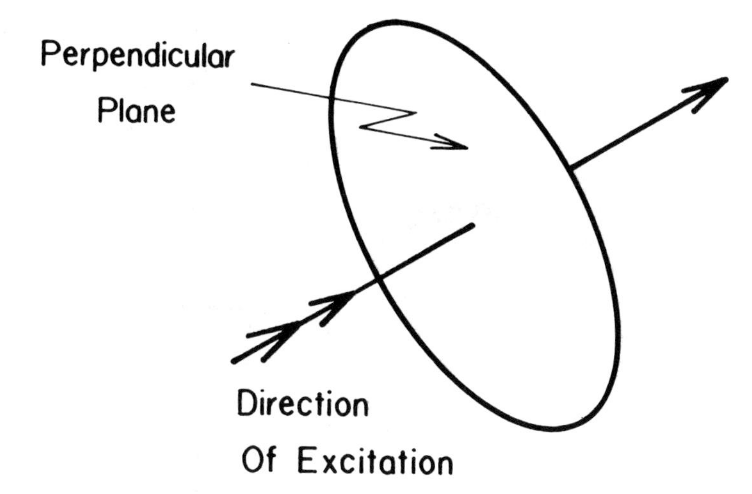

Figure 7-11. Illustration of the Limitation of a Single Rectilinear Test

perpendicular plane, the single excitation might not excite that component. Since we deal with a black box, we do not know the equipment dynamics beforehand. Hence, there is no way of identifying the existence of such components. When the equipment is put into service, a seismic disturbance of sufficient intensity can easily overstress this component along its direction of sensitivity and bring about component failure. It is apparent that at least three tests, performed in three orthogonal directions, are necessary to guarantee excitation of all components, regardless of their direction of sensitivity.

A second example is given in figure 7-12. Consider a dual-arm component with one arm sensitive in the *O-O* direction and the second arm sensitive in the *P-P* direction. If component failure occurs when the two arms are in contact, a single excitation in either the *O-O* direction or the *P-P* direction will not bring about component failure. If the component is located inside a black box, such that either the *O-O* direction or the *P-P* direction is very close to the line of symmetry of the principal axes of the mainframe, a single symmetrical rectilinear test will not result in system malfunction. This may very well be true, because we do not have a knowledge of component dynamics in such cases. Again, under service conditions, a seismic disturbance of sufficient intensity can produce an excitation along the *A-A* direction, subsquently causing system malfunction.

A further consideration in using rectilinear testing is dynamic coupling between the directions of excitation. In the presence of dynamic coupling, the sum of individual responses of the test object resulting from four symmetrical rectilinear tests is not equal to the response obtained when the excitations are

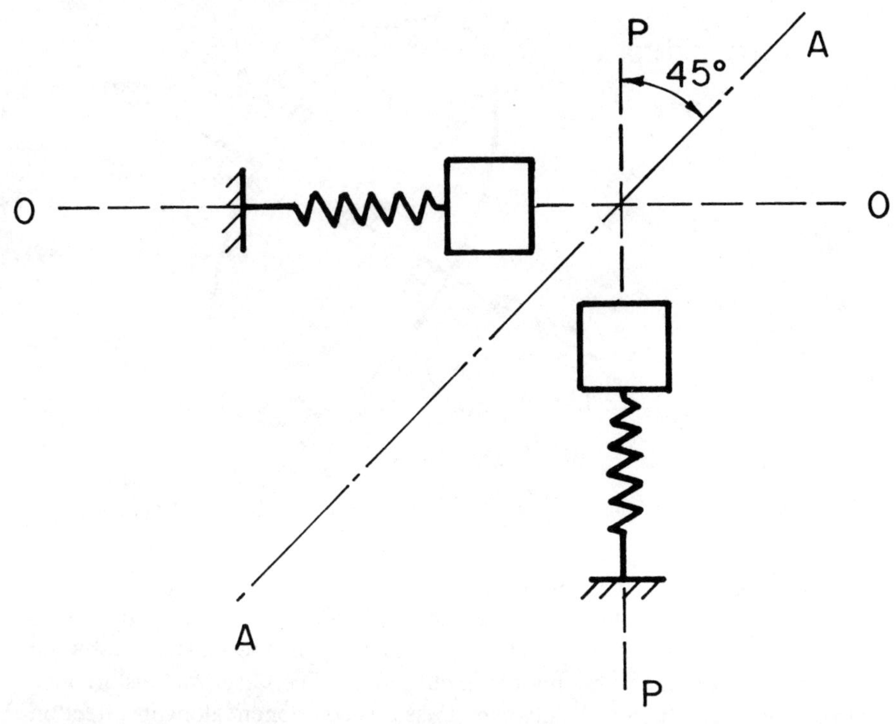

Figure 7-12. Illustrative Example of the Limitation of Several Rectilinear Tests

applied simultaneously in the four directions. Some conservatism should be applied when employing rectilinear testing for objects having a high level of dynamic coupling between the test directions. If the test-object dynamics are restrained to only one direction under normal operating conditions, however, then rectilinear testing could be used without applying any conservatism.

Testing Black Boxes. When the equipment dynamics are unknown, a single rectilinear test does not guarantee proper seismic qualification. To ensure excitation of every component within the test object that has directional intensities, three tests should be carried out along three independent directions. The first test is carried out with a single horizontal excitation, for example. The second test is performed with the equipment rotated through 90° about its vertical axes for the same horizontal input. The last test is performed with a vertical excitation.

Alternatively, if symmetrical rectilinear tests are preferred, four such tests should be performed for four equipment orientations (for example, an

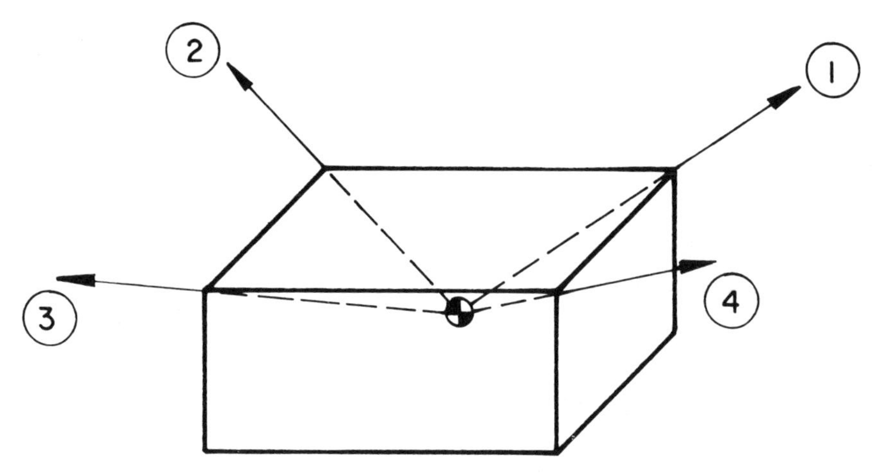

Figure 7-13. Directions of Excitation in a Sequence of Four Rectilinear Tests

original test, a 90° rotation, a 180° rotation, and a 270° rotation about the vertical axis). These tests also ensure excitation of all components that have directional intensities. This procedure might not be very efficient, however. The main shortcoming of this series of four tests is that some of the components would be overtested. It is clear from figure 7-13, for example, that the vertical direction is excited by all four tests. The method has the advantage, however, of simplicity of performance.

Phasing of Excitations. The main purpose of rotating the test orientation in rectilinear testing is to ensure that all components within the equipment are excited. Phasing of different excitations also plays an important role, however, when several excitations are used simultaneously. To explore this concept further, it should be noted that a random input applied in the *A-B* direction or in the *B-A* direction has the same frequency and amplitude (spectral) characteristics. This is clear because the psd of u = psd of $(-u)$ and the autocorrelation of u = autocorrelation of $(-u)$. Hence, it is seen that, if the test is performed along the *A-B* direction, it is of no use to repeat the test in the *B-A* direction. It should be understood, however, that the situation is different when several excitations are applied simultaneously.

The simultaneous action of u and v is not the same as the simultaneous action of $-u$ and v (see figure 7-14). The simultaneous action of u and v is the same, however, as the simultaneous action of $-u$ and $-v$. Obviously, this type of situation does not arise when there are no simultaneous excitations, as in rectilinear testing.

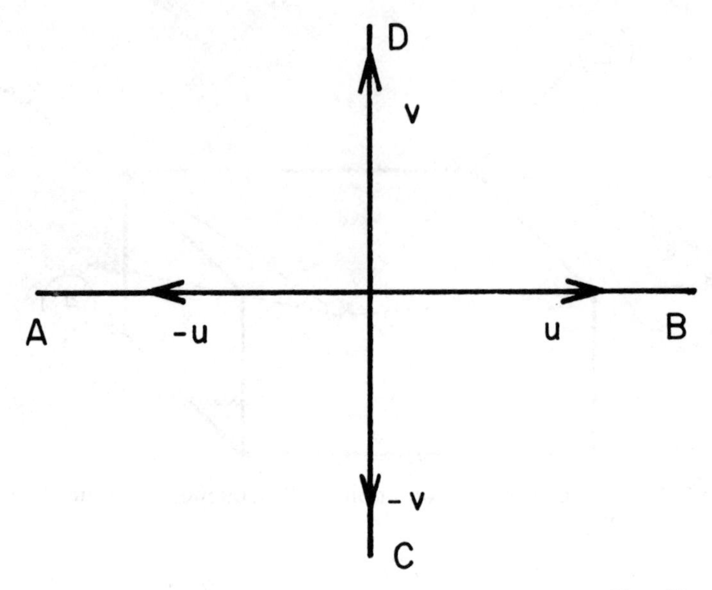

Figure 7-14. Significance of Excitation Phasing in Two-Degree-of-Freedom Testing

Testing a Gray or White Box. When some information regarding the true dynamics of the test object is available, it is possible to reduce the number of necessary tests. In particular, if the equipment dynamics are completely known, then a single test would be adequate (see chapter 10). The best direction for the system in figure 7-12 to be excited, for example, is *A-A*. (Note that *A-A* may be lined up in any arbitrary direction inside the equipment. In such a situation, a knowledge of the equipment dynamics is crucial.) This also indicates that it is very important to accumulate and use any past experience and data on the dynamic behavior of similar equipment. Any test that does not use some previously known information regarding the equipment is a blind test, and it cannot be optimal in any respect. As more and more information is available, better and better tests can be conducted.

Overtesting in Multitest Sequences. It is well known that increasing the test duration increases aging of the test object, because of prolonged stressing and load cycling of various components. This is the case when the test is repeated one or more times at the same intensity as that prescribed for a single test. The symmetrical rectilinear test requires four separate tests at the same excitation intensity as that prescribed for a single test. As a result, the equipment becomes subjected to overtesting, at least in certain directions. The degree of overtesting is small if the tests are performed in only three orthogonal directions. In any event, a certain amount of dynamic coupling is present in the test-object structure, and, to minimize overtesting in these sequential tests, a smaller intensity than that prescribed for a single test should be employed. The value of the intensity-reduction factor clearly depends on the characteristics of the test object, the degree of reliability expected, and the intensity value itself. More research is necessary to develop expressions for intensity-reduction factors for various test objects. For electric capacitors, for example, the test voltage intensity k is related to the duration of the test T through the relationship[2]

$$T \propto \frac{1}{k^p} \tag{7.6}$$

in which p is a parameter that depends on such factors as the particular capacitor used and the environmental conditions. Thus, if the intensity for a single test procedure has been prescribed as k_s, the intensity for a test sequence involving four tests is given by

$$k_m = \frac{k_s}{4^{1/p}} \tag{7.7}$$

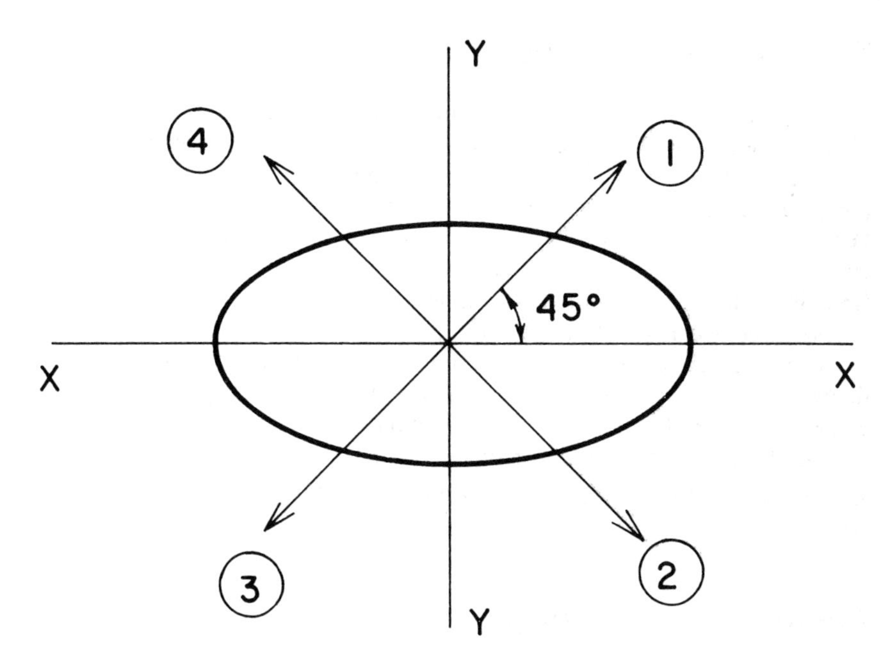

Figure 7-15. An Object That Has Two Orthogonal Planes of Symmetry

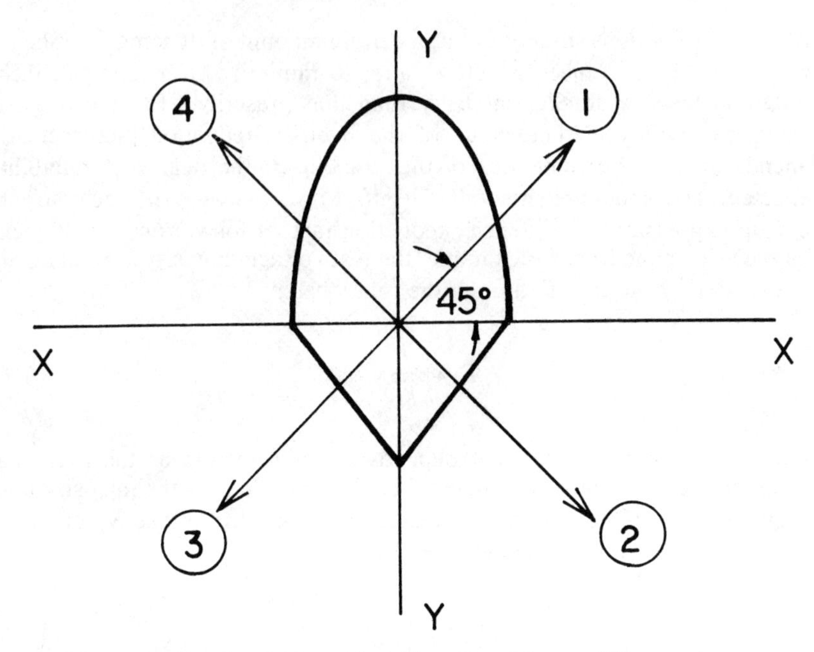

Figure 7-16. An Object That Has One Plane of Symmetry

Test-Object Symmetry Considerations

Consider a test object that has symmetry (dynamic as well as geometric) about vertical planes through the two horizontal principal axes (the x-axis and y-axis in figure 7-15). In this case, symmetrical rectilinear testing in directions 1, 2, 3, or 4 produces identical results. This should be clear because, when the object is looked at in one of these four directions, the same configuration is observed. For such objects, one symmetrical rectilinear test (in one of the four directions) is sufficient.

Next, consider a test object that has dynamic and geometric symmetry about the vertical plane through one horizontal principal axis (the y-axis in figure 7-16). In this case, the orientation of the test object appears the same when viewed in directions 1 and 4. Similarly, directions 2 and 3 appear similar. It follows that, for test objects with one plane of symmetry, two symmetrical rectilinear tests (for example, 1 and 3 or 2 and 4 in figure 7-16) are adequate.

Two-Degree-of-Freedom Testing

Rectilinear testing is the most widespread method employed in seismic qualification. Two-degree-of-freedom testing is employed in some situations, however, depending on the availability of appropriate shaker tables. In this type of testing, if the two excitations are random and statistically independent (or at least uncorrelated), then two tests are adequate. In this case, the two excitations are normally applied along one principal horizontal axis and the vertical principal axis (two-degree-of-freedom biaxial testing) of the test object. Then, the test object is rotated about the vertical axis through 90°, and the test is repeated.

If the two excitations are almost identical but a phase difference can be introduced, still only one rotation of the test object is needed, but four repetitions of the test are required. First, the test is conducted as in the previous case, with the two excitations in phase. Then the test is repeated with the two excitations out of phase. Next, the test object is rotated about the vertical axis through 90°, and the in-phase and out-of-phase tests are repeated as before. It should be clear from figure 7-17 that the first two tests are equivalent to the zero-rotation rectilinear test and the 180° rotation rectilinear rest. The last two tests are equivalent to the 90° rotation rectilinear test and the 270° rotation rectilinear test.

Single-Frequency Testing

Seismic ground motions usually pass through various support structures before they eventually are transmitted to equipment. In seismic qualification of that equipment by testing, we theoretically should apply to it the actual

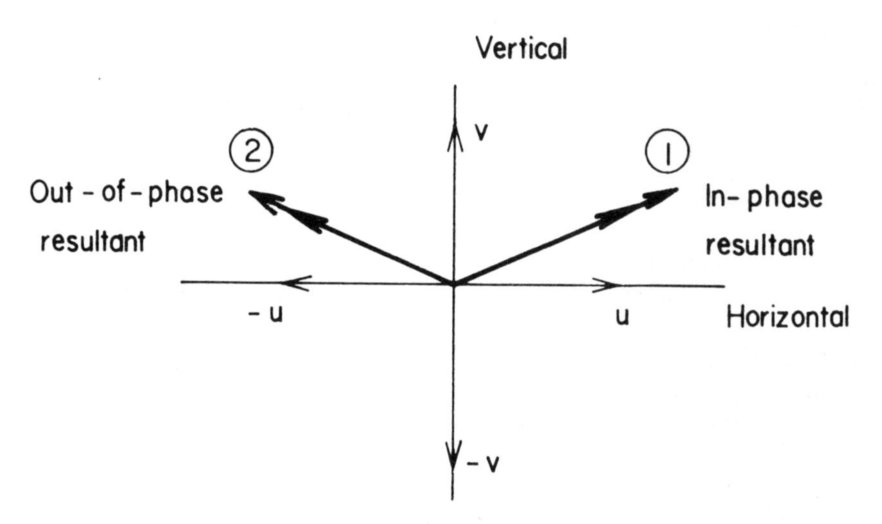

Figure 7-17. Use of Excitation Phasing in Place of Test-Object Rotation in Two-Degree-of-Freedom Testing

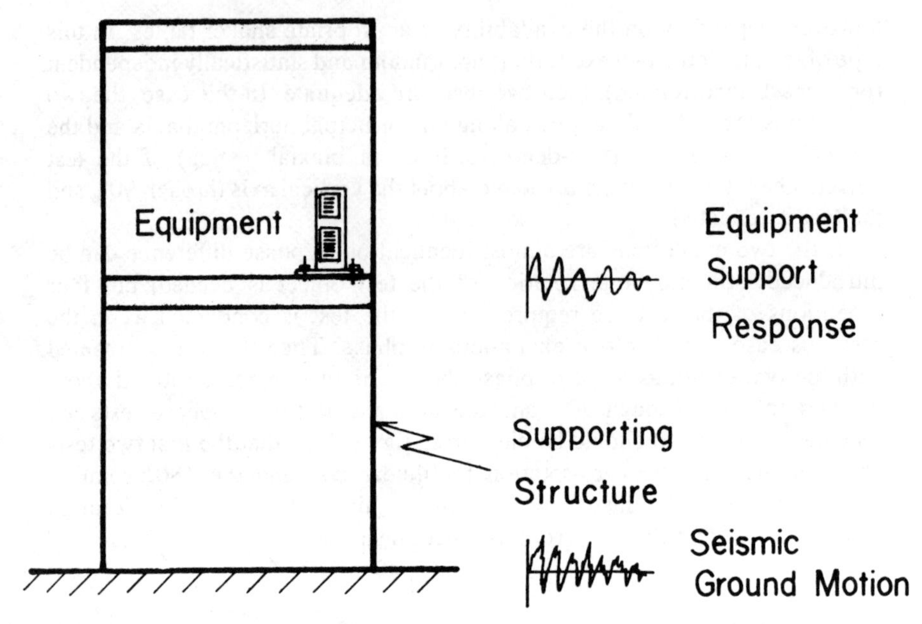

Figure 7-18. Schematic Representation of the Filtering of Seismic Ground Motions by a Supporting Structure

excitations felt by it, not the seismic ground motions. In an ideal case, the shaker-table motion should be equivalent to the seismic response of the supporting structure at the point of attachment of the equipment.

The supporting structure would have a certain frequency-response function between the ground location and the equipment-support location (see figure 7-18). Consequently, it could be considered a filter that modifies seismic ground motions before they reach the equipment mounts. In particular, the components of the ground motion that have frequencies close to the resonance frequency of the supporting structure will be felt by the equipment at a relatively higher intensity. Furthermore, the ground motion components at very high frequencies will be almost entirely filtered out by the structure. If the frequency response of the supporting structure is approximated by a lightly damped simple oscillator, then the response felt by the equipment will be almost sinusoidal, with a frequency equal to the resonant frequency of the structure (see figure 7-19).

When the equipment supporting structure has a very sharp resonance in the significant frequency range of the dynamic environment (for example, 1–35 Hz for seismic ground motions), it follows, from the previous discussion, that it is desirable to use a short-duration single-frequency test in seismic qualification of the equipment. Equipment supported on pipelines (valves, valve actuators, guages, and so forth) falls into this category. Such equipment is termed line-mounted equipment.

Resonant frequency of the supporting structure is usually not known at the time of the seismic qualification test. Consequently, single-frequency testing must be performed over the entire frequency range of interest for that particular dynamic environment.

Another situation in which single-frequency testing is appropriate arises when the test object (equipment) itself does not have more than one sharp resonance in the frequency range of interest. In this case, the most prominent response of the test object occurs at its resonant frequency, even when the dynamic environment is an arbitrary excitation. Consequently, a single-frequency excitation would yield conservative test results. Equipment that has more than one predominant resonance may employ single-frequency testing, provided that each resonance corresponds to a dynamic degree of freedom (for example, one resonance along each dynamic principal axis) and that cross coupling between these degrees of freedom is negligible.

In summary, single-frequency testing may be used if one or more of the following conditions are satisfied:

1. The supporting structure has one sharp resonance in the frequency range of interest (line-mounted equipment is included).
2. The test object does not have more than one sharp resonance in the frequency range of interest.
3. The test object has a resonance in each degree of freedom, but the degrees of freedom are uncoupled (for which adequate verification should be provided in the test procedure).
4. The test object can be modeled as a simple dynamic system (such as a simple oscillator), for which adequate justification or verification should be provided.

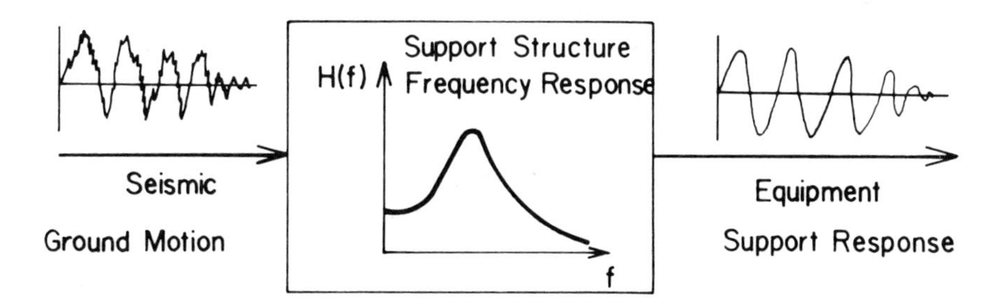

Figure 7-19. Illustration of the Validity of Using Single-Frequency Excitations in Testing Line-Mounted Equipment

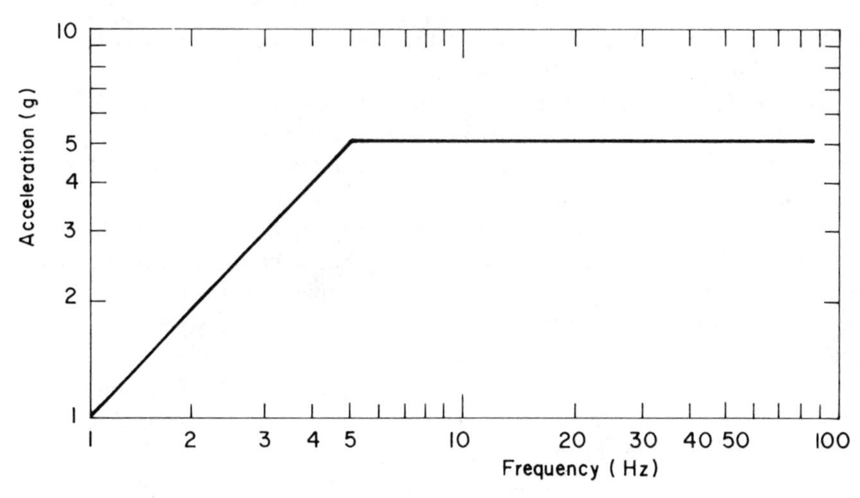

Figure 7-20. RIM Specification for a Single Frequency Test (SSE)

The commonly used single-frequency test excitation signals, as discussed in chapter 5, are sine sweep, sine dwell, sine decay, sine beat, and sine beat with pause. Usually, the required SSE excitation-input level for different frequencies for a single-frequency test is specified by a curve such as the one shown in figure 7-20. This curve is known as the required input motion (RIM) magnitude curve. The OBE excitation level is usually taken as a fraction (typically, 0.5 or 0.7) of the RIM values given for the SSE. For a sine-sweep test, the sweep rate and the number of sweeps in the test should also be specified. Typically, the sweep rate for seismic qualification tests is less than 1 octave/min. One sweep, from the state of rest to the maximum frequency in the range and back to the state of rest, is normally carried out in SSE test (for example, 1 to 35 to 1 Hz). Several sweeps (typically, five) are performed in an OBE test.

In sine-dwell tests, the dwell times for each dwell frequency should be specified for an SSE test. The dwell-frequency intervals should not be high (typically, a half-octive or less). For an OBE test, the dwell times are longer (typically, five times longer) than those specified for an SSE test.

For an SSE test using sine beats, the minimum number of beats and the minimum duration of excitation (with or without pauses) at each test frequency should be specified. In addition, pause time for each test frequency should be specified when sine beats with pauses are employed. For an OBE test, the duration of excitation should be increased (as in a sine-dwell test).

The dwell time at each test frequency should be adequate to perform at least one functional-operability test. Furthermore, a dwell should be carried out at each resonant frequency of the test object as well as at those

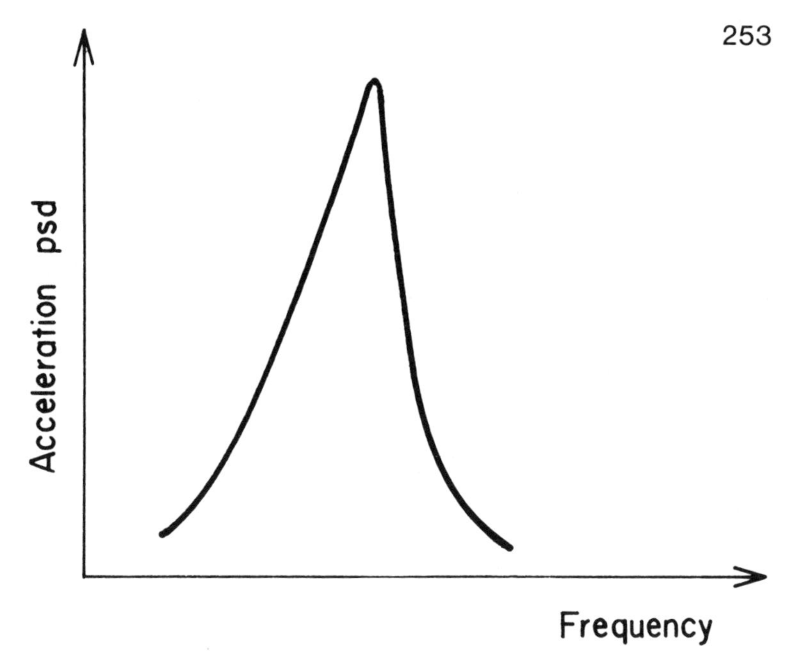

Figure 7-21. Power Spectral Density of a Typical Narrow-Band Random Signal

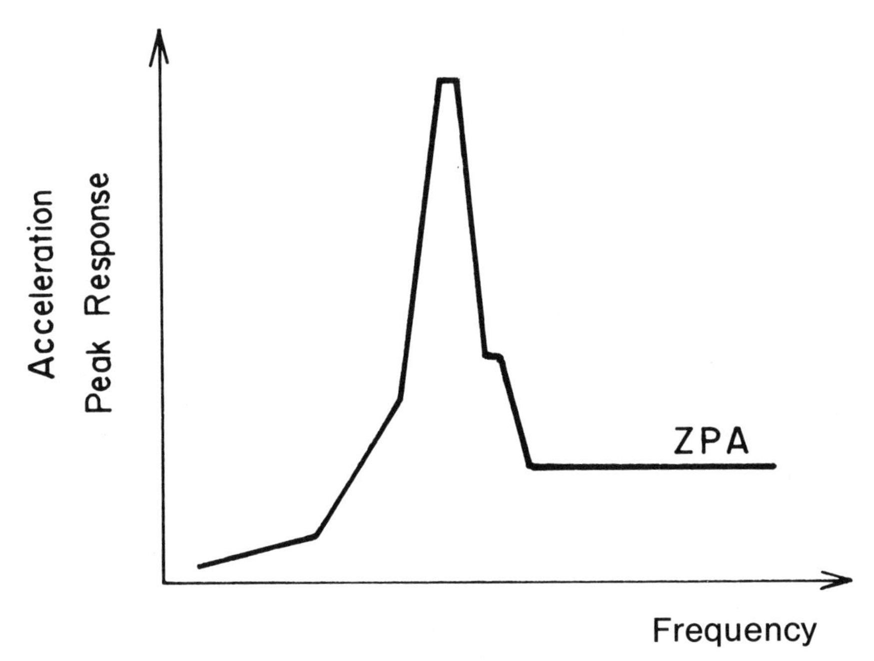

Figure 7-22. A Typical RRS for a Narrow-Band Excitation Test

frequencies that are specified. Total duration of an SSE test should be representative of the duration of the strong-motion part of a standard safe-shutdown earthquake.

Sometimes narrow-band random excitations may be used in cases where single-frequency testing is recommended. Narrow-band random signals are those that have power concentrated over a narrow frequency band. The psd curve for a typical narrow-band random signal is shown in figure 7-21. Such a signal can be generated for test-excitation purposes by passing a random signal through a narrow-band-pass filter. By tuning the filter to different center frequencies in narrow bands, the test-excitation frequency can be varied during testing. This center frequency of the filter should be swept up and down over the desired frequency range at a reasonably slow rate (for example, 1 octave/min) during the test. Thus, a multifrequency test with a sharp frequency-response spectrum (the RRS), as typified in figure 7-22, is adequate in cases where single-frequency testing is recommended. A requirement that has to be satisfied by the test-excitation input signal in this case is that its amplitude should be equal to or greater than the zero-period acceleration (defined in chapter 5) of the RRS for the test.

Multifrequency Testing

When equipment is mounted very close to the ground under its normal operating conditions, or if its supporting structure and mounting can be considered rigid, then seismic ground motions will not be filtered significantly before they reach the equipment mounts. In this case, the seismic excitations felt by the equipment will retain broadband characteristics. Multifrequency testing is recommended for seismic qualification of such equipment.

Whereas single-frequency tests are specified by means of an RIM curve and corresponding durations at each frequency (or sweep rates), multifrequency tests are specified by means of an RRS curve. The test requirement in multifrequency testing is that the response spectrum of the test-excitation input (the TRS), which is felt by the equipment mounts, should envelop the RRS. Note that all frequency components of the test excitation are applied simultaneously to the test object, in contrast to single-frequency testing, in which, at a given instant, only one significant frequency component is applied.

When random excitation inputs are employed in multifrequency testing, enveloping of the RRS by the TRS may be achieved by passing the random signal produced by a signal generator through a spectrum shaper. As the analyzing frequency bandwidth (for example, one-third-octave bands, one-sixth-octave bands) decreases, the flexibility of shaping the TRS improves. A real-time spectrum analyzer (or a microcomputer) may be used to compute and display the TRS curve corresponding to the control accelerometer signal

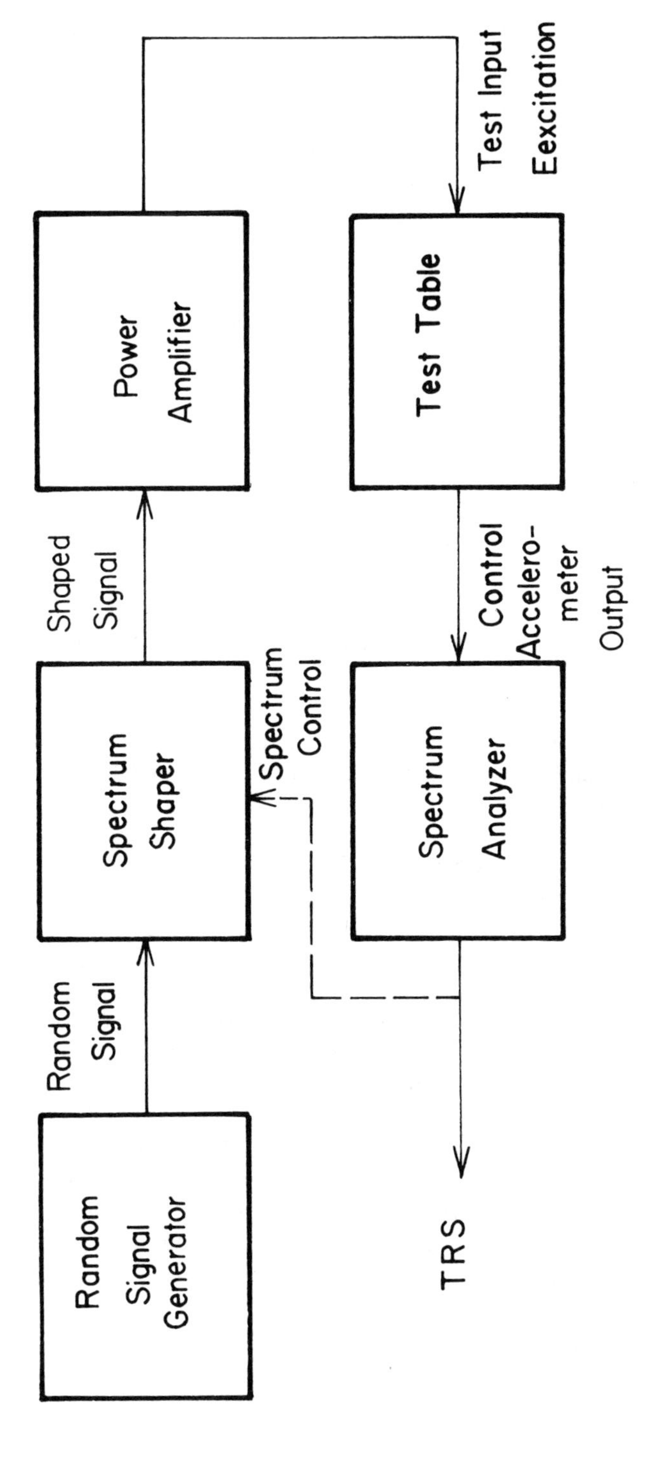

Figure 7-23. Matching of the TRS with the RRS in Multifrequency Testing

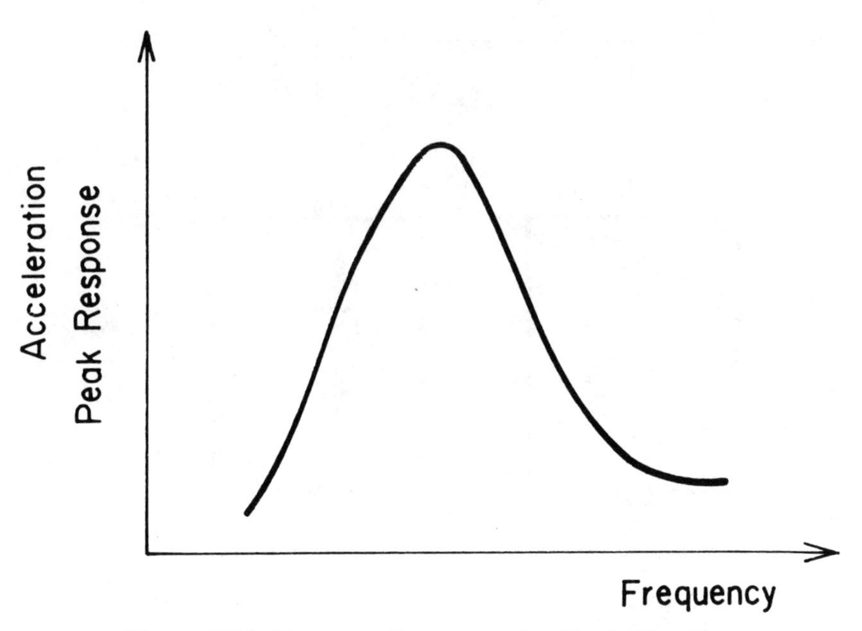

Figure 7-24. Response Spectrum of a Single Sine Beat

(see figure 7-23). By monitoring the displayed TRS, it is possible to adjust the spectrum-shaper filter gains so as to obtain the desired TRS, which would envelop the RRS.

Most test laboratories generate their multifrequency excitation-input signals by combining a series of sine beats that have different peak amplitudes and frequencies. The principle behind this procedure can be explained with reference to the response spectrum curve of a single sine-beat signal (see figure 7-24). By mixing an appropriate set of sine beats, a given response spectrum can be made to envelop (by the composite test signal), as shown in figure 7-25. Care should be taken in this case to phase each sine-beat component properly so that enveloping can be achieved more efficiently.

Using the same method, many other signal types, such as decaying sinusoids, may be superimposed to generate a required multifrequency test-input signal. A combination of signals of different types also could be employed to produce a desired test input. A commonly used combination is a broad-band random signal and a series of sine beats. In this combination, the random signal is adjusted to have a response spectrum that will envelop the broad-band portion of the RRS without much conservatism. The narrow-band peaks of the RRS that generally will not be enveloped by such a broad-band response spectrum will be covered by a suitable combination of sine beats.

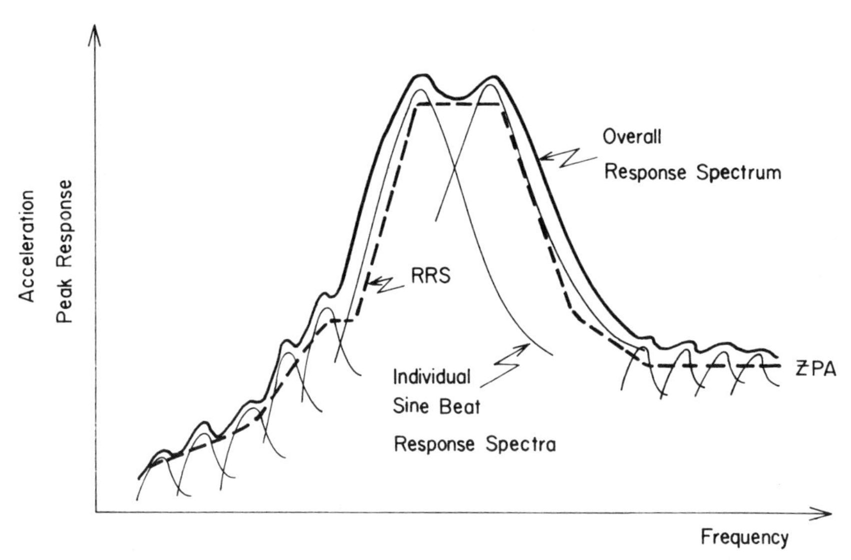

Figure 7-25. TRS Generation by Combining Sine Beats to Envelop the RRS

By employing such mixed composite signals, it is possible to envelop the entire RRS without having to increase the amplitude of the test excitation to be substantially higher than the ZPA of the RRS. One important requirement in multifrequency testing is that the amplitudes of the test-excitation input be equal to or greater than the ZPA of the RRS.

Seismic Qualification by Analysis

Qualification by analysis is usually preferred to dynamic testing for dynamically simple objects. Analysis is often less time-consuming and more economical. Well-developed computer programs are currently available to perform extensive dynamic analysis of relatively complex systems. Seismic qualification by analysis is generally recommended, however, only if the following requirements are satisfied:

1. The object is not very complex, so that it can be represented by a relatively simple analytical model (linear or simple nonlinear) of reasonable order.
2. Structural integrity alone governs the seismic qualification decision, or functional operability can be represented in terms of simple mechanical motions.

Sometimes, analysis is used by test laboratories in the seismic qualification of test objects that are too large for the available test tables. Unfortunately, this is not a valid justification for using analysis, and, in such circumstances, justification should be provided for using analysis in terms of the two requirements listed above. Analysis alone usually is not adequate in seismic qualification of complex systems, such as control equipment and reactor internals.

Analysis Procedures

Once the decision is made to employ analysis in seismic qualification of an object, it is customary to perform a resonance-search test on the object if frequency-response data are not available from a previous test. If the object can be represented by a simple linear model, its natural frequencies can be determined by analysis rather than by testing.

Since the strong-motion portion of a typical seismic ground motion is concentrated in a frequency range below 33 Hz, it is unlikely that any significant resonance of the object would be excited by seismic motions if the object does not have resonances below about 35 Hz. In that case, the object can be assumed to be rigid, and a static analysis alone is adequate for seismic qualification of the object. If there are resonances or natural frequencies below 35 Hz, then the object is assumed to be flexible, and a dynamic analysis should be employed in seismic qualification of the object. This initial decision-making stage is shown in figure 7-26.

Typical instances of static analysis include the seismic qualification of rigid valves and pumps. Dynamic analysis is usually used in the seismic qualification of flexible objects, such as flexible valves, tanks, and heat exchangers. Dynamic testing is also required for the seismic qualification of complex objects, such as valve operators, limit switches, solenoid valves, and relays.

Static Analysis. If an object is rigid, all points in the object will move at the same acceleration as the point to which seismic excitation is applied (that is, the mounting location). In this case, the distributed mass of the object can be represented by an equivalent set of lumped masses, concentrated at several key locations of the object.

Static-analysis methods used in seismic qualification can be divided into two categories: specified static analysis and response-spectra static analysis. The second method is also called static oefficient analysis.

In specified static analysis, the support acceleration under SSE conditions is specified. An acceleration value of 4 g, for example, may be used for objects that are line-mounted, and a value of 2 g for objects that are floor-mounted.

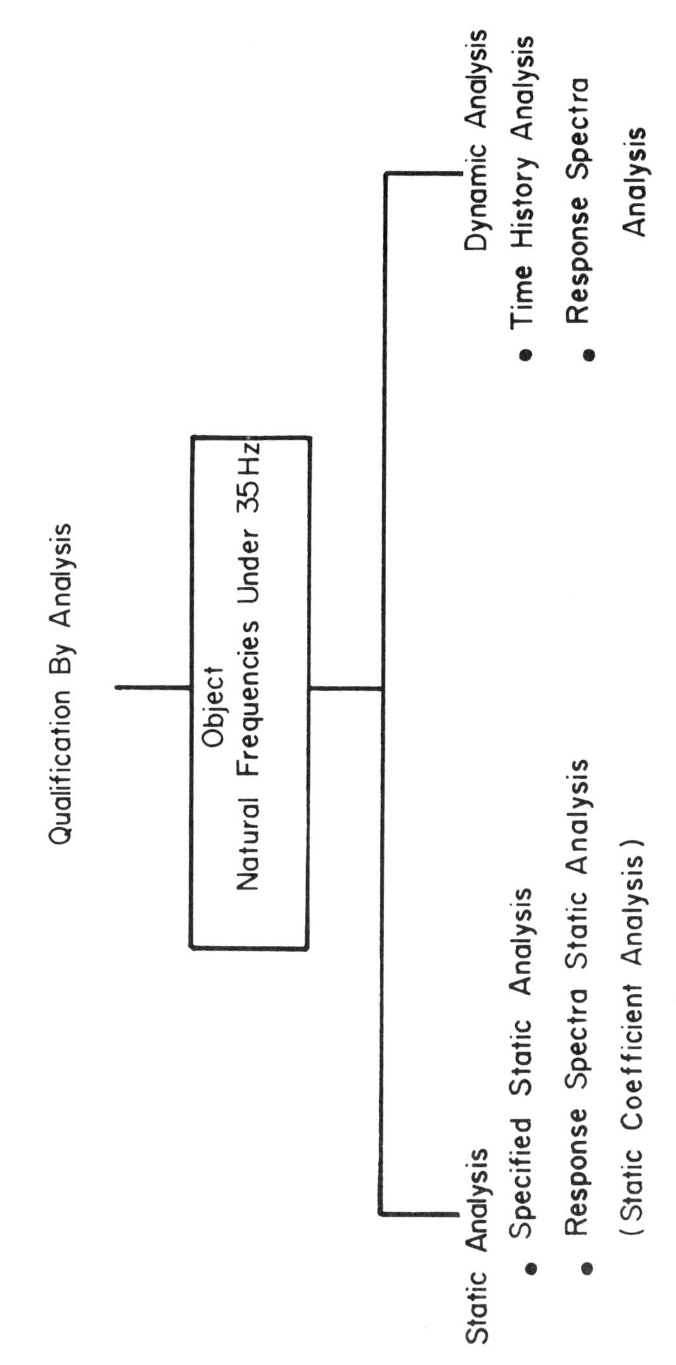

Figure 7-26. Selection of the Analysis Method in Seismic Qualification

In response-spectra static analysis, the support acceleration is taken as the peak value of the RRS curve corresponding to a justifiable value of damping. Sometimes, particularly when the assumption of rigidity is not quite valid, the peak acceleration thus obtained is multiplied by a coefficient that is greater than unity (a static coefficient) to account for multimode response effects. Typically, an SSE required response spectrum is specified. Then the OBE peak acceleration is taken as a fraction (typically, 0.5 or 0.7) of the SSE value.

In both methods, the support acceleration is multiplied by the lumped-mass values in the model to obtain the inertia-force values at the locations of these masses. Three orthogonal components of inertia force should be applied at each mass to account for the three-dimensional nature of earthquake disturbances. If the RRS curves for three orthogonal directions (for example, vertical, east-west, and north-south) are available, the largest peak acceleration of the three should be used in computing the inertia-force components. In addition, usual design loads should be applied at proper locations of the model.

A stress analysis and a static-deflection analysis should be performed using combined seismic inertia loads and design loads. The stress results should be checked against allowable stress values to gurantee structural integrity. Static-deflection values can be employed to determine functional-operability problems that could arise from such causes as misalignment, component rubbing, and short-circuiting.

In stress analysis, it is customary to follow the ASME Boiler and Pressure Vessel Code. For OBE loading conditions, an elastic analysis is usually sufficient. Since the object has to be qualified for a number of OBEs, however, a fatigue analysis might be required. This is normally done by considering the application of a specified number of peak stress cycles (for example, 100 to 1,000 cycles). For SSE loading, however, an inelastic analysis (considering plastic material behavior) might be required. A fatique analysis is not required in SSE analysis, however, because the object is qualified for the application of only one SSE.

Dynamic Analysis. The major steps in seismic qualification by analysis are outlined in figure 7-27. The first crucial step in seismic qualification of an object by analysis is the development of a suitable analytical model for the object. The contractor accomplishes this task using information provided by the vendor or the customer (see figure 7-1). The primary steps of this task are as follows:

1. Identify the system and all interfaces.
2. Identify response and operability parameters.
3. Identify mass, stiffness, and damping elements in the system.

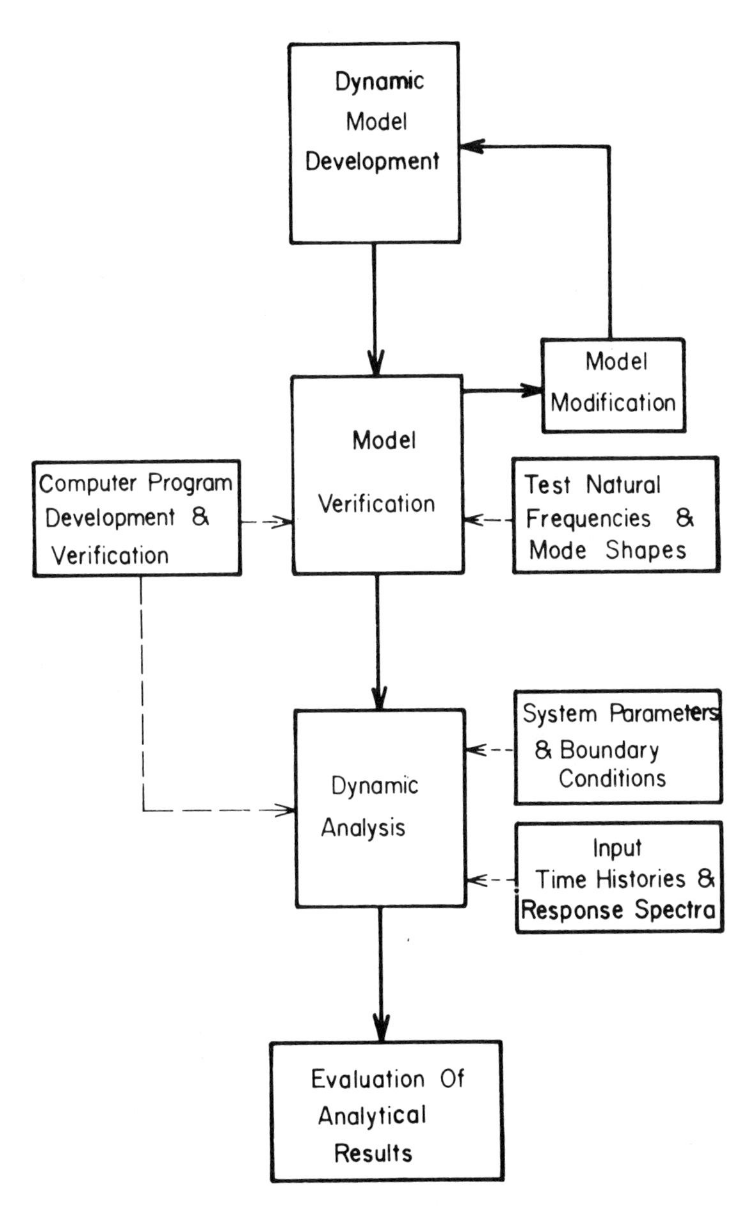

Figure 7-27. Major Steps in Dynamic Analysis for Seismic Qualification

4. Model the system elements by their constitutive relations, incorporating nonlinear relations if necessary.
5. Identify boundary conditions and loading.

The system selected should be uncoupled as much as possible from any interfacing systems. Otherwise, it is necessary to perform a coupled analysis of all system that have a high degree of dynamic coupling. As an approximation to this procedure, an interfacing system could be replaced by interface loads that satisfactorily represent the interacting loads of the coupled system at the interface.

Identification of the operability requirements and associated response parameters should be done before deciding on the final model of the system. The main reason for this is that the model elements in regions where operability conditions are determined must be chosen more carefully than those in the rest of the system. Operability parameters include deflections, clearances, loads, degree of alignment, and strains. Operability limits for these parameters should be available from the vendor or purchaser.

The number of mass, stiffness, and damping elements to be included in the system model should be realistic. When high accuracy is desired, a large number of these basic elements are used. A too-complex analytical model could decrease computational accuracy, however, and would increase the computational effort required. Economics and time restrictions will influence these decisions.

When using constitutive relations to represent the basic elements (mass, stiffness, damping), any significant nonlinearities that may be present in the system should be carefully accounted for. The origin of major nonlinearity could be material (for example, plastic stress-strain relations) or geometric (for example, presence of clearances in moving parts). One of the important considerations in obtaining system equations would be modeling of energy dissipation caused by impact between adjacent parts (impact damping). This modeling could be done by using either a coefficient of restitution or an equivalent viscous damping coefficient. If a finite-element method is used in modeling, an impact element could be incorporated in the model.

Mounting features and mounting loads have a great influence on system boundary conditions. The loads include those generated by relative anchor movements, external structural restraints, pipes, cables, and coupled subsystems. The locations, directions, and nature of such loading should be identified. In addition to three-dimensional seismic loading at mounting, other loads, such as nozzle loads and slush loads resulting from contained liquid, should be included in the model. Boundary conditions incorporated in the model should be consistent with these loads.

Model verification is usually done by performing a natural-frequency and mode-shape analysis, using the developed model, and comparing the results with frequency-response test results. The results from an earlier resonance-search test (for example, the test to determine rigidity for the static or dynamic analysis decision) would be adequate for this purpose. Realistic system parameters should be used in all simulations. Stiffness and damping parameters for the model are either directly measured or obtained from information already available to the contractor. It is customary to use low (conservative) values of equivalent viscous damping (for example, 2 percent of critical) in the simulations. If the analytical natural frequencies and mode shapes are not agreeable with the test results (for example, a maximum of 10 percent deviation), then the model should be refined until satisfactory results are obtained. The error could originate from the computer program used. It is therefore advisable to verify the computer program before using it in the system simulation. The program verification and the model simulation should be done on the same computer to avoid any coding discrepancies. Program verification is done by using a baseline problem for which a standard and verified solution is available. Hand calculations, results from previous tests, or results available in prestigious technical literature may be used in verification of the computer program. Minimal verification is needed if the computer programs are those used extensively in industry. Sometimes, new programs must be developed to perform specialized analysis tasks. Thorough verification is required under such circumstances.

Dynamic analysis may be performed according to the standard guidelines given in the ASME Boiler and Pressure Vessel Code, Section III, Appendix N, "Dynamic Analysis Methods." There are two commonly used dynamic-analysis methods for seismic qualification: time-history analysis and response-spectrum analysis. In time-history analysis, a suitable set of time histories is used as seismic excitation inputs at the mounting locations, and the system response usually is determined by numerical integration of the system equations. For simple systems, direct analytical solution might be possible. For complex time histories and for large (high-order) and complex (nonlinear and the like) systems, numerical integration is used exclusively. Various numerical integration schemes could be used for this purpose. The time-history inputs normally are generated by the contractor so that their response spectra envelop the RRS that is usually provided by the purchaser. Since three-dimensional seismic effects must be considered, two orthogonal horizontal inputs and a vertical input should be used at each mounting point. Their response spectra should envelop the resultant horizontal RRS and the vertical RRS, respectively. For objects supported at more than one point, different RRSs may be applicable at different support locations if the supporting structure is relatively flexible. Then, corresponding time-history inputs should be used at various mounting points. It often is acceptable to use the same time-history input for all support points and in all three directions. Its response spectrum, however, should envelop all RRS curves for various support points and directions.

At times, it may not be convenient to apply time history inputs in all three orthogonal directions simultaneously when performing the numerical integration. In such cases, it is acceptable to apply the three time histories one at a time and compute the system response separately. The system response at a given location and in a given direction is then determined by combining the three individual responses for that location and that direction. This combination can be done as follows:

1. If the three orthogonal input time histories are statistically independent (verification should be provided), then the combination is done by algebraic addition at each time step.
2. Otherwise, the maximum value of each response is squared and added, and the square root of the sum is determined.

The second method is termed the square root of the sum of the squares (SRSS) method. Only an estimate for the maximum value of the system response is obtained by this method. Evaluation of the object under the specified seismic loading is done using the response results (stresses, strains, deflections, and so on). When determining the system response, all loads present under operating conditions should be included, in addition to the seismic loads.

Response-spectrum analysis is more convenient and economical than the time-history method. In the response-spectrum method, only an estimate for the peak response of the system is obtained. First, the natural frequencies of the system that fall below the ZPA asymptote point of the response spectrum (typically, 33 Hz for seismic qualification analysis) and the corresponding mode shapes are determined by analysis. Then, using the RRS curve, the peak response (of a simple oscillator) corresponding to each natural frequency is determined. Finally, these peak responses are combined in accordance with U.S. NRC Regulatory Guide 1.92. The corresponding mode-shape factor and the modal-participation factor should be included in each peak-response value obtained from the RRS before combining them. In essence, the combination is done by the SRSS method if there are no closely spaced modes (that is, modes that have natural frequencies very close to each other). Closely spaced modal responses are combined by absolute addition.

As in the time-history analysis method, three-dimensional seismic motion is considered in the spectrum analysis method by using horizontal RRS curves in two orthogonal directions and a vertical RRS curve. Usually, the three orthogonal directions of input are considered separately, and the net peak response finally is obtained by combining the individual peak responses, using the SRSS method. If there is more than one support location, it is acceptable to use a single RRS that envelops the RRS curves at all suport locations and in all directions (as in the time-history method). Final seismic evaluation is done using peak-response results (stresses, strains, deflections, and so on).

Notes

1. C. W. Roberts, "Seismic Testing Capabilities of a Typical Commercial Laboratory," *Earthquake Environment Simulation,* Report No. PB-240 404 Washington, D.C.: National Academy of Engineering, pp. 213–217, 1974.

2. J. Kimmel, "Accelerated Life Testing of Paper Dielectric Capacitors," *Proceedings of the Fourth National Symposium on Reliability and Quality Control*, pp. 120–134, January 1958.

8 Excitation System and Instrumentation

To perform dynamic testing on an object, we require various apparatus and instruments. A typical shaker-test setup can be divided into several subsystems, depending on the basic function of the various equipment and instruments employed. The main subsystems are as follows:

1. The signal-generating system
2. The excitation system
3. The response-sensing system
4. The signal-conditioning system
5. The response-signal-recording system

A specified excitation signal is generated by the signal generator. This signal is converted into a dynamic motion (or excitation) by the excitation system and applied to the test object. Dynamic response of the test object must be monitored, using various sensors and transducers. The response signals must be conditioned, using signal-conditioning devices, before they are recorded for subsequent analysis and processing. Even the input signal must be conditioned before it is used for actuating the shaker.

In addition to these five subsystems, we might need special instruments for functional-operability monitoring if it is part of the test. The nature of such instruments depends on the features of the particular test object. Hence, a general discussion of the functional-operability monitoring system is not undertaken in this book.

The data-acquisiton and processing system could be considered a separate subsystem. Since control decisions on the excitation input are based on the results obtained from data acquisition and processing, however, it is considered a consistuent of the excitation system.

Excitation System

In dynamic testing, the excitation system typically consists of three basic subsystems: the shaker (or actuator or drive unit), the control system, and the data-acquisition and processing system.

Other essential components in a dynamic-testing setup include such auxiliary equipment as the excitation-input-signal source (for example, tape players, random-signal generators), test-object mounting fixtures, signal-conditioning equipment (for example, amplifiers, filters), signal-sensing equipment (for example, accelerometers, strian gauges, LVDTs), and test-monitoring equipment (for monitoring functional operability and structural integrity). These components, however, may be considered interfacing devices that are introduced through the interaction of the three main subsystems.

The interactions that exist between the shaker and the data-acquisition and processing system are shown schematically in figure 8-1. Although the term *shaker* sometimes is used loosely to denote the entire excitation system, in the present context it denotes the actuator that is the driving unit of the test object. Shaker and test object are interfaced through a test table and the required mounting fixtures. The control system uses feedback signals, which could be response signals from the test table (including mounting fixtures) and from the test object itself, to control the test. The controlling is done by modifying the excitation-input signals to the actuator. The objective of control is generally twofold: (1) to control the excitation-input motion felt by the test object from its mounting, and (2) to stabilize or limit motions in various directions of freedom (for example, three translational motions and three rotational motions—pitch, roll, and yaw) of the test object. A combination of these objectives might be achieved by a complex control system. The control system itself can be either analog or digital, but the input signal to the actuator should always be analog. Interfacing between the control system and the actuator usually consists of a digital-to-analog converter if the control system is digital, filter circuits, and power amplifiers.

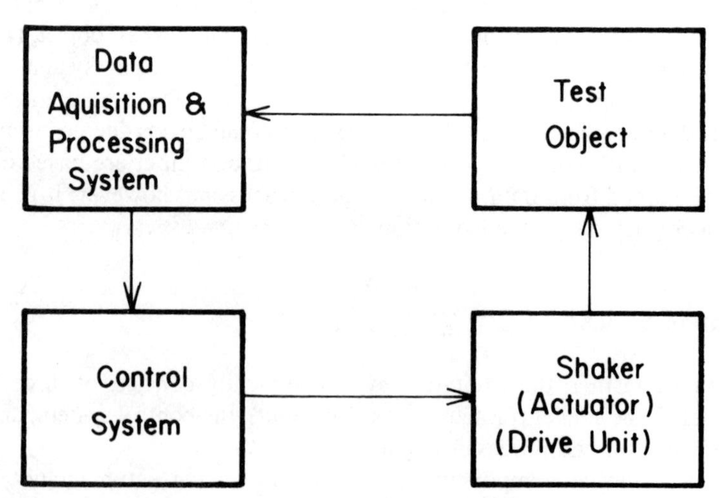

Figure 8-1. Interactions between Major Subsystems of an Excitation System in Dynamic Testing

The data-acquisition and processing system receives various response signals from the test object. It processes these data and, based on the results, generates command signals. It also can analyze the results and compile them. A suitable minicomputer (or microcomputer) is the heart of the data-acquisition and processing system. It is interfaced with the test object through various transducers and sensors attached to the test object, the signal-conditioning circuitry and amplifiers, and analog-to-digital converters (ADCs). Automatic monitoring of the test-object response may be realized in this manner. Command signals from the data-acquisiton and processing system are fed into the control system. Processed test results are usually displayed using oscilloscopes, hard-copy units, X-Y plotters, strip-chart recorders, level recorders, and the like, or they are stored on magnetic tapes or discs.

Shakers

Three basic types of shakers are widely used in dynamic testing applications: hydraulic shakers, inertial shakers, and electromagnetic shakers. The operation-capability ranges of typical shakers in these three categories are summarized in table 8-1. Stroke, or maximum displacement, is the largest displacement the shaker is capable of imparting onto a test object whose weight is assumed to be within its design load limit. Maximum velocity and acceleration are similarly defined. Maximum force is the largest force that could be applied by the shaker to a test object of acceptable weight (within the design load). The values given in table 8-1 should be interpreted with caution. Maximum displacement is achieved only at very low frequencies. Maximum velocity corresponds to intermediate frequencies in the operating-frequency range of the shaker. Maximum acceleration and force ratings are usually achieved at high frequencies. It is not feasible, for example, to operate a shaker at its maximum displacement and its maximum acceleration simultaneously.

Consider a loaded actuator that is executing harmonic motion. Its displacement is given by

$$x = s \sin \omega t \tag{8.1}$$

in which s is the displacement amplitude (or stroke). Corresponding velocity and acceleration are

$$\dot{x} = s\omega \cos \omega t \tag{8.2}$$

$$\ddot{x} = -s\omega^2 \sin \omega t \tag{8.3}$$

Table 8-1
Typical Operation-Capability Ranges for Various Shaker Types

	Typical Operational Capabilities					
Shaker Type	*Frequency*	*Maximum Displacement (Stroke)*	*Maximum Velocity*	*Maximum Acceleration*	*Maximum Force*	*Excitation Waveform*
Hydraulic (electrohydraulic)	Low 0.1–500 Hz	High 12 in 30 cm	Intermediate 50 in/s 125 cm/s	Intermediate 20 g	High 100,000 lbf 450,000 N	Average flexibility (simple to complex and random)
Inertial (counter-rotating mass)	Intermediate 2–50 Hz	Low 1 in 2.5 cm	Intermediate 50 in/s 125 cm/s	Intermediate 20 g	Intermediate 1,000 lbf 4,500 N	Sinusoidal only
Electromagnetic (electrodynamic)	High 2–10,000 Hz	Low 1 in 2.5 cm	Intermediate 50 in/s 125 cm/s	High 100 g	Low to intermediate 450 lbf 2,000 N	High flexibility and accuracy (simple to complex and random)

If the velocity amplitude is denoted by v and the acceleration amplitude by a, it follows from equations (8.2) and (8.3) that

$$v = \omega s \tag{8.4}$$

and

$$a = \omega v \tag{8.5}$$

An idealized peformance curve of a shaker has a constant displacement-amplitude region, a constant velocity-amplitude region, and a constant acceleration-amplitude region for low, intermediate, and high frequencies, respectively, in the operating-frequency range. Such an ideal performance curve is shown in figure 8-2(a) on a frequency-velocity plane and in figure 8-2(b) on a frequency-acceleration plane. In both cases, logarithmic axes are used. In practice, typical shaker-performance curves would be smooth curves, similar to those shown in figure 8-3. The curves depend on the load (weight) that the shaker carries. As the weight increases, the performance curve compresses. To standardize the performance curves, they usually are defined at the rated load of the shaker.

Several general observations can be made from equations (8.4) and (8.5). In the constant-peak displacement region of the performance curve, the peak velocity increases proportionately with the excitation frequency, and the peak acceleration increases with the square of the excitation frequency. In the constant-peak velocity region, the peak displacement varies inversely with the excitation frequency, and the peak acceleration increases proportionately. In the constant-peak acceleration region, the peak displacement varies inversely with the square of the excitation frequency, and the peak velocity varies inversely with the excitation frequency. This further explains why rated stroke, maximum velocity, and maximum acceleration values are not simultaneously realizable.

Hydraulic Shakers. A typical hydraulic shaker consists of a piston-cylinder arrangement (also called a ram), a servo-valve, a fluid pump, and a driving electric motor. Hydraulic fluid (oil) is pressurized (typical operating pressure, 4,000 psi) and pumped into the cylinder through a servo-valve by means of a pump that is driven by an electric motor (typical power, 150 hp). The flow (typical rate, 100 gal/min) that enters the cylinder is controlled (modulated) by the servo-valve, which, in effect, controls the resulting piston (ram) motion. A typical servo-valve consists of a two-stage spool valve, which provides a pressure difference and a controlled (modulated) flow to the piston, which sets it in motion. The servo-valve itself is moved by means of a linear torque motor, which is driven by the excitation-input signal (electri-

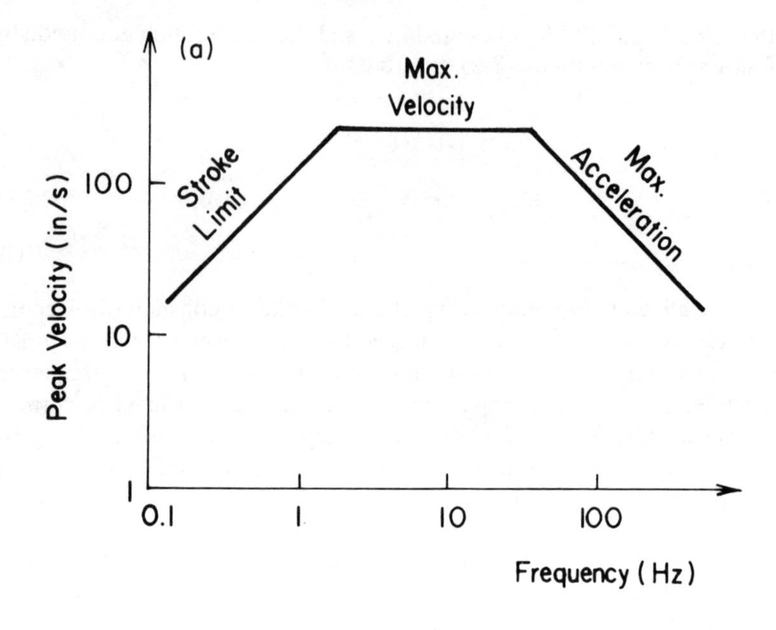

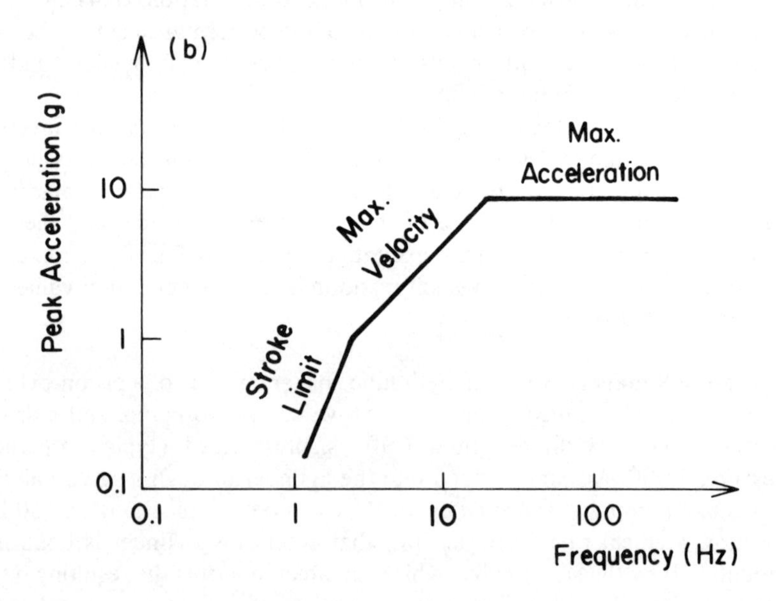

Figure 8-2. Ideal Performance Curve for a Shaker: (a) In the Frequency-Velocity Plane (Log), (b) In the Frequency-Acceleration Plane (Log)

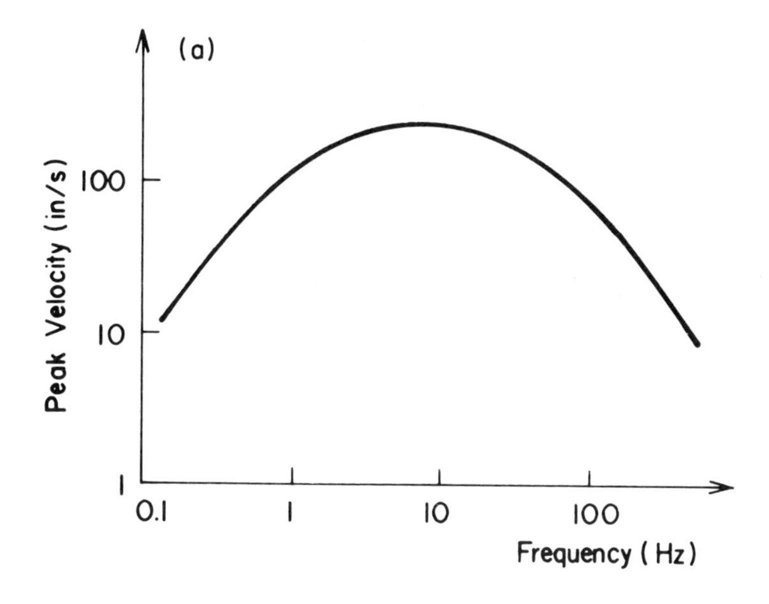

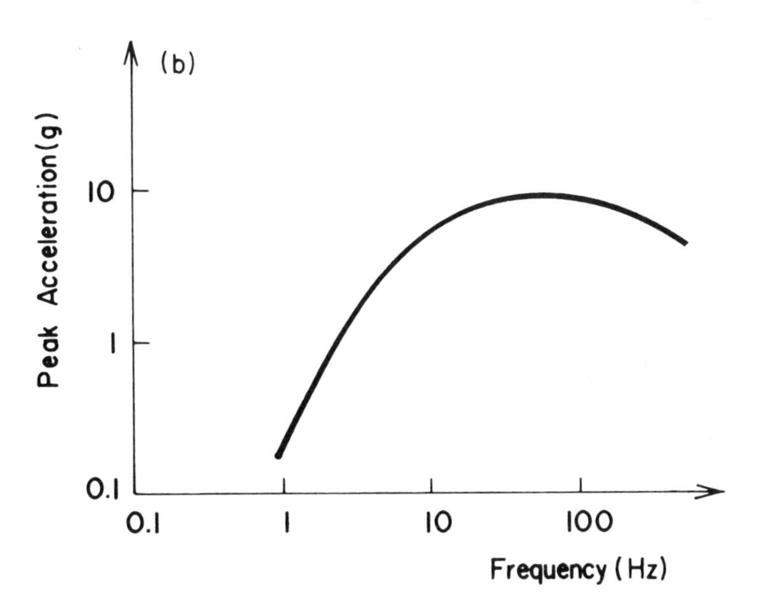

Figure 8-3. A Typical Performance Curve for a Shaker: (a) In the Frequency-Velocity Plane (Log), (b) In the Frequency-Acceleration Plane (Log)

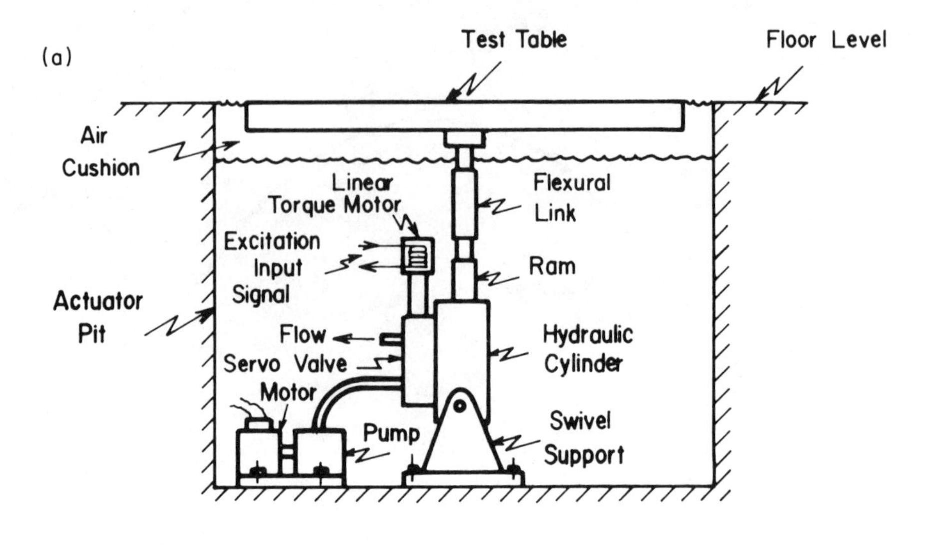

Figure 8-4. A Typical Hydraulic Shaker Arrangement: (a) Schematic, (b) Operational Block Diagram.

cal). A primary function of the servo-valve is to provide a stabilizing feedback to the ram. In this respect, the servo-valve complements the main control system of the test setup. The ram is coupled to the shaker table by means of a flexure link. The cylinder frame is mounted on the support foundation with swivel joints. This allows for some angular and lateral misalignment, which might primarily be caused by test-object dynamics as the table moves.

Two-degree-of-freedom testing requires two independent sets of actuators, and three-degree-of-freedom testing requires three independent actuator sets. Each independent actuator set can consist of several actuators operated in parallel, using the same pump and the same excitation-input signal to the torque motor.

If the test table is directly supported on the vertical actuators, they must withstand the total dead weight (that is, the weight of the test table, the test object, the mounting fixtures, and the instrumentation). This is usually prevented by providing a pressurized air cushion in the gap between the test table and the foundation walls. Air should be pressurized so as to balance the total dead weight exactly (typical required gauge pressure, 3 psi).

Figure 8-4(a) shows the basic components of a typical hydraulic shaker. The corresponding operational block diagram is shown in Figure 8-4(b). It is advisable to locate the actuators in a pit in the test laboratory so that the test table top is flush with the test laboratory floor under no-load conditions. This minimizes the effort required to place the test object on the test table. Otherwise, the test object will have to be lifted onto the test table with a forklift. Also, installation of an aircushion to support the system dead weight would be difficult under these circumstances.

Hydraulic actuators are most suitable for heavy load testing and are widely used in seismic qualification applications. They can be operated at very low frequencies (almost DC), as well as at intermediate frequencies (see table 8-1). Large displacements (stroke) are possible at low frequencies.

Hydraulic shakers have the advantage of providing high flexibility of operation during the test, including a capability of variable-force and constant-force testing and wide-band random-input testing. Velocity and acceleration capabilities of hydraulic shakers are intermediate. Although any general excitation-input motion (for example, sine wave, sine beat, wide-band random) can be used in hydraulic shakers, faithful reproduction of these signals is virtually impossible at high frequencies because of distortion and higher-order harmonics introduced by the high noise levels that are common in hydraulic systems. This is only a minor drawback in heavy-duty, intermediate-frequency applications.

Inertial Shakers. In inertial shakers, the force that causes the shaker-table motion is generated by inertia forces (accelerating masses). Counterrotating-mass inertial shakers are typical in this category. To explain their principle of operation, consider two equal masses rotating in opposite directions at the same angular speed ω and in the same circle of radius r (see figure 8-5). This produces a resultant force equal to $2m\omega^2r \cos \omega t$ in a fixed direction (the direction of symmetry of the two rotating arms). Consequently, a sinusoidal force with a frequency of ω and an amplitude proportional to ω^2 is generated. This reaction force is applied to the shaker table.

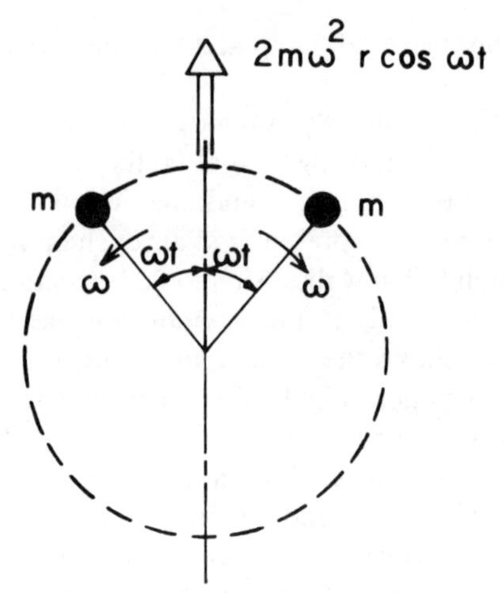

Figure 8-5. Principle of Operation of Counter-Rotating-Mass Inertial Shakers

Figure 8-6 shows a schematic of a typical counterrotating-mass inertial shaker. It consists of two identical rods rotating at the same speed in opposite directions. Each rod has a series of slots to place weights. In this manner, the eccentric mass can be varied to achieve various force capabilities. The rods are driven by an electric motor through a gear mechanism that usually has several speed ratios. A speed ratio is selected depending on the required test-frequency range. The whole system is symmetrically supported on a carriage that is directly connected to the test table. The test object is attached to the test table. The preferred mounting configuration is horizontal, so that the excitation force is applied to the test object in the horizontal direction. In this configuration, there are no variable gravity moments (weight × distance to center of gravity) acting on the drive mechanism. Figure 8-6 shows the vertical configuration. In dynamic testing of large structures, the carriage can be mounted directly on the structure at a location where the excitation force should be applied.

Inertially driven reaction-type shakers are widely used for prototype testing of civil engineering structures. Their first application dates back to 1935.[1] Inertial shakers are capable of producing intermediate excitation forces. The force generated is limited by the strength of the carriage frame. The frequency range of operation and the maximum velocity and acceleration capabilities are also intermediate for inertial shakers, whereas the maximum displacement capability is typically low. A major limitation of inertial shakers is that their excitation force is exclusively sinusoidal and that the force amplitude is directly proportional to the square of the excitation frequency. As a result, complex and random excitation testing, constant-force testing (for example, transmissibility tests and constant-force sine-sweep tests), and flexibility to vary the force amplitude or the displacement amplitude during a test are not feasible with this type of shakers. The sinusoidal excitation generated by inertial shakers is virtually undistorted, however, which is an advantage over the other types of shakers when used in sine-dwell and sine-sweep tests. Small portable shakers with low-force capability are available for use in on-site testing.

Electromagnetic Shakers. In electromagnetic shakers, the motion is generated using the principle of operation of an electric motor. Specifically, the excitation force is produced when a variable excitation signal (electrical) is passed through a moving coil placed in a magnetic field.

A schematic of an electromagnetic shaker is shown in figure 8-7. A steady magnetic field is generated by a stationary electromagnet that consists of field coils wound on a ferromagnetic base that is rigidly attached to a protective shell structure. The shaker head has a coil wound on it. When the excitation electrical signal is passed through this drive coil, the shaker head, which is supported on flexure mounts, will be set in motion. The shaker head

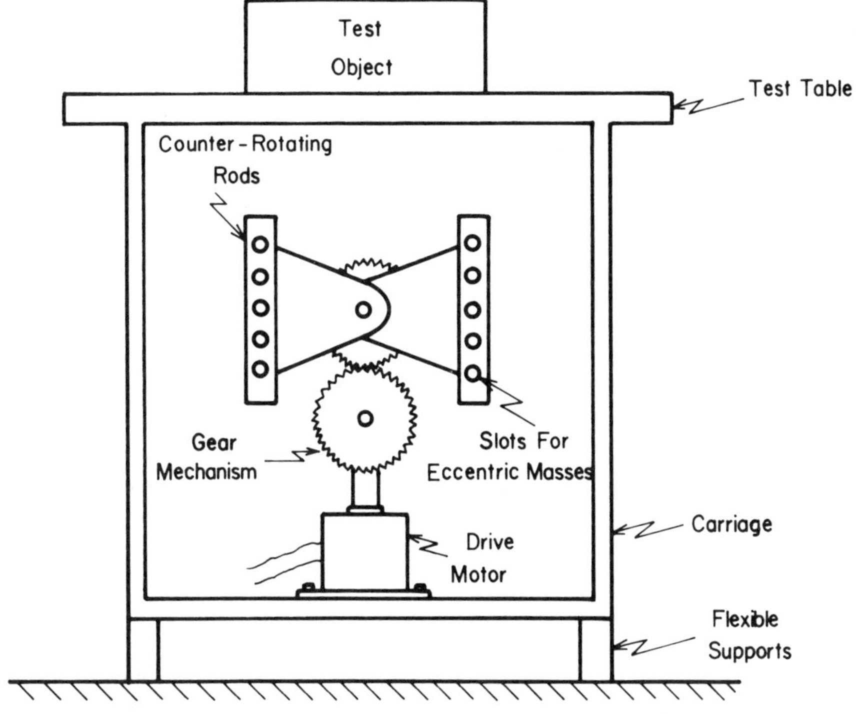

Figure 8-6. Schematic of a Counter-Rotating-Mass Inertial Shaker

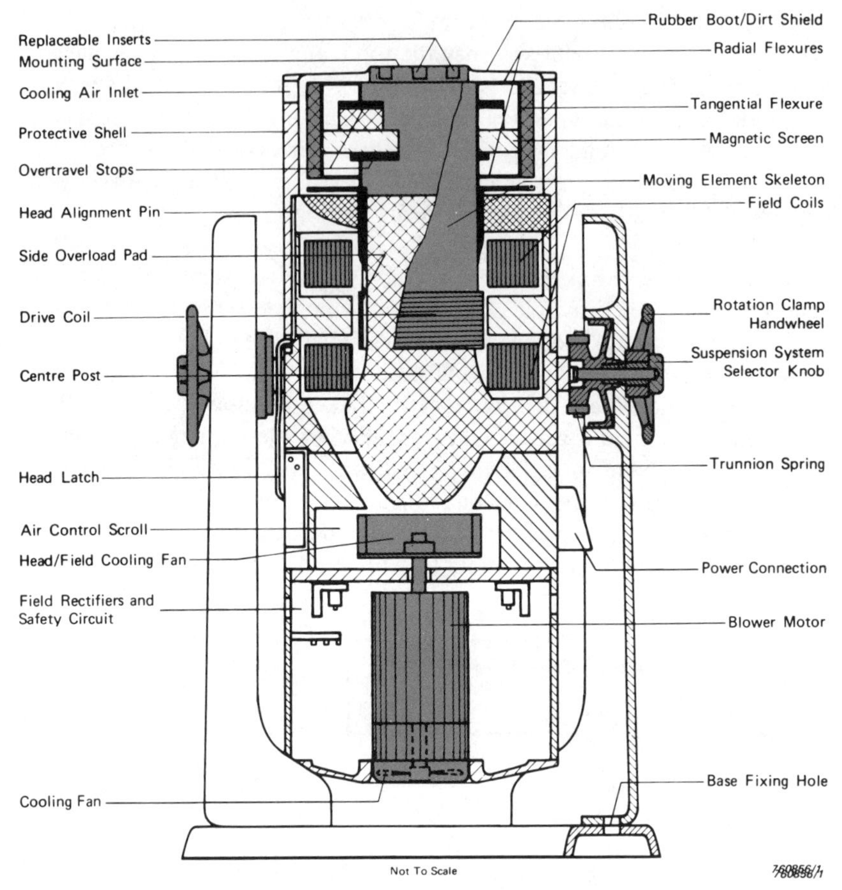

Source: From "Vibration Testing Systems," courtesy of Brüel & Kjaer, Naerum, Denmark.

Figure 8-7. Schematic Sectional View of a Typical Electromagnetic Shaker

consists of the test table on which the test object is mounted. Shakers with interchangeable heads are available. The choice of appropriate shaker head is based on the geometry and mounting features of the test object. The shaker head can be turned to different angles by means of a swivel joint. In this manner, different directions of excitation (in biaxial and triaxial testing) can be obtained.

The main advantages of electromagnetic shakers are their high frequency range of operation, their high degree of operating flexibility, and the high level of accuracy of the generated shaker motion. Faithful reproduction of complex excitations is possible because of the advanced electronic control systems used in this type of shakers. Unfortunately, electromagnetic shakers are not suitable for heavy-duty applications (large test objects). High test-input accelerations are possible at high frequencies, when electromagnetic shakers are used, but displacement and velocity capabilities are limited to low or intermediate values (see table 8-1).

Control System

The two primary functions of the shaker control system in dynamic testing are (1) to guarantee that the specified excitation is applied to the test object, and (2) to ensure that dynamic stability (motion constraints) of the test setup is preserved. An operational block diagram illustrating these control functions is given in figure 8-8. The reference input to the control system represents the desired excitation force that should be applied to the test object. In the absence of any control, however, the force reaching the test object will be distorted, primarily because of (1) dynamic interactions and nonlinearities of the shaker, the test table, the mounting fixtures, the auxiliary instruments, and the test object itself; (2) noise and errors in the signal generator, amplifiers, filters, and other equipment; and (3) external loads and disturbances (for example, aerodynamic forces, friction) acting on the test object and other components. To compensate for these distorting factors, response measurements (displacements, velocities, acceleration, and so on) are made at various locations in the test setup and are used to control the system dynamics. In particular, responses of the shaker, the test table, and the test object are measured. These responses are used to compare the actual forcing input felt by the test object with the desired (specified) input. The command (control) signal to the shaker is modified, depending on the error that is present.

Two types of control are commonly employed in shaker aparatus: simple manual control and complex automatic control. Manual control normally consists of simple, open-loop, trial-and-error methods of manual adjustments (or calibration) of the control equipment to obtain a desired dynamic response. The actual response is usually monitored (on an oscilloscope screen, for example,) during manual-control operations. The pretest adjustments in manual control can be very time-consuming; as a result, the test object might be subjected to overtesting (which could produce cumulative damage), which is undesirable and could defeat the test purpose. Furthermore, the calibration procedure must be repeated for each new test object.

The disadvantages of manual control suggest that automatic control is desirable in complex test schemes in which high accuracy of testing is desired. The first step of automatic control involves automatic measurement of system response, using control sensors. The measurement is then fed back into the control system, which instantaneously computes the best command

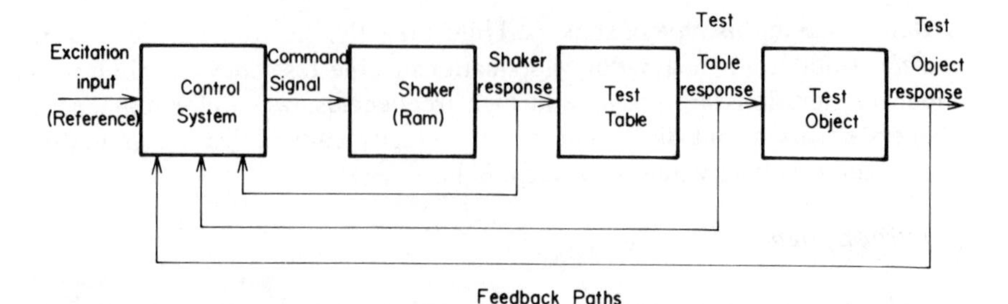

Figure 8-8. Operational Block Diagram Illustrating Shaker Control System Functions

signal to actuate the shaker to get the desired response. This may be done by analog means, by digital methods, or by a combination of the two.

Primitive control systems required an accurate mathematical description of the test object. This dependency of the control system on the test-object dynamics is clearly undesirable. Performance of a good control system should not be sensitive to dynamic interactions and nonlinearities of the test object or to specified excitation. Proper selection of feedback signals and control-system components can reduce such sensitivities.

In the response-spectrum method of dynamic testing, it is customary to use displacement control at low frequencies, velocity control at intermediate frequence, and acceleration control at high frequencies. This necessitates feedback of displacement, velocity, and acceleration responses. Generally, however, the most important feedback is the velocity feedback. In sine-sweep tests, the shaker velocity must change steadily over the frequency band of interest. In particular, the velocity control must be precise near the resonances of the test object. Velocity (speed) feedback has a stabilizing effect on the dynamics, which is desirable. This effect is particularly useful in ensuring stability in motion when testing is done near resonances of lightly damped test objects. On the contrary, displacement (position) feedback can have a destabilizing effect on some systems.

The control-system black box usually consists of various instruments, equipment, and computation hardware and software. Often, the functions of the data-acquisition and processing system overlap with those of the control system to some extent. An example might be a digital-control system for dynamic testing apparatus. First, the responses are measured through sensors (and transducers), filtered, and amplified (conditioned). These data channels are passed through a multiplexer, whose purpose is to select one data channel at a time for processing. The analog data are converted into digital data using analog-to-digital converters (ADCs). The resulting sampled data are stored on a disc or as block data in the computer memory. The reference input signal (typically, a signal recorded on an FM tape) is also sampled (if it is not already digital), using an ADC, and fed into the computer. Digital processing is done on the reference signal and the response data to compute the command signal to drive the shaker. The digital command signal is converted into an analog signal, using a digital-to-analog converter (DAC), and amplified (conditioned) before it is used to drive the actuator. A block diagram of these operations is shown in figure 8-9.

The nature of the control-system components depends to a great extent on the nature and objectives of the particular test to be conducted. Some of the basic components in a control system are described in the following subsections.

Compressor. A compressor circuit is incorporated in automatic excitation-control devices to control the excitation-input level automatically. The level control depends on the feedback signal from a control sensor and the specified (reference) input signal. Usually, the compressor circuit is included in the excitation-input generator (for example, a sine generator). Level control by this means typically is done on the basis of a single-frequency component (usually the fundamental frequency).

Equalizer (Spectrum Shaper). Random-signal equalizers are used to shape the spectrum of a random signal in a desired manner. In essence, an equalizer

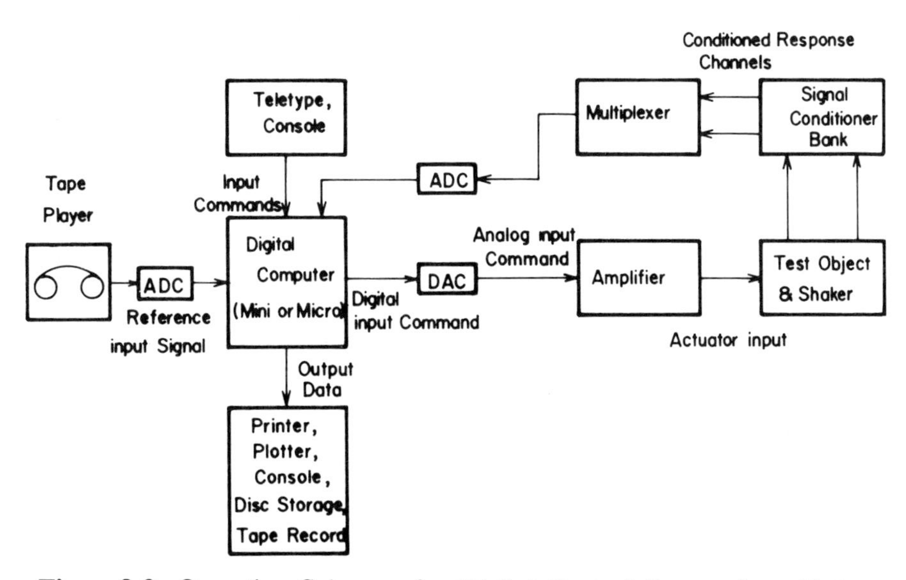

Figure 8-9. Operation Scheme of a Digital Control System in a Shaker Apparatus

consists of a bank of narrow-band filters (for example, 80 filters) in parallel over the operating-frequency range. By passing the signal through each filter, the spectral density (or the mean square value) of the signal in that narrow frequency band (for example, each one-third-octave band) is determined. This is compared with the desired spectral level, and automatic adjustment is made in that filter in case there is an error. In some systems, response-spectrum analysis is made in place of power spectral density analysis. In that case, the equalizer consists of a bank of simple oscillators, whose resonant frequencies are distributed over the operating-frequency range of the equalizer. The feedback signal is passed through each oscillator, and the peak value of its output is determined. This value is compared with the desired RRS value at that frequency. If there is an error, automatic gain adjustment is made in the appropriate excitation-signal components.

Random-noise equalizers are used in conjunction with random-signal generators. They recieve feedback signals from the control sensors. In digital control systems, there are algorithms (software) that are used to iteratively converge the spectrum of the excitation signal felt by the test object into the desired spectrum.

Tracking Filter. Many dynamic tests are based on single-frequency excitations. In such cases, the control functions should be performed on the basis of amplitudes of the fundamental-frequency component of the signal. A tracking filter is simply a frequency-tuned band-pass filter. It automatically tunes its very narrow band-pass filter center frequency to the frequency of a carrier signal. Then, the noisy input signal is passed through the tuned filter, whose output is the required fundamental-frequency component in the signal. Tracking filters also are useful in obtaining amplitude-frequency plots using an X-Y plotter. In such cases, the frequency value comes from the signal generator (sweep oscillator), which produces the carrier signal to the tracking filter. The tracking filter then determines the corresponding amplitude of a response signal that is fed into it. Most tracking filters have dual channels, so that two signals can be handled (tracked) simultaneously.

Excitation Controller (Amplitude Servo-Monitor). An excitation controller is typically an integral part of the signal generator. It can be set so that automatic sweep between two frequency limits can be performed at a selected sweep rate (linear or logarithmic). More advanced excitation controllers have the capability of automatic switch-over between constant-displacement, constant-velocity, and constant-acceleration excitation-input control at specified frequencies over the sweep-frequency interval. Consequently, integrator circuits, to determine velocities and displacements from acceleration signals, should be present within the excitation controller unit. Sometimes, integration is performed by a separate unit called a vibration meter. This unit also offers the operator the capability of selecting the desired level of each signal (acceleration, velocity, or displacement). There is an automatic cutoff level for large displacement values that could result from noise in acceleration signals. A compressor is also a subcomponent of the excitation controller. The complete unit is sometimes known as an amplitude servo-monitor.

Data-Acquisition and Processing System

The data-acquisition and processing system performs several important functions, in addition to augmenting the control system. In figure 8-9, the data-acquisition and processing system consists of the response sensors (and transducers), the signal conditioners, the multiplexer, the ADCs, and the digital computer, with associated input-output devices. The main functions of a digital data-acquisition and processing system are as follows:

1. Measuring, conditioning, sampling, and storing the response signals and test-object operability data (using input commands if necessary);
2. Digital processing of the measured data according to the test objectives (and using input commands when necessary);
3. Generation of command signals for the control system;
4. Generation and recording of test results (outputs) in a required format.

The capacity and the capabilities of a data-acquisition and processing system are determined by such factors as:

1. The number of response data channels that can be handled simultaneously;
2. The data-sampling rate (samples per second) for each data channel;
3. Computer memory size;
4. Computer processing speed;
5. External storage capability (floppy discs, tapes, and so forth)
6. The nature of the input and output devices;
7. Software (computer program) features.

Commercial data-acquisition and processing systems with a wide range of processing capabilities are available for use in dynamic testing. Some of the standard processing capabilities are the following:

1. Response-spectrum analysis
2. FFT analysis (spectral densities, correlations, coherence, Fourier spectra, and so on)

3. Frequency-response function, transmissibility, and mechanical-impedance analysis;
4. Natural-frequency and mode-shape analysis;
5. System-parameter identification (for example, damping parameters).

Most processing is done in real time, which means that the signals are analyzed as they are being measured. The advantage of this is that outputs and command signals are available simultaneously as the monitoring is done, so that any changes can be detected as they occur (for example, degradation in the test object or deviations in the excitation signal from the desired form) and automatic feedback control can be effected. For real-time processing to be feasible, the data-acquisition rate (sampling rate) and the processing speed of the computer should be sufficiently fast. In real-time frequency analysis, the entire frequency range (not narrow bands separately) is analyzed at a given instant. Results are presented as Fourier spectra, power spectral densities, cross-spectral densities, coherence functions, correlation functions, and response-spectra curves. Averaging of frequency plots can be done over small frequency bands (for example, one-third-octave analysis), or the running average of each instantaneous plot can be determined.

Signal-Generating Equipment

Shakers are force-generating devices that are operated using excitation signals generated from a source. The excitation-signal source is known as the signal generator. Three major types of signal generators are used in dynamic testing applications: (1) oscillators, (2) random-signal generators, and (3) tape players. In some units, oscillators and random-signal generators are combined. We shall discuss these two generators separately, however, because of their difference in functions. It also should be noted that almost

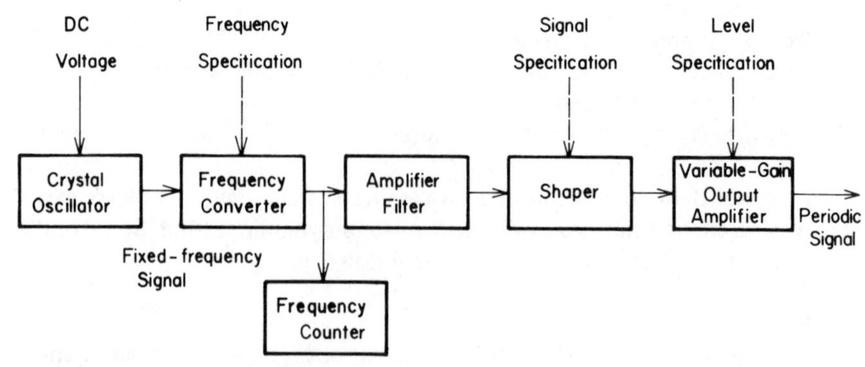

Figure 8-10. Block Diagram of an Oscillator-Type Signal Generator

any digital signal (deterministic or random) can be generated by a digital computer using a suitable computer program; it eventually can be passed through a DAC to obtain the corresponding analog signal. Such digital techniques are a separate subject (digital programming and software) with which the reader should attempt to become familiar to some degree.

The dynamic range of equipment is the ratio of the maximum and minimum output levels (expressed in decibels) at which it is capable of operating without significant error. This is an important specification for signal-generating equipment. The output level of the signal generator should be set to a value within its dynamic range.

Oscillators

Oscillators are essentially single-frequency generators. Typically, sine signals are generated, but other waveforms (such as rectangular and triangular pulses) are also available in most oscillators. Normally, an oscillator has two modes of operation: (1) up-and-down sweep between two frequency limits and (2) dwell at a specified frequency. In the sweep operation, the sweep rate should be specified. This can be done either on a linear scale (Hz/min) or on a logarithmic scale (octaves/min). In the dwell operation, the frequency points (or intervals) should be specified. In either case, a desired signal level can be chosen using the gain-control knob. An oscillator that is operated exclusively in the sweep mode is called a sweep oscillator.

Primitive oscillators employed variable inductor-capacitor types of electronic circuits to generate signals oscillating at a desired frequency. The oscillator is tuned to the required frequency by varying the capacitance or inductance parameters. A DC-voltage initial condition is applied to the capacitor and released to obtain the desired oscillating voltage signal, which subsequently is amplified and conditioned. Modern oscillators use crystal parallel-resonance oscillators to generate voltage signals accurately at a fixed frequency. The crystal is activated using a DC-voltage source. Other frequencies of interest are obtained by passing this high-frequency signal through a frequency converter. The signal is then conditioned (amplified and filtered). Required shaping (for example, rectangular pulse) is obtained using a shaper circuit. Finally, the required signal level is obtained by passing the resulting signal through a variable-gain amplifier. A block diagram of an oscillator, illustrating various stages in the generation of a periodic signal, is given in figure 8-10.

A typical oscillator offers a choice of several (typically six) linear and logarithmic frequency ranges and a sizable level of control capability (for example, 80 db). Upper and lower frequency limits in a sweep can be preset on the front panel to any of the available frequency ranges. Sweep-rate settings are continuously variable (typically, 0 to 10 octaves/min in the logarithmic range, and 0 to 60 kHz/min in the linear range), but one value must be selected for a given test or part of a test. Most oscillators have a repetitive-sweep capability, which allows the execution of more than one sweep continuously (for example, for mechanical aging and in OBE single-frequency tests). Some oscillators have the capability of also varying the signal level (amplitude) during each test cycle (sweep or dwell). This is known as level programming. Also, automatic switching between acceleration, velocity, and displacement excitations at specified frequency points in each test cycle can be implemented with some oscillators. A frequency counter, which is capable of recording the fundamental frequency of the output signal, is usually an integral component of the oscillator.

Random Signal Generators

In modern random-signal generators, zener diodes are used to generate a random signal that has Gaussian distribution. This is accomplished by applying a suitable DC voltage to a zener diode. The resulting signal is then amplified and passed through a bank of conditioning filters, which effectively acts as a spectrum shaper. In this manner, the bandwidth of the signal can be varied. Extremely wideband signals (white noise), for example, can be generated for random-excitation dynamic testing in this manner. The block diagram in figure 8-11 shows the essential steps in a random-signal generation process. A typical random-signal generator has several (typically eight) bandwidth selections over a wide frequency range (for example, 1 Hz to 100 kHz). A level-control capability (typically 80 db) is also available.

Tape Players

Dynamic testing for seismic qualification is usually performed using a tape player as the signal source. A tape player is essentially a signal reproducer.

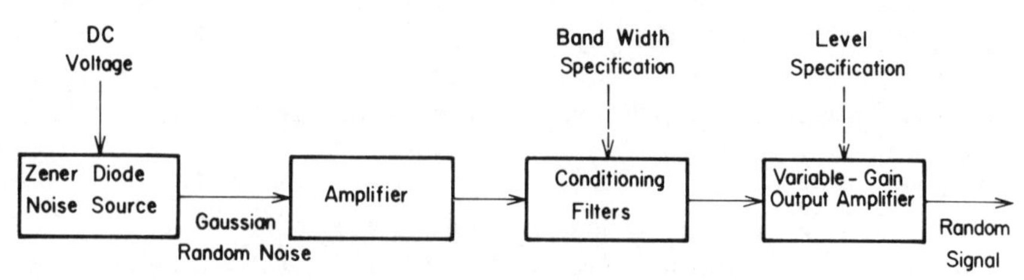

Figure 8-11. Block Diagram of a Random Signal Generator

The test-input signal that has a certain specified response spectrum (the RRS) is obtained by playing a magnetic tape and mixing the contents in the several tracks of the tape in a desirable ratio. Typically, each track contains a sine-beat signal (with a particular beat frequency, amplitude, and number of cycles per beat) or a random-signal component (with a desired spectral characteristic).

Two types of tape records are commonly available for replay: frequency modulation (FM) tapes and direct-record tapes. In FM case, the signal amplitude is proportional to the frequency of a carrier (modulating) signal. The carrier signal is the one that is recorded on the tape. When played back, the actual signal is reproduced, based on detecting the frequency content of the carrier signal in different segments of it. The FM method is usually favorable to direct recording and play-back of a signal, particularly for low-frequency signals (below 100 Hz), which are typical for seismic qualification test inputs).

Performance of a tape player is determined by several factors, including tape type and quality, signal reproduction (and recording) circuitry, characteristics of the magnetic heads, and the tape-transport mechanism. Some important specifications for tape players are (1) the number of tracks per tape (for example, 14 or 28); (2) the available tape speeds (for example, 3.75, 7.5, 15, or 30 in/sec); (3) reproduction filter-amplifier capabilities (for example, 0.5 percent third-harmonic distortion in a 1 kHz signal recorded at 15 in/sec tape speed, peak-to-peak output voltage of 5 V at 100 ohm load, signal-to-noise ratio of 45 db, output impedance of 50 ohms); and (4) the available control options and their capabilities (for example, stop, play, reverse, fast-forward, record, speed selection, channel selection). Tape-player specifications for dynamic testing are usually governed by the Communication and Telemetry Standard of the Intermediate Range Instrumentation Group (IRIG Standard 106-66).

A common practice in dynamic testing is to generate the test-input signal by repetitively playing a closed tape loop. In this manner, the input signal becomes periodic but has the desired frequency content. Frequency-modulation players can be fitted with special loop adaptors for playing tape loops. In spectral (Fourier) analysis of such signals, the analyzing-filter bandwidth should be more than the repetition frequency (tape speed/loop length). Extraneous noise is caused by discontinuities at the tape joint. This can be suppressed by using suitable filters or gating circuits.

A technique that can be employed to generate low-frequency signals with high accuracy is recording the signal first at a very low tape speed and then playing it back at a high tape speed (for example, r times higher). This has the effect of multiplying all frequency components in the signal by the speed ratio (r). Consequently, the filter circuits in the tape player will allow some low-frequency components in the signal that would normally be cut off, and will cut off some high-frequency components that would normally be allowed. Hence, this process is a way of emphasizing the low-frequency components in a signal.

Signal-Sensing Equipment

The response parameter that is being measured (for example, acceleration) is termed the measurand. A measuring device passes through two stages in making a measurement. First, the measurand is sensed. Then, the measured signal is transduced (converted) into a form that is particularly suitable for signal conditioning, processing, or recording. Often, the output from the transducer stage is an electrical signal. The two stages are illustrated by the block diagram in figure 8-12. It is common practice to identify the combined sensor-transducer unit as either a sensor or a transducer.

The measuring device itself might contain some of the signal-conditioning circuitry and recording (or display) devices or meters. These are components of an overall measuring system. For our purposes, we shall consider these components separately.

In most applications, the following four variables are particularly useful in determining the dynamic response and structural integrity of the test object:

1. Displacement (potentiometer or LVDT)
2. Velocity (tachometer)
3. Acceleration (accelerometer)
4. Strain (strain gauge)

In each case the usual measuring devices are indicated in parentheses. It is somewhat common practice to measure acceleration first and then determine velocity and displacement by direct integration. Any noise and DC components in the measurement, however, could give rise to erroneous results in such cases. Consequently, it is good practice to measure displacement, velocity, and acceleration by using separate sensors, particularly when the measurements are employed in feedback control of the vibratory system. It is not recommended to differentiate a displacement (or velocity) signal to

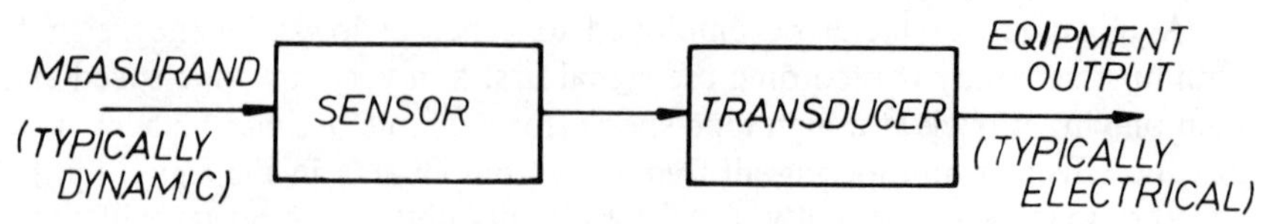

Figure 8-12. Block Diagram of a Measuring Device

obtain velocity (or acceleration), because this process would multiply any noise present in the measured signal. Consider, for example, a sinusoidal signal given by $A \sin \omega t$. Since $d/dt\,(A \sin \omega t) = A\omega \cos \omega t$, it follows that any high-frequency noise would be amplified by a factor proportional to its frequency. Also, any discontinuities in noise components would produce large deviations in the results. Using the same argument, it may be concluded that the acceleration measurements are desirable for high-frequency signals and the displacement measurements are desirable for low-frequency signals. It follows that the selection of a particular measurement transducer should depend on the frequency content of the useful portion of the measured signal.

Transducers are divided into two broad categories: active transducers and passive transducers. Active transducers do not require an external electric source for activation. Some examples are electromagnetic, piezoelectric, and photovoltaic transducers. Passive transducers, however, do not possess self-contained energy sources and thus need external activation. Examples are resistive, inductive, and capacitive transducers.

In selecting a particular transducer (measuring device) for a specific dynamic-testing application, special attention should be given to the following characteristics, which usually are provided as instrument ratings by the manufacturer:

1. Sensitivity
2. Dynamic range
3. Linearity
4. Useful frequency range
5. Resolution

Sensitivity of a transducer is measured by the magnitude (peak, RMS value, and so on) of signal output corresponding to a unit input of the measurand (for example, displacement, velocity, acceleration) along its direction of sensitivity. Cross-sensitivity is the sensitivity along directions that are orthogonal to the direction sensitivity; it is expressed as a percentage of the direct sensitivity. It should be clear that high sensitivity and low cross-sensitivity are desirable.

Dynamic range of a transducer is measured by the allowed lower and upper limits of its output, so as to maintain its measurement accuracy. The range is often expressed as a ratio (in decibels).

Linearity of a transducer is measured by the curve of output amplitude (peak or RMS value) versus input amplitude in its dynamic range. This input-output curve's closeness to a straight line measures the linearity of the transducer.

Useful frequency range is the frequency range within which the frequency-response function of the transducer is flat. The upper frequency of

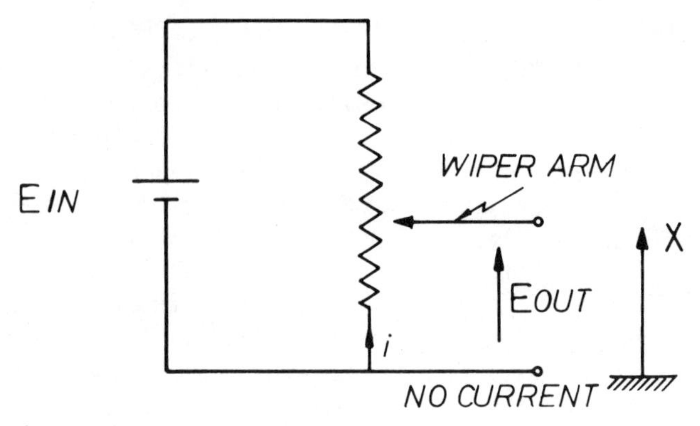

Figure 8-13. Schematic of a Potentiometer.

this range typically is several times smaller than the lowest resonant frequency of the transducer.

Resolution is the smallest change in the measurand that can be detected by the transducer. It is usually expressed as a percentage of the dynamic range.

In the following sections, we shall give a brief discussion of four types of measuring devices commonly used in dynamic testing.

Displacement Measurement (Potentiometer, Linear-Variable Differential Transformer)

A potentiometer is a simple displacement transducer. This passive transducer consists of a uniform coil of wire whose resistance is proportional to the length. A fixed voltage E_{in} is applied to the potentiometer coil, using an external, constant, DC-voltage source. The transducer output signal E_{out} is the voltage between the movable contact (wiper arm) sliding on the coil and one terminal of the coil (see figure 8-13). Slider displacement x is proportional to E_{out}:

$$E_{out} = kx \tag{8.6}$$

This relationship assumes that the output terminals are open-circuit (that is, infinite impedance, or resistance in the present DC case, is connected across the output), so that the output current is zero. In actual practice, however, the voltmeter that measures the output voltage has a finite impedance. Consequently, the output current (current through the voltmeter) is nonzero (see figure 8-14). The output voltage thus drops to $\widetilde{E}_{out}$ even if the source voltage E_{in} is assumed to remain constant with load variations. Then the linear relationship given by equation (8.6) would no longer be valid. This causes an error in the displacement reading. To reduce the error, a voltage source that is not seriously affected by load variations and a voltmeter that has high impedance (resistance in this case) should be used. Similarly, any signal-conditioning equipment used for such potentiometer devices should have high input impedance. Otherwise, parallel shunting effects would cause a further drop in output voltage, producing additional nonlinearity errors. Other error sources of a simple potentiometer are inertia and friction effects of the wiper arm and rapid wearout of sliding contact.

The linear-variable differential transformer (LVDT) is a displacement measuring device that overcomes most of the shortcomings of the simple potentiometer. It is a passive transducer that operates on the mutual-induction principle. The operational principle of the LVDT is illustrated in figure 8-15. A cylindrical insulating form has a central primary coil and a symmetrically placed secondary coil, as in figure 8-15(a). The primary coil is activated by an AC excitation of RMS voltage E_{in}. This will induce (by mutual induction) an AC of RMS voltage E_{out} in the secondary coil. If a core made of ferromagnetic material is inserted coaxially into the cylindrical form, as shown, the inductance will increase. Nevertheless, since the two end coils in the secondary circuit are connected so that the potentials induced in the two segments oppose each other, $E_{out} = 0$ when the core is centered between the end coils. If the core is moved from this position, a nonzero E_{out} will be generated, which is proportional to the core displacement x. Consequently, E_{out} may be used as a measure of the displacement. For the form geometry that is shown in the figure, the relationship between x and E_{out} is linear. The LVDT has other advantages over the simple potentiometer. Since the moving core is not in contact with the coils, for instance, there is minimal friction and

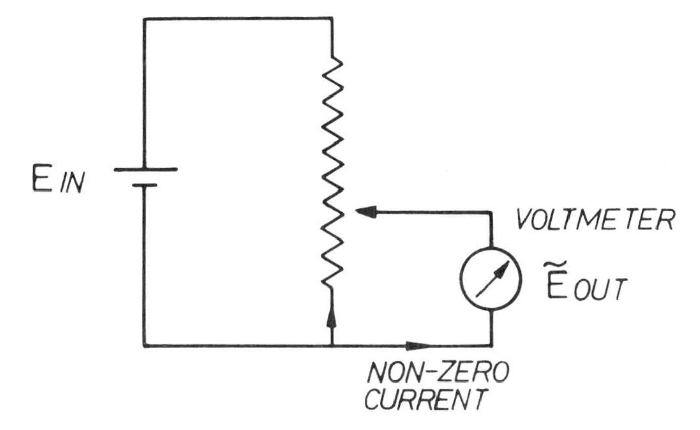

Figure 8-14. Loading Effect on a Potentiometer

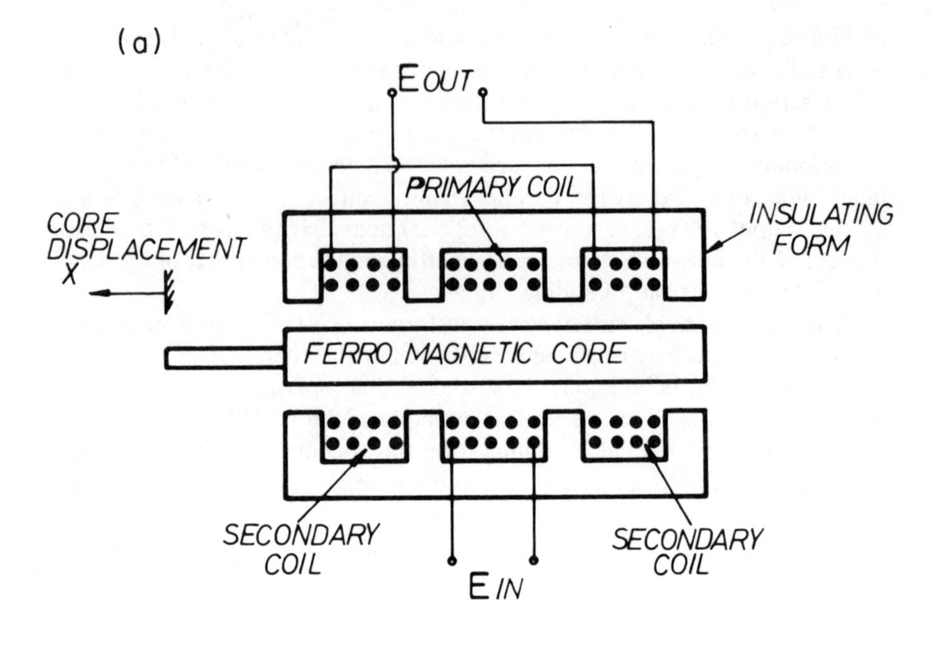

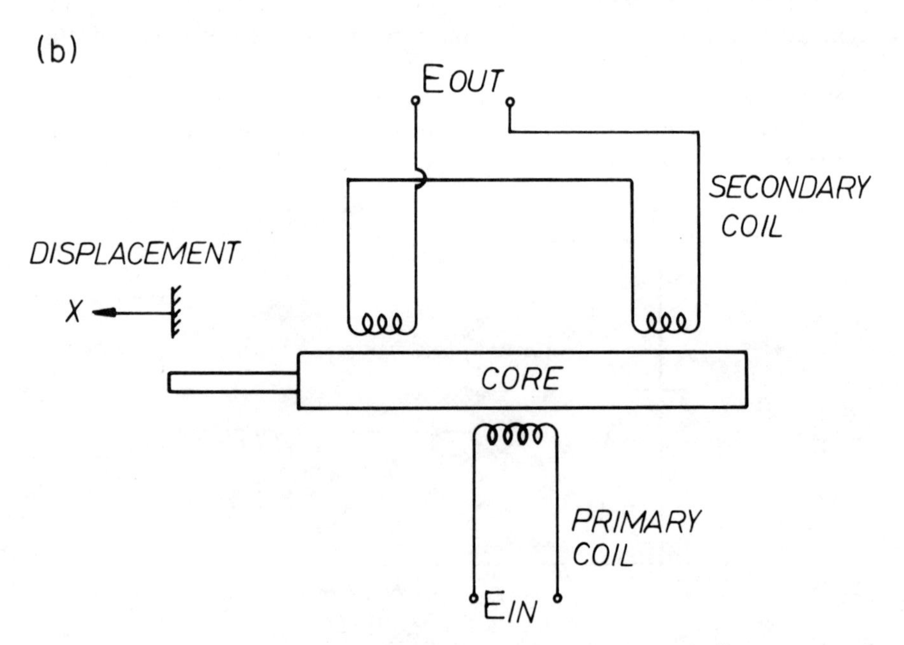

Figure 8-15. Linear-Variable Differential Transformer: (a) Constructional Geometry, (b) Operation Schematic

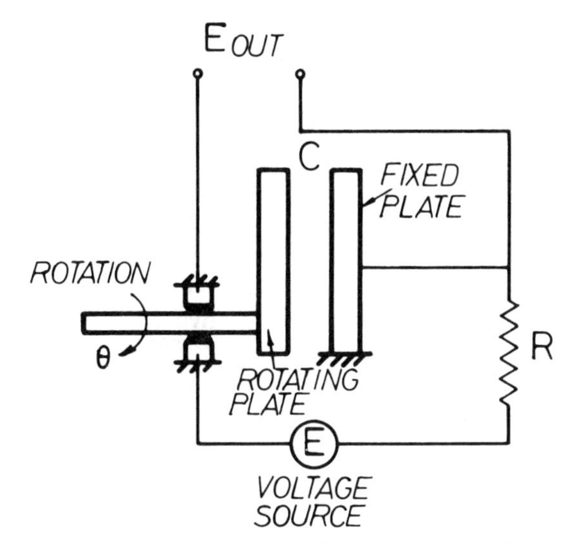

Figure 8-16. Operation Schematic of a Capacitance-Type Displacement Transducer

wear. Inertia can be reduced by making the mass of the core as small as practical.

Configurations such as those in figures 8-13 and 8-15 correspond to linear (translatory) displacement measuring. The potentiometer and the variable differential transformer also may be employed in measuring angular displacements by straightforward modification of their geometry of construction. Such a transformer is known as a rotary-variable differential transformer (RVDT).

Capacitance-type displacement transducers are also available. They operate on the principle that the capacitance is proportional to the common facing area of two capacitor plates. If one plate is rotated eccentrically with respect to the other, the capacitance will change, and this may be used as a measure of the angular displacement (see figure 8-16). If the voltage source is DC, then output voltage gives a direct measure of capacitance, assuming that the electrical charge in the capacitor is constant. If the voltage source is AC, however, impedance measurement (voltage across capacitor or current through capacitor in complex representation) is necessary to determine the capacitance.

Velocity Measurement (Tachometer)

When the magnetic flux crossing an electrical conductor varies, a voltage is generated in the conductor. This is the principle of electromagnetic induction.

In practice, the flux variation can be effected by relative motion between the magnetic field and the conductor. The induced voltage is proportional to the relative velocity. This principle is applied widely in velocity transducers. They are active transducers because an external electricity source is not employed (although electromagnets, rather than permanent magnets, are sometimes used to generate the magnetic field).

The operating principle of a magnetic-induction linear-velocity transducer is illustrated in figure 8-17. The conducter coil is wrapped on a core and placed centrally between two magnetic poles (which produce a cross-magnetic field). The core is attached to the moving part whose velocity must be measured. Its velocity v is proportional to the induced voltage E_{out} and is a measure of the velocity. Sometimes a moving-magnet and fixed-coil arrangement is also used, thus eliminating the need for any sliding contacts for the output leads. The tachometer is a device that uses the same principle to sense angular velocities.

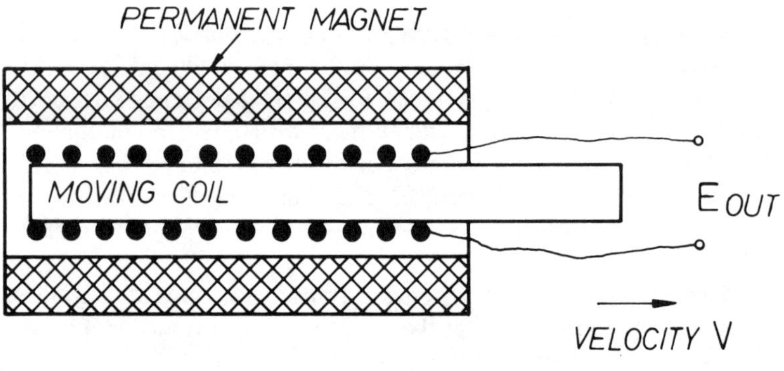

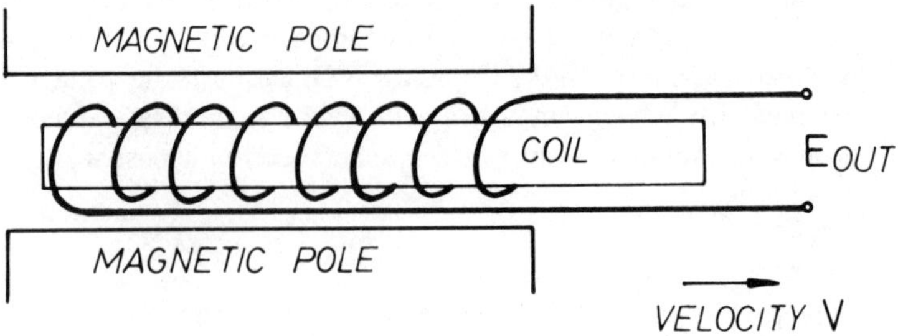

Figure 8-17. Linear-Velocity Transducer: (a) Constructional Geometry, (b) Operation Schematic

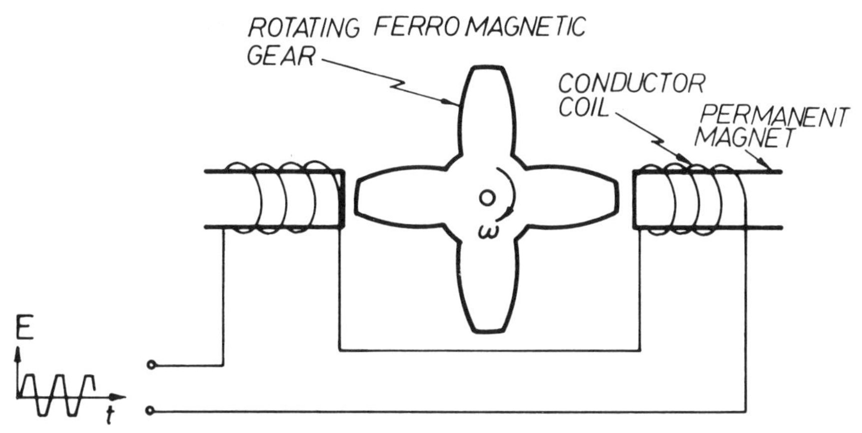

Figure 8-18. Principle of Operation of Variable-Reluctance Velocity Transducer (Also Refers to Eddy-Current Tachometer and Accelerometer)

A somewhat different principle is also used to measure angular velocities. When a ferromagnetic material is placed in a magnetic flux path, its reluctance (reluctance in a magnetic circuit is analogous to resistance in electrical circuits) is reduced, causing an increase in flux linkage through the circuit. Consider a gear wheel made of ferromagnetic material rotating between the poles of a permanent magnet (figure 8-18). The flux linkage through a coil wound on this permanent magnet will fluctuate with the gear-wheel motion, causing a fluctuating voltage E_{out} to be induced in the coil. The number of pulses per second in the output voltage is proportional to the number of gear teeth and the angular speed of the gear. Using this arrangement, an analog signal or a digital signal proportional to the pulse rate can be produced. If a rack gear is employed, this principle can be used to measure linear velocities.

Other types of velocity transducers include the eddy-current tachometer, the Hall-effect tachometer, and the microwave tachometer.[2] It should be observed that the LVDT (see figure 8-15) may be used for velocity measurement if E_{in} is a constant (DC) voltage. In that case, E_{out} will be proportional to the velocity of the ferromagnetic core (which affects a variation in the magnetic flux linkage).

If, in figure 8-18, a nonferrous gear wheel that has high electrical conductivity is used, an eddy-current velocity transducer (tachometer) results. In this case, when the conducting gear wheel rotates at constant speed in the original magnetic field, eddy currents are generated in the gear wheel. Since the gap changes because of the presence of gear teeth, the eddy currents themselves will change with time. The magnetic field created by eddy currents opposes the original magnetic field. This produces a fluctuation in the resultant magnetic field. Hence, the voltage induced in the coil will fluctuate. The rate of the voltage pulses induced in the coil determines the rotating speed of the gear wheel.

Suppose that a uniform disc of high electrical conductivity is used in place of the gear wheel in figure 8-18. In this case, if the rotating speed is constant, the eddy currents will not vary, because the gap does not change (no teeth). Hence, the resultant magnetic field will not fluctuate and so the induced voltage in the coil will be zero. If the conducting disc rotates with a constant acceleration, however, the eddy currents will increase uniformly, and the resultant magnetic field will decrease uniformly (because the eddy-current magnetic field opposes the original magnetic field. This will induce a voltage in the coil that is proportional to the angular acceleration of the disc. This arrangement can be employed as an eddy-current accelerometer.

Finally, consider two coils in self-induction (that is, a primary coil carrying a fluctuating current that induces a voltage in the secondary coil). If a metal plate of high electrical conductivity is moved toward the pair of coils, eddy currents will be generated in the plate, creating a magnetic field that opposes the original mutual-induction magnetic field. Consequently, the mutual inductance in the system will decrease, resulting in a drop in the induced voltage in the secondary coil. The voltage is proportional to the proximity of the plate to the coils. This principle is used in eddy-current displacement transducers, or proximity probes.

Acceleration Measurement (Accelerometer)

It follows from Newton's second law that a force, f, is necessary to accelerate a mass (inertia element), and its magnitude is given by the product of mass, m, and acceleration, a. This product ma is usually termed inertia force. The logic is that, if a force whose magnitude equals ma were applied to the accelerating mass in the direction opposing the acceleration, then the system could be analyzed using static-equilibrium considerations (see figure 8-19). This is known as D'Alembert's principle.

The force that causes acceleration is itself a measure of the acceleration. This is the principle used in acceleration transducers (accelerometers). Two types of accelerometers are widely used in dynamic-testing applications: strain-gauge accelerometers and piezoelectric (crystal) accelerometers. Strain-gauge accelerometers have one or more strain members of negligible mass that are supported by the strain-gauge housing at one end and support an inertia element (mass) at the other end. When the housing is accelerated, the force that is necessary to accelerate the mass (which is equal to the inertia force) is applied by the strain members. The reaction on the strain members

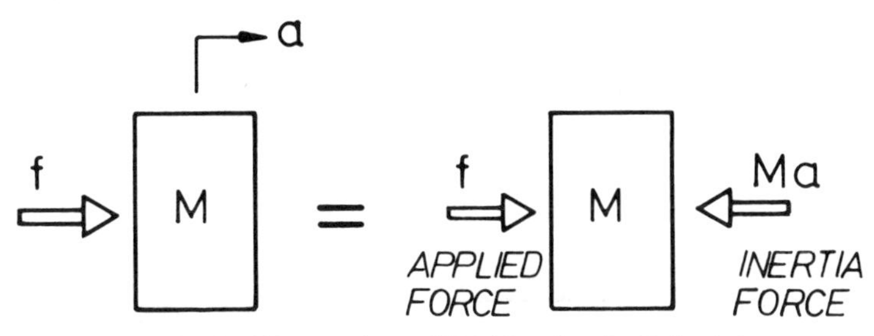

Figure 8-19. Illustration of D'Alembert's Principle

will cause them to deform. The amount of deformation (strain) is a measure of the acceleration. The direction of motion of the mass to which the strain gauges inside the accelerometer housing are sensitive is the direction of sensitivity of the accelerometer. This is determined by the strain-member geometry and the strain-gauge configuration.

Figure 8-20 shows the operating principle of a strain-gauge accelerometer whose strain member is a cantilever. When the mass accelerates in the direction shown, the cantilever bends, thus generating an output signal in the strain gauge. How this signal is generated is discussed in the next section.

Piezoelectric accelerometers (also known as crystal accelerometers) use the piezoelectric effect present in certain materials (for example, barium titanite). These substances generate electrical potentials when they are stressed (in tension, compression, or shear). This principle may be used in accelermoeters by substituting a piezoelectric element for a strain gauge. A

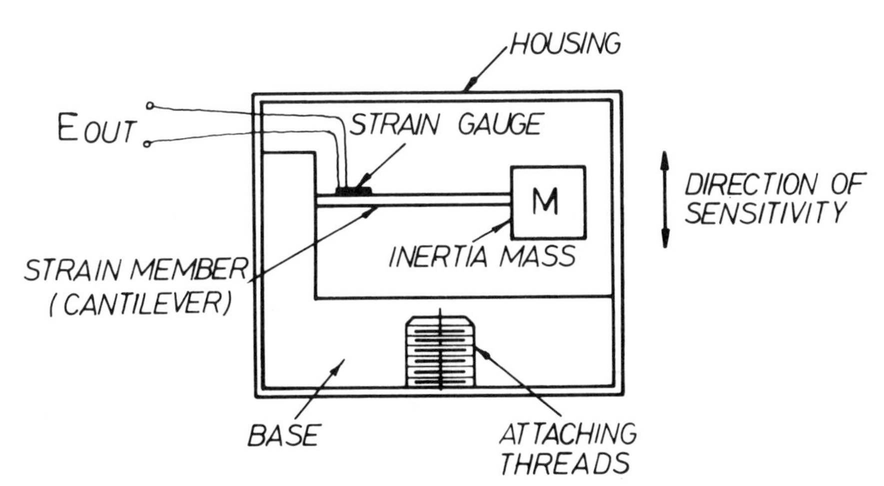

Figure 8-20. Schematic of a Strain-Gauge Accelerometer (Cantilever Type)

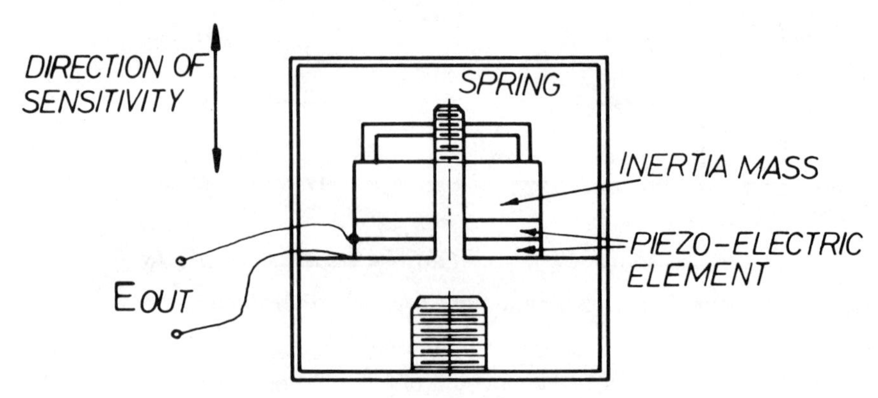

Figure 8-21. Schematic of a Piezoelectric Accelerometer (Compressive Type)

schematic of a compression-type crystal accelerometer is shown in figure 8-21. In this case, the flexural resistance to the accelerating mass is provided by a spring that has high stiffness. When the mass is accelerated in the direction of sensitivity of the accelerometer, a variation in the compressive force on the piezoelectric element results. This produces the output-voltage signal E_{out}.

Sensitivity of an accelerometer is a measure of the size of its output signal when a unity acceleration is applied in its direction of sensitivity. It is desirable to have as high a sensitivity as possible, so that the output signal is not obscured by a high noise content.

The piezoelectric element in a crystal accelerometer actually acts as an electrical-charge generator with a certain capacitance. This produces the voltage output (*Note: Charge* = capacitance × *voltage.*) Because of this, accelerometer sensitivity is commonly expressed in terms of charge per unit acceleration and voltage per unit acceleration. The former is termed charge sensitivity and the latter voltage sensitivity. Electrical charge is measured in coulombs [1 coulomb (C) = 10^{12} picocoulombs (pC)]. Acceleration is measured in units of acceleration due to gravity (g). Typical accelerometer sensitivites for dynamic-testing applications are 10 pC/g and 5 mV/g.

Sensitivity of an accelerometer depends on the piezoelectric properties (or strain-gauge factors) and on the mass of the inertia element. If a large mass is used, the reaction inertia force on the strain member will be large (for a given acceleration), thus generating a relatively large output signal. Large accelerometer mass could have several disadvantages, however. In particular, (1) the accelerometer mass could modify test-object dynamics when installed, and (2) heavy accelerometers have a lower useful frequency range. As a rule, the accelerometer mass should be less than 10 percent of the mass of the component in the test object to which it is attached. The useful frequency range of an accelerometer is determined by its resonant frequency. To explain this concept, consider a typical frequency-response curve (figure 8-22) for the inertia element of an accelerometer. At frequencies several times lower than the resonant frequency, the response curve is flat, and the accelerometer base and the inertia element act as a rigid body. The base acceleration equals the acceleration of the inertia element in this flat region. This is the useful frequency range. Since the resonant frequency is proportional to the square root of stiffness/mass, heavy accelerometers have a lower useful frequency range. This range can be increased somewhat by increasing the stiffness of the spring element.

The upper acceleration limit in the useful range of a given accelerometer primarily depends on its structural strength. The lower acceleration limit depends on its sensitivity, because, if the sensitivity is low, the output signal at low accelerations might be so small that the electrical noise from connecting cables and signal-conditioning equipment (amplifiers, and the like) could significantly distort it, giving erroneous readings. In dynamic-testing applications, the typical upper acceleration limit is 20 g. The lower limit is about 0.001 g.

Another factor that should be considered in selecting an accelerometer is its cross-sensitivity. This is defined as its sensitivity in an axis that is orthogonal to its direction of sensitivity and is usually expressed as a percentage of the direct sensitivity. Cross-sensitivity primarily results from

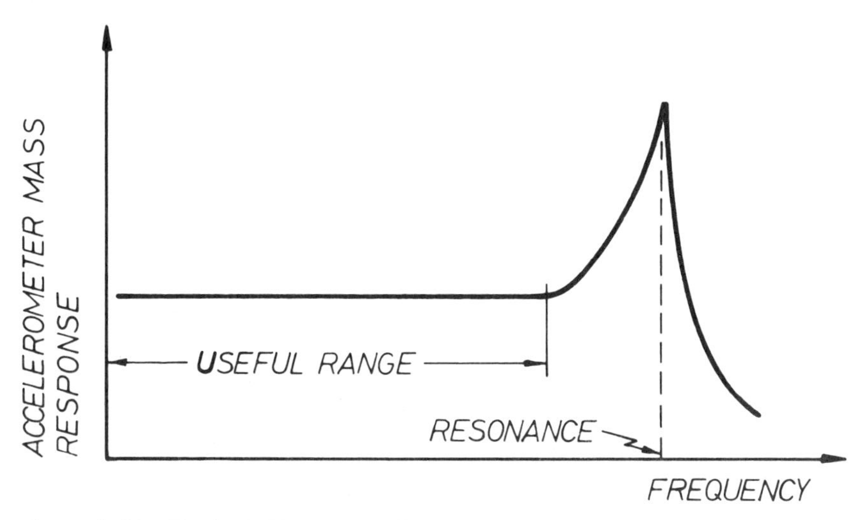

Figure 8-22. Typical Frequency-Response Curve for an Accelerometer Inertia Element

manufacturing irregularities of the piezoelectric (or strain-gauge) element. Material unevenness and incorrect orientation of sensing element are such irregularities. Cross-sensitivity should be less than 5 percent for most dynamic-testing applications.

Accurate calibration of accelerometers is a very important consideration in determining the accuracy of dynamic-testing results. An accelerometer could be calibrated by first holding its direction of sensitivity in a horizontal direction and then turning it to vertical direction (90° rotation). In the first orientation, the weight of the inertia element does not act on the piezoelectric element, and it corresponds to zero acceleration. In the second orientation, the weight of the inertia element acts directly on the piezoelectric element, and it corresponds to an acceleration of 1 g. An arm rotating about a horizontal axis, with the accelerometer attached to the free end, and an accurate angle gauge may be used to facilitate such a calibration test. An alternative calibration procedure is to mount the accelerometer on an arm rotating at a constant angular velocity and about vertical axis such that its direction of sensitivity is oriented toward the center of rotation. The acceleration to which the accelerometer is subjected under these conditions is $\omega^2 r$, in which r is the distance of the accelerometer from the center of rotation. Accelerometer readings when the system is at rest and when it is rotating at a known angular velocity are used to calibrate the accelerometer.

Although the foregoing discussion exclusively considered linear accelerometers, the same principle can be extended to angular accelerometers. One method of construction is to use a simple pendulum as the inertia element.

An accelerometer can be attached to the test object in many ways. Some commonly employed means are the following:

1. Screw-in base
2. Magnetic base
3. Glue, cement, or wax
4. Hand-held probe

Drilling holes in the test object can be avoided by using methods 2 through 4, but the useful range of the acceleration can decrease significantly when hand-held probes are used and, to some extent, when the magnetic-attachment method is employed. This is because of the high compliance between the accelerometer and the test object introduced by these two methods, which can reduce the overall natural frequency of the accelerometer and the attachment significantly.

In dynamic testing, we might need to monitor force or impedance at various component locations in the test object. This is done with force sensors and impedance sensors. A crystal-type force sensor consists of a

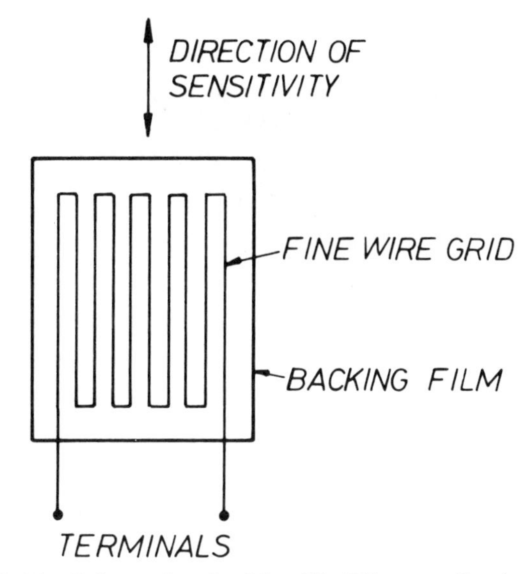

Figure 8-23. Schematic of a Metallic-Filament Strain Gauge

piezoelectric element. An electrical signal proportional to the force (or pressure) transmitted through the piezoelectric element is generated when the sensor is attached to the test object.

An impedance sensor (impedance head) usually consists of a force sensor and a velocity sensor (or an accelometer with an integrator). The impedance at a location of the test object is obtained as the ratio of force to velocity in the complex frequency domain (see chapter 2). Since the signals are generated by a piezoelectric element in these two types of sensors, the same signal conditioning methods are used as are used for crystal accelerometers.

Strain Gauge

As noted in the previous section, the strain gauge responds to strains in its direction of sensitivity. It senses strains of the member to which it is attached and produces a proportional voltage at its output terminals. Strain gauges are passive transducers and require external voltage sources for their operation. Operation of a typical strain gauge depends on the change in electrical resistance associated with elongation (or shortening) of a conducting wire. Such strain gauges are known as resistance strain gauges. The sensor element could be a metallic foil or a piezoresistive (semiconductor) element.

A metallic-filament strain gauge is shown schematically in figure 8-23. A fine grid of conductive wire is attached to a backing film made of high

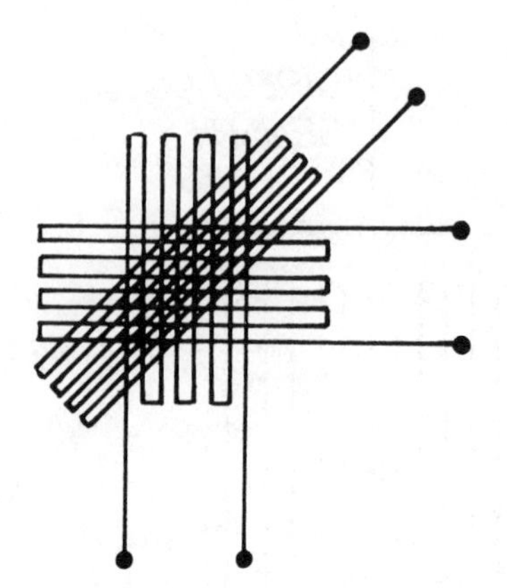

Figure 8-24. Three-Arm Rosette Element

insulating material, which is used to attach the strain gauge to the test object by means of a suitable adhesive. The direction of sensitivity of a strain gauge is the major elongation direction of the wire grid. To measure strains in more than one direction, multiple strain gauges are available in single units (for example, the rosette arrangement shown in figure 8-24). These units have more than one direction of sensitivity. Principal strains in a given plane can be obtained by using such multiple strain-gauge units.

One way of using a strain gauge is to apply a constant voltage to it (through a resistor) and measure the output voltage across the strain gauge under open-circuit conditions. As the test member is strained, the resistance of the strain-gauge element will change, and a corresponding change in the output voltage will result. This arrangement, shown in figure 8-25, is known as the potentiometer circuit, for obvious reasons. The method has several

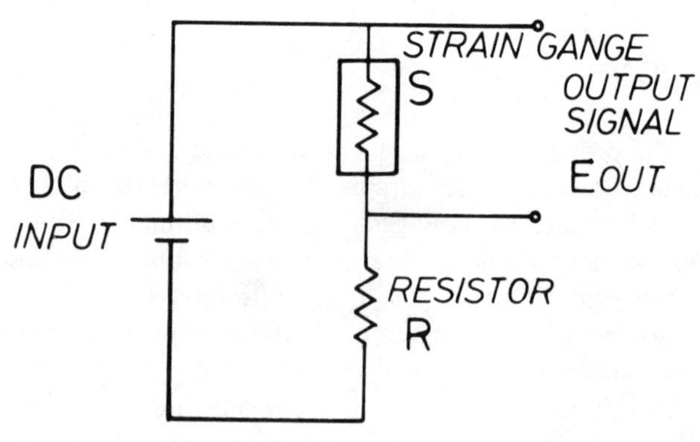

Figure 8-25. Potentiometer Circuit for Strain-Gauge Measurements

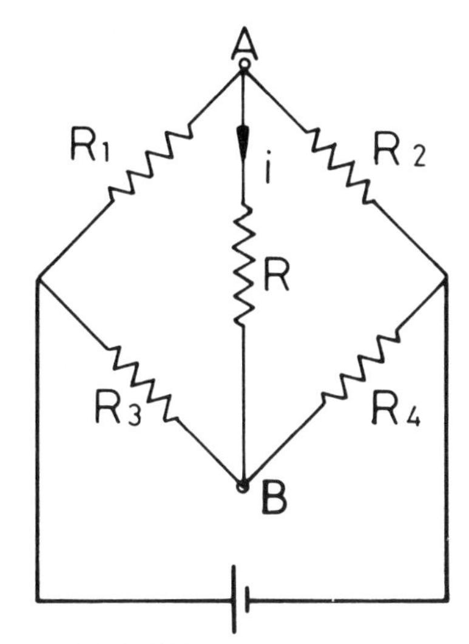

Figure 8-26. Wheatstone Bridge Circuit

weaknesses. Any ambient temperature variations will cause some measurement error because of associated changes in strain-gauge resistance and the resistance of the connecting circuitry. Also, the measurement accuracy will be affected by possible variations in the supply voltage. Furthermore, when a voltmeter is connected to measure E_{out}, the open-circuit condition is violated.

A more favorable arrangement is a bridge circuit that has two or four strain gauges. A wheatstone bridge is shown in figure 8-26. If

$$\frac{R_1}{R_3} = \frac{R_2}{R_4} \tag{8.7}$$

then the bridge is said to be balanced, and the potential difference between points A and B will be zero. Consequently, the current i through the resistance R will be zero. A slight unbalance—that is, a variation in any of the resistances R_1, R_2, R_3, or R_4, violating equation (8.7)—will produce a potential difference across R. In strain-gauge bridges, either two or all four resistances may be replaced by identical strain gauges. Consider the two-strain-gauge bridge shown in figure 8-27. Two strain gauges S_1 and S_2 form two resistors of the bridge. When both strain gauges are not active (that is, when they are not strained), a balanced bridge results, and $E_{out} = 0$. Any ambient changes will affect all four resistors equally, and thus the balance will be maintained. If S_1 is made active (by straining the member to which it

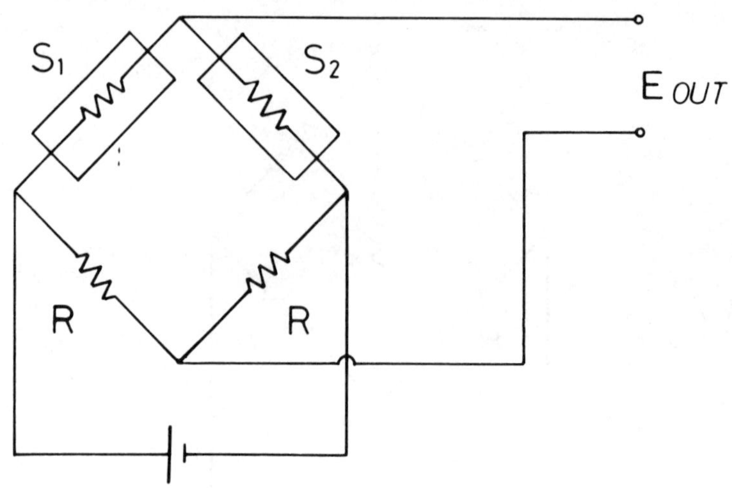

Figure 8-27. Two-Strain-Gauge Bridge

is attached), then a nonzero E_{out} will result. It is a measure of the strain at the active gauge. Note that S_2 acts as a dummy gauge in this situation. Numerous other activating combinations are possible (for example, tension on S_1 and compression on S_2). Since there could be a slight discrepancy in strain-gauge and circuit resistances, it is desirable to have one of the resistors R_i variable, with fine-adjustment capability. In this way, initial balance can be obtained accurately when both strain gauges are inactive before making strain measurements.

We have seen how strain gauges can be employed in acceleration measurement. An inertia element is used to generate a force that is proportional to applied acceleration. Strain gauges also may be used in displacement and velocity measurements. In displacement measurement, a spring element (for example, a cantilever) that produces a strain (or force) proportional to displacement is used. In velocity measurement, a damping element (for example, a viscous damper) that produces a strain (or force) proportional to velocity is used. In conclusion, it should be noted that piezoelectric and strain-gauge elements also can be used in force, pressure, and torque transducers.

Signal-Conditioning Equipment

In dynamic testing, response signals often must be conditioned in order to transform them into a suitable form for recording, processing (digital and analog), and use as feedback-control signals or actuator inputs. In most cases, the transducer output signal (the measurement) is a very weak voltage. Its power might have to be stepped up several times, while preserving the original shape and characteristics (particularly frequency content), so that it can be recorded, processed, or used as an actuating or control signal. In short, such signals must be conditioned. A main reason for this is that the recording, processing, control, and actuating equipment must extract a sizable amount of power from the signal in order to function. Furthermore, noise from these circuitry and connecting cables could distort weak signals almost to the point that the original signal would be undetectable (that is, a low signal to noise ratio, or SNR). Also, the transducer itself can introduce noise, which must be filtered out from the authentic signal. Reasons such as these have made signal conditioning an essential part of dynamic-testing procedures.

Three primary types of signal conditioning are important in dynamic testing: amplification, filtering, and signal modification. Amplification consists of using circuitry powered by an external electricity source to step up the level (typically, voltage, current, or power) of the signal. Filtering is the procedure whereby unwanted frequency components (noise) in the signal are extracted. Amplification and filtering also may be considered signal modification. For our purposes, however, signal modification is defined as changing the basic characteristics of the signal. Some examples of signal modification are (1) converting a series of pulses into a continuous signal that is proportional to its pulse rate (pulse-rate integration), (2) converting an analog signal into a digital signal (by an ADC), and (3) converting a digital signal into an analog signal (by a DAC). Before dealing with these methods of signal conditioning, it is useful to discuss the importance of the impedance characteristics of electrical equipment.

Impedance Characteristics of Equipment

Dynamic-testing systems use various devices, such as transducers and signal-generating, conditioning, and recording equipment. When these devices are interconnected, it is necessary to properly match the impedances to realize their rated performance level.

Consider a standard input-output electrical device, presented by the operational block diagram in figure 8-28. The output impedance of such a

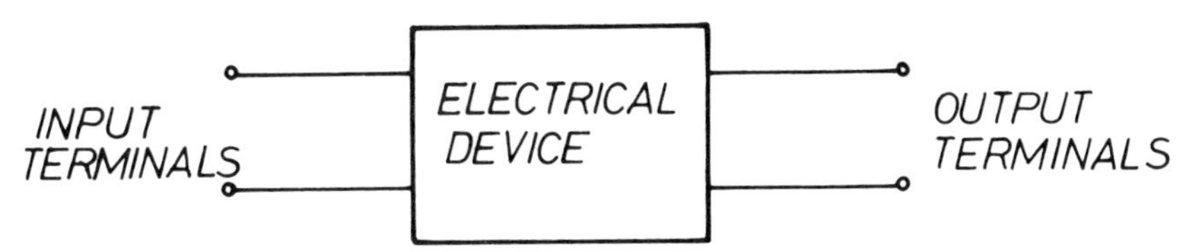

Figure 8-28. Block Diagram of a Standard Input-Output Device

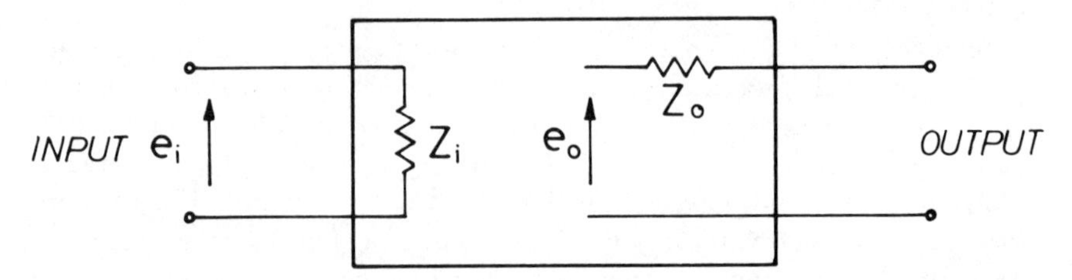

Figure 8-29. Representation of Input Impedance and Output Impedance

device is defined as the ratio of the open-circuit (that is, no-load) voltage at the output terminals to the short-circuit current at the output. Open-circuit voltage is the output voltage present when there is no current flowing across the output terminals. This is the case if the two output terminals are not connected through a load (impedance). As soon as a load is connected across the terminals, a current will flow through it, and the output voltage will drop to a value less than that of the open-circuit voltage. The following procedure may be adopted to determine the open-circuit voltage. First, the rated input voltage is applied at the input terminals and maintained constant. Next, the output voltage is measured using a voltmeter that has a very high impedance. To measure the short-circuit current, a very low impedance ammeter should be connected across the output terminals. Input impedance is defined as the ratio of the rated input voltage to the corresponding current through the input terminals.

Using the foregoing definitions, input impedance Z_i and output impedance Z_o may be represented in the input-output device block diagram as in figure 8-29. This representation satisfies the definitions of the two parameters. Note that e_o is the open-circuit output voltage. When a load is connected across the output terminals, the voltage across the load will be less than e_o. This is caused by the presence of Z_o.

When input e_i and output e_o are represented by their respective Fourier integral transforms, their relationship can be expressed in terms of H, the

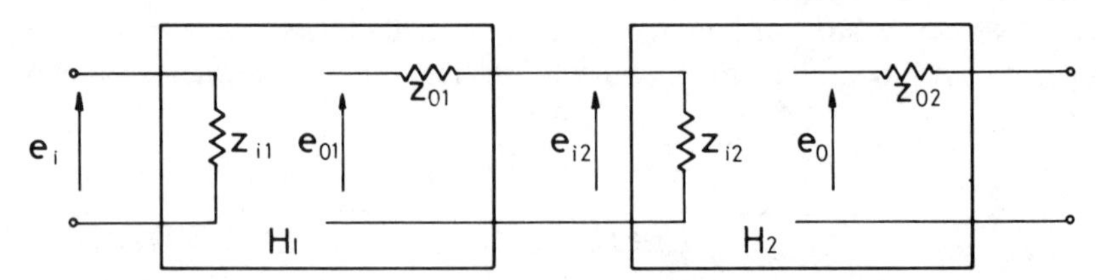

Figure 8-30. Effect of Cascade Connection of Devices on the Overall Frequency Response

frequency-transfer (response) function of the input-output device under open-circuit (no-load) conditions. Specifically,

$$e_o = He_i \tag{8.8}$$

Consider two input-output devices connected in cascade, as shown in figure 8-30. It is easy to see that the following relations hold:

$$e_{o1} = H_1 e_i \tag{8.9}$$

$$e_{i2} = \left(\frac{Z_{i2}}{Z_{o1} + Z_{i2}}\right) e_{o1} \tag{8.10}$$

$$e_o = H_2 e_{i2} \tag{8.11}$$

These can be combined to give the overall input-output relation,

$$e_o = \left(\frac{Z_{i2}}{Z_{o1} + Z_{i2}}\right) H_2 H_1 e_i \tag{8.12}$$

We see from equation (8.12) that the overall frequency-transfer function differs from the expected H_2H_1 by the factor

$$\left(\frac{Z_{i2}}{Z_{o1} + Z_{i2}}\right) = \left(\frac{1}{\dfrac{Z_{o1}}{Z_{i2}} + 1}\right)$$

In other words, cascading has affected the frequency-response characteristics of the two devices. It is noticed, however, that, if $Z_{o1}/Z_{i2} \ll 1$, the effect becomes insignificant. From this we can conclude that, whenever frequency-response characteristics are important, cascading of two devices should be done, such that the output impedance of the first device is much smaller than the input impedance of the second device. Using the same argument, we can see that, when a load is connected to a piece of equipment, the output voltage varies depending on the size of the load. This variation is known as the loading error. It is measured as the difference between the output voltage corresponding to the minimum load and that corresponding to the maximum load, and is expressed as a percentage of the no-load (open-circuit) voltage. From the previous discussion, it follows that the loading error becomes small if the output impedance of the equipment is small in comparison to the maximum load impedance.

Transducers that have high output impedance (for example, piezoelectric accelerometers) are devices that produce low output current. Furthermore, because of high output impedance, loading error becomes high. For these reasons, the output signals from high-output-impedance transducers need conditioning before they are used in dynamic-testing applications (that is, recording, processing, control, and actuating). Impedance-matching amplifiers that have higher input impedance and lower output impedance should be used to correct the situation.

Amplification

Amplification is the process of increasing the level of a signal. This is achieved by a device called an amplifier, which needs an external energy source to achieve the required amplification. The three primary types of amplifiers are the voltage amplifier, the current amplifier, and the power amplifier. This nomenclature clearly indicates the parameter (voltage, current, or power) that is being stepped up during the amplification process. The charge amplifier is a signal-conditioning device that is used primarily to condition accelerometer output signals. It produces an output voltage proportional to the input charge (from a charge-generating element such as a piezoelectric crystal). Certain amplifiers characteristically distort the input signal. In dynamic testing, we normally do not employ such amplifiers, but a phase difference of 180° between input and output signals is allowed in most applications.

A detailed discussion of the electronics of the various types of amplifiers does not fall within the scope of this book. A preliminary understanding of the basic principle of operation, however, can be gained by reference to the transistor circuit shown in figure 8-31. This is a common-base amplifier that uses a negative-positive-negative (NPN) transistor. It is useful as a voltage amplifier. The DC voltages V_1 and V_2 are the external energy sources to the amplifier. They provide bias voltages to the transistor. The input bias V_e is

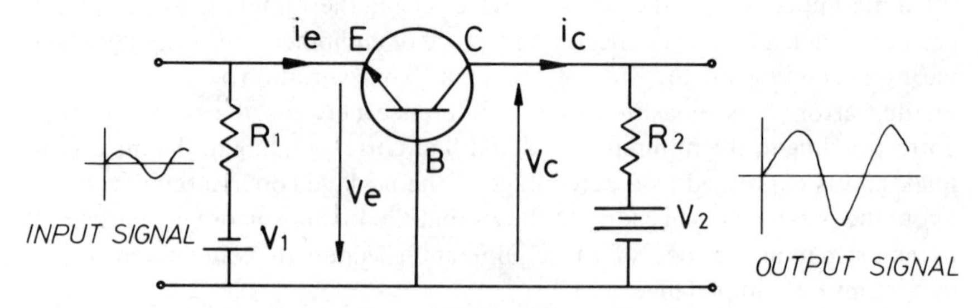

Figure 8-31. Common-Base Amplifier Circuit Using an NPN Transistor

the voltage between emitter E and base B. The output bias V_c is the voltage between collector C and base B. The emitter current i_e is slightly higher but approximately equal to the collector current i_c. When there is no input signal, these currents are AC. When an input signal is applied, the output bias will fluctuate. This will cause significant fluctuations at the output that forms the output signal (voltage, current, power). Figure 8-31 shows only one amplifier stage, but many such circuits (stages) are normally used in a single amplifier unit.

Stability is necessary for operation of an amplifier. If any disturbance at the amplifier output (load) tends to grow, the amplifier is said to be unstable. Low stability will distort the output signal. Amplifier stability can be improved by means of negative feedback. This amounts to sensing a fraction of the amplifier output and feeding it back to the input in an opposing (subtractive) direction. Any unwanted load fluctuation will change the input so as to correct the output. The feedback process changes the impedance of an amplifier, depending on whether the feedback signal is current or voltage. If the feedback signal is current, it tends to decrease the output current, thereby increasing the output impedance. If the feedback signal is voltage, it tends to decrease the output voltage, thereby decreasing the output impedance.

Characteristics of an amplifier are specified in such quantities as voltage gain and accuracy, current gain and accuracy, power gain and accuracy, phase shift between input and output, dynamic range, maximum input voltage, maximum output voltage, input impedance, and output impedance. Gain refers to the maximum possible amplification factor for the amplifying parameter. Input impedance and output impedance were defined in the previous section. Voltage amplifiers generally have high input impedance and low output impedance. In contrast, current amplifiers have low input impedance and high output impedance.

Accelerometers have high-output-impedance characteristics, and their output is not usually sufficient for direct recording, processing, control, or actuating. To correct this situation, accelerometer outputs are conditioned by using preamplifiers. The two basic requirements of these amplifiers are high gain (to amplify the signal) and low output impedance (to reduce the combined output impedance). These requirements can be achieved by using a high-gain (typically, 1,000) voltage amplifier. Voltage amplifiers have high input impedance, however, which affects low-frequency inputs. Also, when a voltage amplifier is used in cascade with an accelerometer, the system becomes very sensitive to the length of the cable connecting the accelerometer to the amplifier. Output voltage decreases with any increase in the capacitance of the connecting cable. A solution to this is to use a charge amplifier. In principle, a charge amplifier is a high-gain amplifier with capacitance negative feedback (figure 8-32). The charge amplifier provides

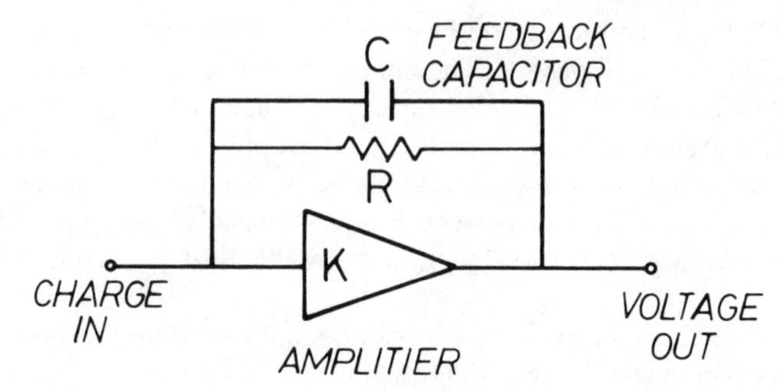

Figure 8-32. Operational Schematic of a Charge Amplifier

an output voltage signal that is proportional to an input electrical charge. A feedback capacitor of high capacitance (for example, 1,000 pF) would reduce the input impedance of the amplifier and also would lower its sensitivity to the capacitance of connecting cables and circuitry. Obviously, charge amplifiers are more costly than voltage amplifiers.

Modern amplifiers use integrated-circuit (IC) construction on printed-circuit boards (PCB). This results in compact, reliable devices that consume less power.

Filtering

The main function of filtering is extraction of unwanted frequency components in a signal. In this sense, any dynamic system can be considered a filter. The range of the frequency components that are allowed through a filter depends on the frequency-response function of the filter. For an analog circuit, the frequency-response function is normally a smooth function, and therefore a sharp cutoff is not feasible. As an example, consider the circuit shown in figure 8-33. Its transfer function between the input voltage V_1 and the output voltage V_2 is given by

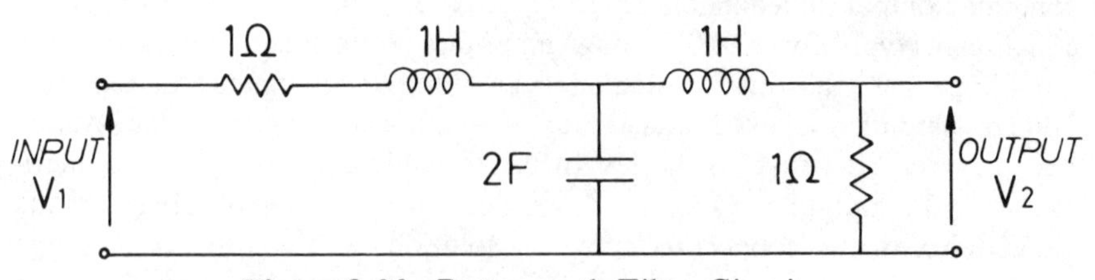

Figure 8-33. Butterworth Filter Circuit

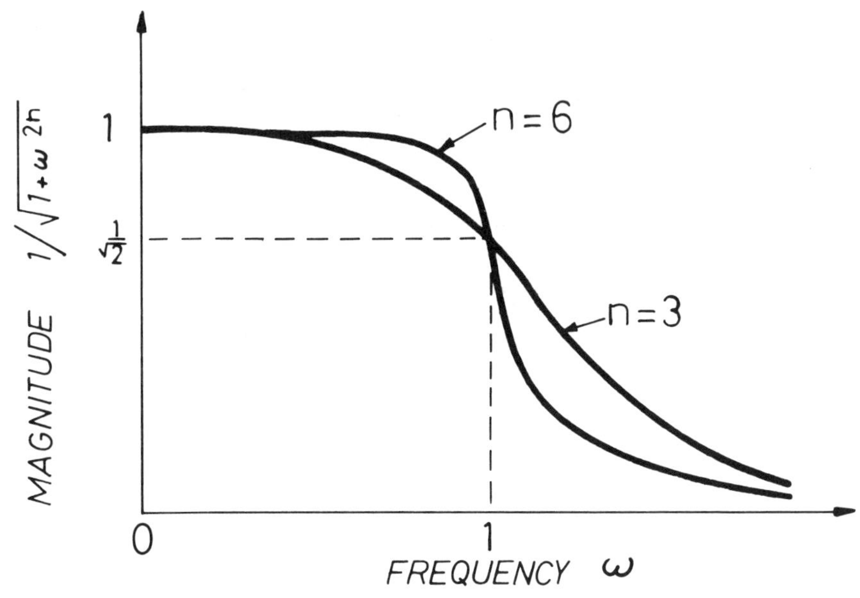

Figure 8-34. Frequency Response of a Butterworth Filter of Order *n*

$$H(s) = \frac{1}{2(s^3 + 2s^2 + 2s + 1)} \tag{8.13}$$

The corresponding frequency-response function is

$$H(j\omega) = \frac{1}{2[1 - 2\omega^2 + j(2\omega - \omega^3)]} \tag{8.14}$$

which has the magnitude

$$|H(j\omega)| = \frac{1}{2\sqrt{1 + \omega^6}} \tag{8.15}$$

This is of the form $1/\sqrt{1 + \omega^{2n}}$, which is known as the Butterworth filter of order n. It is a low-pass filter, which filters out a large fraction of high-frequency components in the input signal. This is clear from the frequency-response magnitude plots shown in figure 8-34. It is seen that the high-frequency cutoff performance, as given by the slope at $\omega = 1$, improves as n increases.

We can idealize filters into several categories. The main categories are

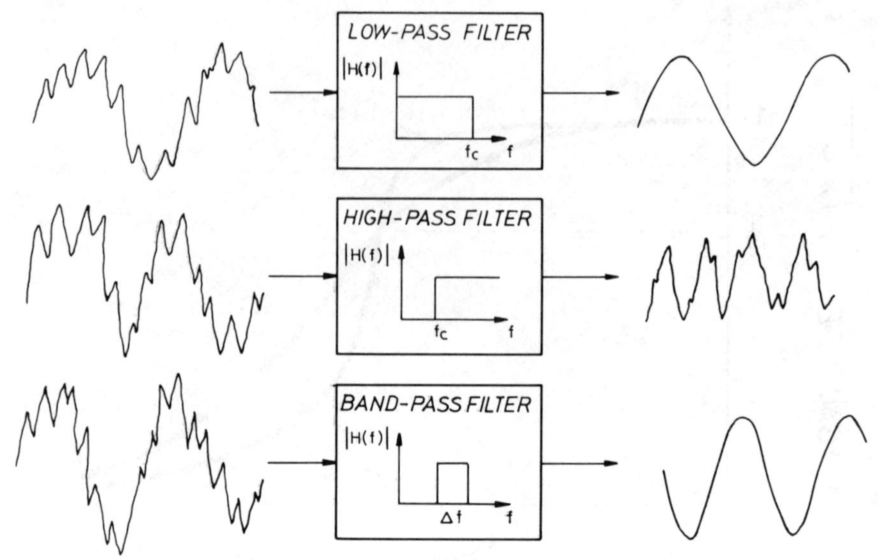

Figure 8-35. Functions of Some Ideal Filters

low-pass filters, high-pass filters, and band-pass filters. The characteristic function of each of these filter types is illustrated in figure 8-35. Low-pass filters do not allow high-frequency input-signal components above the cutoff frequency f_c. High-pass filters do not allow low-frequency input-signal components below f_c. Band-pass filters allow frequency components within the frequency band Δf and reject the remaining components.

A tracking filter is a device that extracts a single frequency from a complex input signal. In that sense, it is a band-pass filter that has very narrow band and variable center (tuned) frequency. The tuning-frequency value is specified by using a high-frequency carrier input that contains the required frequency information. More than one input channel is usually available (multiple tracking). Its operation is illustrated in figure 8-36. In certain tracking filters, a sine-reject output (that is, a tracked sinusoidal output signal subtracted from the input signal) is also available.

Industrial filter units normally have built-in amplifiers. Consequently, they need external power to operate. Performance of a filter is specified in terms of such factors as the following:

1. Number of frequency bands available for operation and their bandwidths
2. Sharpness of the cutoff slope (db/octave)
3. Input and output impedance
4. Maximum input-voltage amplitude
5. Maximum power of the output signal

6. Linearity over the operating output-amplitude range (expressed in db as a maximum variation from a straight line)
7. External power requirement
8. Internal noise generation

Filter frequency-response curves are normally available with the operating data. Control knobs and dials on the panel are used to vary the frequency band of operation and the maximum input-voltage amplitude.

Common industrial filters are analog devices that use analog circuitry to condition analog input signals. Their outputs are also analog signals. Digital filters are used to process digital input data resulting in digital outputs. In essence, they are digital computers. A suitably programmed digital computer can function as a digital filter. Digital filters have the usual advantages of digital systems and are widely used in signal-processing applications.

Signal Modification

Most signal modifications are performed by means of suitable electronic circuitry, using semiconductor (solid-state) elements, integrator networks, and the like. Operations such as rectification of an AC signal, clipping, DC to AC conversion, and various modulation and demodulation procedures may be achieved in this manner.

Sometimes information signals have to travel very large distances (for example, in radio transmission), which will weaken and distort them and

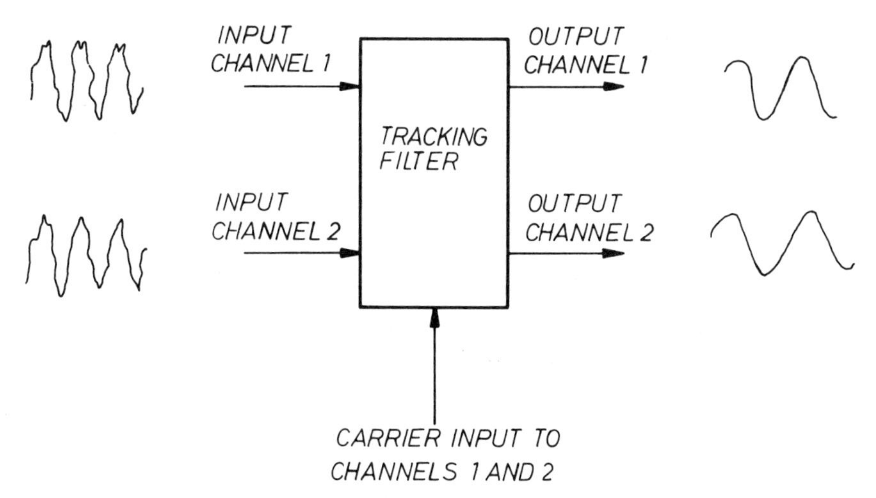

Figure 8-36. Operational Schematic of a Tracking Filter

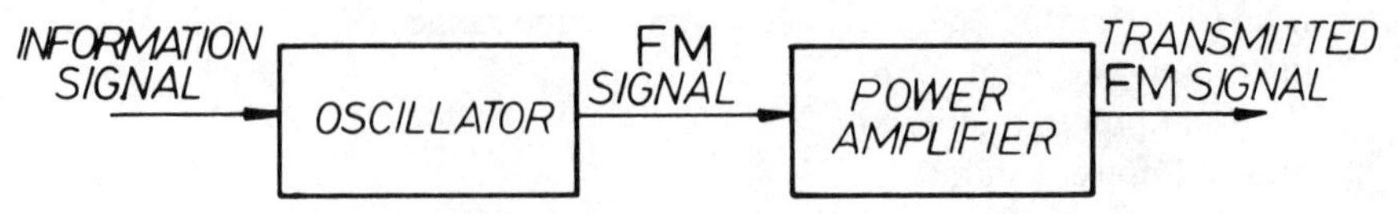

Figure 8-37. Generation of an FM Signal

corrupt them with extraneous noise. Modulation techniques may be used to overcome this problem. Information is carried by a powerful carrier wave in this method. To extract the information from the carrier wave, a demodulation process has to be employed.

In frequency modulation (FM), the amplitude of the carrier wave is kept constant and its frequency is varied according to the information-signal level. The carrier wave is generated by an oscillator, which uses the information signal as the input. The oscillator output is the FM signal, which usually must be amplified (see figure 8-37). Often, other conditioning stages are required before the signal is transmitted. Demodulation of an FM signal is achieved by using a frequency discriminator that can detect frequency changes in signals.

In data acquisition and processing associated with dynamic testing, analog-to-digital converters (ADCs) and digital-to-analog converters (DACs) are necessary if a digital computer is used as the processor. The ADC is a signal-modification process by which a digital signal (a series of binary numbers) is generated from an analog signal. One way of achieving this is by using a comparison circuit. A voltage source is associated with each significant figure of a suitable digital number system, up to a certain number of significant figures. Starting from the most significant voltage, successive comparison is made with the analog signal until a source that is below the level of the analog signal is obtained for the first time. The next weaker source is added to it and again compared. If the level is still below that of the analog signal, the next source is also added and again compared. If

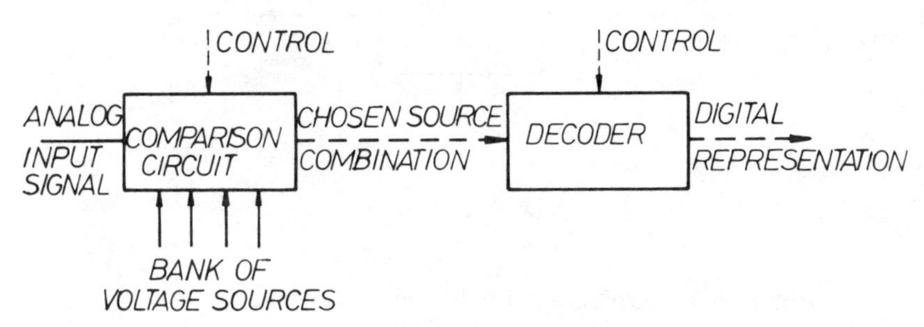

Figure 8-38. Typical Analog-to-Digital Conversion (ADC) Process

the level is higher, then the second source is removed before the next source is added and compared. In this manner, a digital signal (expressed as a number) that is very close to but slightly weaker than the input analog signal is obtained. The digital number corresponding to the set of voltage sources that is active is determined simply by using a decoder (see figure 8-38). The reverse process would be used in digital-to-analog conversion.

Signal-Recording Equipment

Signal-recording equipment commonly employed in dynamic testing includes tape recorders, oscilloscopes, strip-chart recorders, and *X-Y* plotters. Tape recorders are used to record dynamic-testing data (transducer outputs) that is subsequently reproduced for processing or examination. Often, tape-recorded waveforms are also used to generate (by replay) signals that drive dynamic-test actuators. Tape recorders use tapes made of a plastic material that has a thin coating of a specially treated ferromagnetic substance. During the recording process, magnetic flux proportional to the recorded signal is produced by the recording head (essentially an electromagnet), which magnetizes the tape surface in proportion to the signal variation. Reproduction is the reverse process, whereby an electrical signal is generated at the reproduction head by electromagnetic induction in accordance with the magnetic flux of the magnetized (recorded) tape. Several signal-conditioning circuitries are involved in the recording and reproducing stages. Recording by FM is very common in dynamic testing. Some important performance specifications for tape recorders are outlined in the earlier section on signal-generating equipment.

Oscilloscopes may be used to observe (monitor) test data such as test-object-response time histories obtained from the transducer outputs. They are also useful in observing and examining test results, such as frequency-response plots, psd curves, and response spectra. Typically, only temporary records are available on an oscilloscope screen. The main component of an oscilloscope is the cathode-ray tube (CRT), which consists of an electron gun (cathode) that deflects an electron ray according to the input-signal level. The oscilloscope screen has a coating of electron-sensitive material, so that the electron ray that impinges on the screen leaves a temporary trace on it. The electron ray sweeps across the screen horizontally, so that waveform traces can be recorded and observed. Usually, two input channels are available. Each input may be observed separately, or the variations in one input may be observed against those of the other. In this manner, signal phasing can be examined. Several sensitivity settings for the input-signal-amplitude scale (in the vertical direction) and sweep-speed selections are available on the panel. Normal and auto triggering modes generally are also available.

Strip-chart recorders are usually employed to plot time histories (that is, quantities that vary with time), although they also may be used to plot such data as frequency-response functions and response spectra. In these recorders, a paper roll unwinds at a constant linear speed, and the writing head moves aross the paper (perpendicular to the paper motion) proportionally to the signal level. There are many kinds of strip-chart recorders, which are grouped according to the type of writing head employed. Graphic-level recorders, which use ordinary paper, employ such heads as ink pens or brushes, fiber pens, and sapphire styli. Visicoders are simply oscilloscopes that are capable of producing permanent records; they employ light-sensitive paper for this. Several channels of input data can be incorporated with a visicoder. Obviously, graphic-level recorders are generally limited by the number of writing heads available (typically, one or two), but visicoders can have many more input channels (typically, twenty-four). Performance specifications of these devices include paper speed, frequency range of operation, dynamic range, and power requirements.

In dynamic-testing applications, *X-Y* plotters are generally employed to plot frequency data (for example, psd, frequency-response functions, response spectra, transmissibility curves), although they also can be used to plot time-history data. Many types of *X-Y* plotters are available, most of them using ink pens on ordinary paper. There are also hard-copy units that use heat-sensitive paper in conjunction with a heating element as the writing head. The writing head in a *X-Y* plotter is moved in the *X* and *Y* directions on the paper by two input signals that form the coordinates for the plot. In this manner, a trace is made on stationary plotting paper. Performance specifications of *X-Y* plotters are governed by such factors as paper size; writing speed (in/sec, cm/sec); dead band (expressed as a percentage of the full scale), which measures the resolution of the plotter head; linearity (expressed as a percentage of the full scale), which measures the accuracy of the plot; minimum trace separation (in., cm) for multiple plots on the same axes; dynamic range; input impedance; and maximum input (mV/in., mV/cm).

Notes

1. J. A. Blume, "A Machine for Setting Structures and Ground into Forced Vibration," *Bulletin of the Seismological Society of America* 25, October 1935.

2. F. J. Oliver, *Practical Instrumentation Transducers* (London: Pitman Publishing, 1972).

Seismic Qualification Practice

Generation of required response spectra (RRS) for a particular seismic qualification project is a factor that needs careful consideration in practice. Actual generation of the RRS may be done by the test laboratory, based on the requirements of the customer. Alternatively, the test laboratory might be supplied with a complete set of RRS for various pieces of equipment that must be qualified.

If the test laboratory has access to a wide selection of shaker tables, it is desirable to have a method to quickly select the one that is most suitable for a particular test. This decision is made on the basis of test-object characteristics and the limitations of the available shakers.

The report of data and results from a seismic qualification test or analysis should be done in a suitable format. The primary objective of a seismic qualification report is to include the test or analysis information and results as completely and comprehensively as possible, without confusing the reviewers and others who will be making the final qualification decision. The report should address the qualification objective.

This chapter discusses generation of the RRS specifications and selection of shaker specifications. In addition, appendix 9A is a sample of a typical seismic qualification report, and appendix 9B is a sample of a typical report review.

Generation of RRS Specifications

Seismic qualification of an object usually is specified in terms of a required response spectrum (RRS). The excitation input used in seismic qualification analysis and testing should conservatively satisfy the RRS; that is, the response spectrum of the actual excitation input should envelop the RRS (without excessive conservatism, of course,).

For equipment to be installed in a building or on some other supporting structure, the RRS generally cannot be obtained as the response spectrum of a modified seismic ground-motion time history. The supporting structure usually introduces an amplification effect and a filtering effect on seismic ground motions. This amplification factor alone could be as high as 3. Some of the major factors that determine the RRS for a particular seismic qualification test are as follows:

1. Nature of the building to be qualified
2. Dynamic characteristics of the building or structure and the location (elevation and the like) where the object is to be installed
3. In-service mounting orientation and support characteristics of the object
4. Nature of the seismic ground motions in the geographic region where the object is to be installed
5. Test severity and conservatism required by the purchaser or regulatory agency

The basic steps in developing the RRS for a specific seismic qualification application include the following:

1. Development of representative safe-shutdown earthquake (SSE) ground-motion time histories for the building (or support structure) location
2. Development of a suitable building (or support structure) model
3. Response analysis of the building model, using the time histories obtained in step 1
4. Development of response spectra for various critical locations in the building (or support structure), using the response time histories obtained in step 3
5. Normalization of the response spectra obtained in step 4 to unity ZPA (that is, dividing by their individual ZPA values)
6. Identification of the similarities in the set of normalized response spectra obtained in step 5 and grouping them into a small number of groups
7. Representation of each similar group by a response spectrum consisting of straight-line segments that envelop all members in the group, giving normalized RRS for each group.
8. Determination of scale factors for various locations in the building for use in conjunction with the corresponding normalized RRS

Representative strong-motion earthquake time histories (SSEs) are developed by suitably modifying actual seismic ground-motion time histories observed in that geographic location (or a similar one) or by using a random-signal-generation (simulation) technique or any other appropriate method. These time histories may be available as either digital or analog records, depending on the way in which they are generated. If computer simulation is used in their development, a statistical representation of the expected seismic disturbances in the particular geographic region (using geological features in the region, seismic activity data, and the like) should be incorporated in the algorithm. The intensity of the time histories can be adjusted, depending on the required test severity and conservatism.

A structural dynamic model of the building in which the object would be installed following its seismic qualifications (for example, the reactor building) is developed, using available data for the actual (or expected) nature of the building. This could be a dynamic finite-element model. Flexibility, mass distribution, dissipation (damping), and geometric-size characteristics of various critical structural members of the building, effects of local soil conditions, and various dynamic coupling effects present should be included adequately in the model. Boundary conditions and model parameters employed in the dynamic simulation should be realistic.

Dynamic analysis of the building model is performed using the representative SSE ground-motion time histories as the foundation excitation inputs. Proper phasing or correlation of inputs at various locations of the building foundation should be included in the analysis whenever possible. Building response time histories (for example, floor response at various elevations in a multistory building) in three orthogonal directions (typically, vertical, east-west, and north-south) are determined by computer simulation. The corresponding response spectra are computed for a range of damping values. These analytical response spectra are normalized with respect to their individual ZPA values. To achieve this, the asymptote is drawn parallel to the frequency axis for high frequencies (low periods) of each response-spectrum curve. The corresponding asymptotic value gives the ZPA (see chapter 5). Then, the entire response spectrum is divided by this ZPA value.

The normalized response spectra are grouped so that those spectra that have roughly the same shape are put in the same group. In this manner, relatively few groups of analytical response spectra (normalized) are obtained. Then, the response spectra that belong to each group are plotted on the same graph paper. Next, straight-line segments are drawn to envelop each group of response spectra. This procedure results in a normalized RRS for each group of analytical response spectra, as shown in figure 9-1.

The RRS used for a particular seismic qualification scheme is obtained as follows. First, the normalized RRS corresponding to the location in the building where the object would be installed is selected. The normalized RRS curve is then multiplied by the appropriate scale factor. The scale factor normally consists of the product of the actual ZPA value under SSE conditions at that location (as obtained previously from the analytical response spectrum at that location, for example) and a factor of safety that depends on the required test severity and conservatism.

Actually, three RRS curves corresponding to vertical, east-west, and north-south directions might be needed, even for single-degree-of-freedom seismic qualification tests, because, by mounting three control accelerometers in these three directions, triaxial monitoring could be accomplished. If

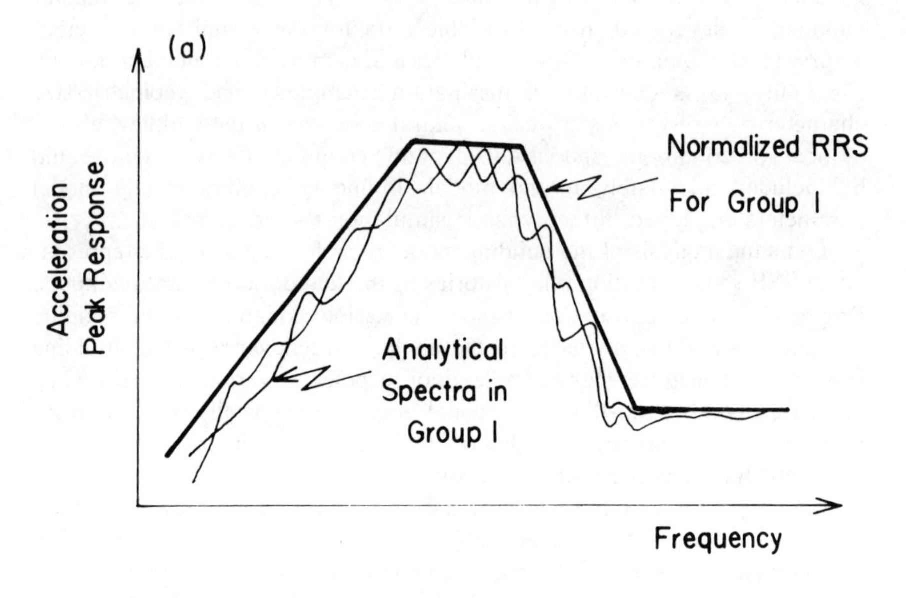

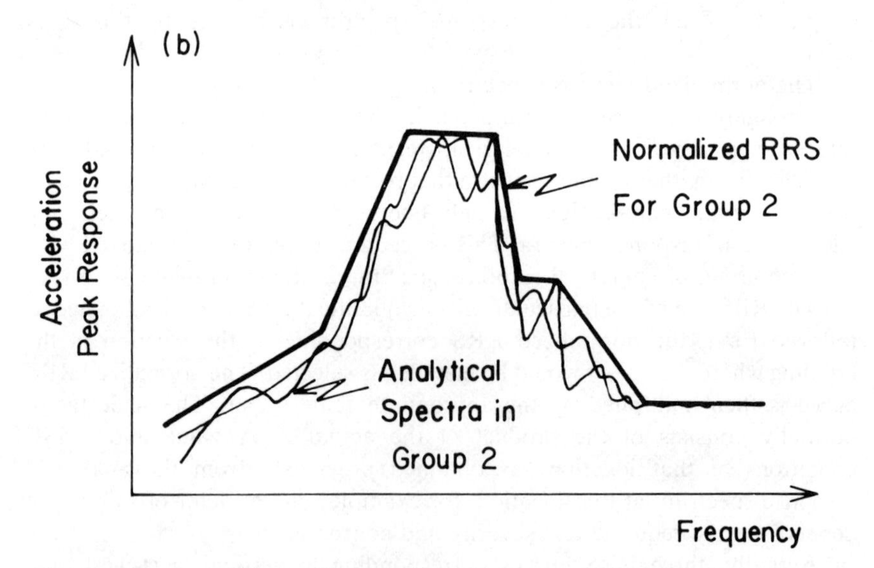

Figure 9-1. Illustration of the RRS Generation Procedure for a Complex Qualification Project

only one control accelerometer is used in the test, then only one RRS curve is used. In this case, the resultant of the three orthogonal RRS curves should be used. One way to obtain the resultant RRS curve is by applying the SRSS method to the three orthogonal components. Alternatively, the envelope of the three orthogonal RRS curves is obtained and multiplied by a safety factor (greater than unity).

Note that more than one building or even many different geographic locations could be included in the described procedure for developing RRS curves. The resulting RRS curves are then valid for the collection of buildings or geographic locations considered. When the generality of a RRS curve is extended in this manner, the test conservatism increases. This also will result in a RRS curve with a much broader band.

In a particular seismic qualification project, in practice, only a few normalized RRS curves are employed. In conjunction with these RRS curves, a table of data is provided that identifies the proper RRS curve and the scale factor that should be used for different locations (for example, elevation) in various buildings situated at several geographic locations.

Selection of Shaker Specifications

A major step in planning a seismic qualification test program is selection of a proper shaker system for a given test package. The three specifications that are of primary importance in selecting a shaker are the force rating, the power rating, and the stroke (maximum displacement) rating.

The force and power ratings are particularly useful in moderate-frequency excitations, and the stroke rating is the determining factor for low-frequency excitations. In this section, a procedure is developed to determine the minimum allowable values of these parameters in a specified test for a given test package. It should be noted that the shaker selection is normally done by referring to manufacturers' specifications. If the required information is not available, however, the selection method outlined in this section would be useful.

Theoretical Considerations

Force Rating. The full-load maximum force of the shaker head is given by

$$F = m(a + g) \tag{9.1}$$

in which m is the total mass of the test package, including mounting fixture and attachments, a is the acceleration amplitude of the input excitation, and g is the acceleration due to gravity. It should be noted that equation (9.1) takes into account the static load caused by the weight of the test package as well as the dynamic amplitude caused by the external excitation. If the test package cannot be assumed rigid in the frequency range of interest, the maximum apparent mass defined by $\max|F(\omega)/a(\omega)|$ should be used in place of m.

Equation (9.1) is written,

$$F = m(\omega v + g) \tag{9.2}$$

in which v is the velocity amplitude of the input excitation and ω is the corresponding frequency of excitation. In the present derivation, it is convenient to assume that the excitation is harmonic. For nonharmonic excitation, each Fourier frequency must be considered separately, or only the fundamental frequency is considered.

It follows that the force rating is given by the highest constant-acceleration line touching the RRS curve. Equation (9.1) should be used if the acceleration RRS is supplied, and equation (9.2) is the desired form when the velocity RRS is available.

Output Power Rating. The full-load output maximum power is given by the magnitude (absolute value) of

$$p = Fv = F\frac{a}{\omega} \tag{9.3}$$

in which the maximum value occurs when F and v are in phase. By substituting equations (9.1) and (9.2), respectively, in equation (9.3), two versions of the power equation are obtained:

$$p = m\frac{a}{\omega}(a + g) \tag{9.4}$$

$$p = mv(\omega v + g) \tag{9.5}$$

Again, equation (9.4) is the appropriate version for use in conjunction with the acceleration RRS, and equation (9.5) is the desired form when the velocity RRS is provided.

Representative segments of typical acceleration RRS curves are shown in figure 9-2. Each segment represents a relation of the form

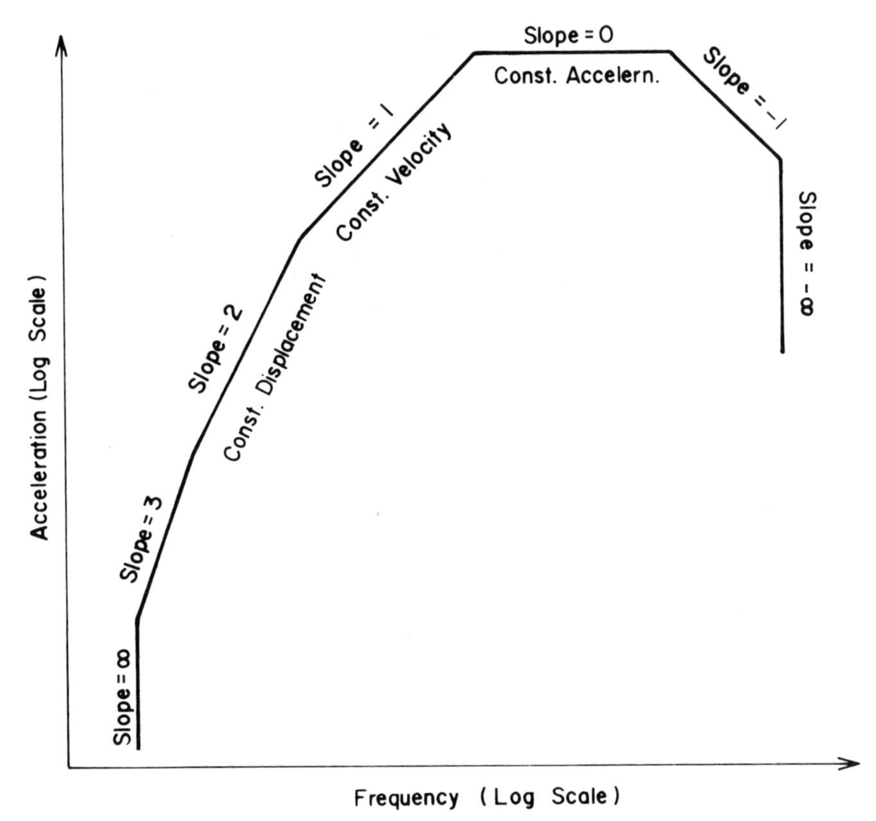

Figure 9-2. Representative Segments in Typical Acceleration RRS Curves

$$a = k\omega^n \tag{9.6}$$

in which k is a constant and n is an integer constant. Equation (9.6) is substituted in equation (9.4) to obtain

$$p = k(k\omega^{2n-1} + g\omega^{n-1}) \tag{9.7}$$

which is an increasing function of ω for $n = 1, 2, 3, \ldots$ and a decreasing function of ω for $n = 0, -1, -2, \ldots$ Since $n = 0$ corresponds to constant acceleration and $n = 1$ corresponds to constant velocity, it follows that the power rating corresponds to the leftmost point of contact between the highest constant-acceleration line and the RRS curve, or the rightmost point of contact between the highest constant-velocity line and the RRS curve.

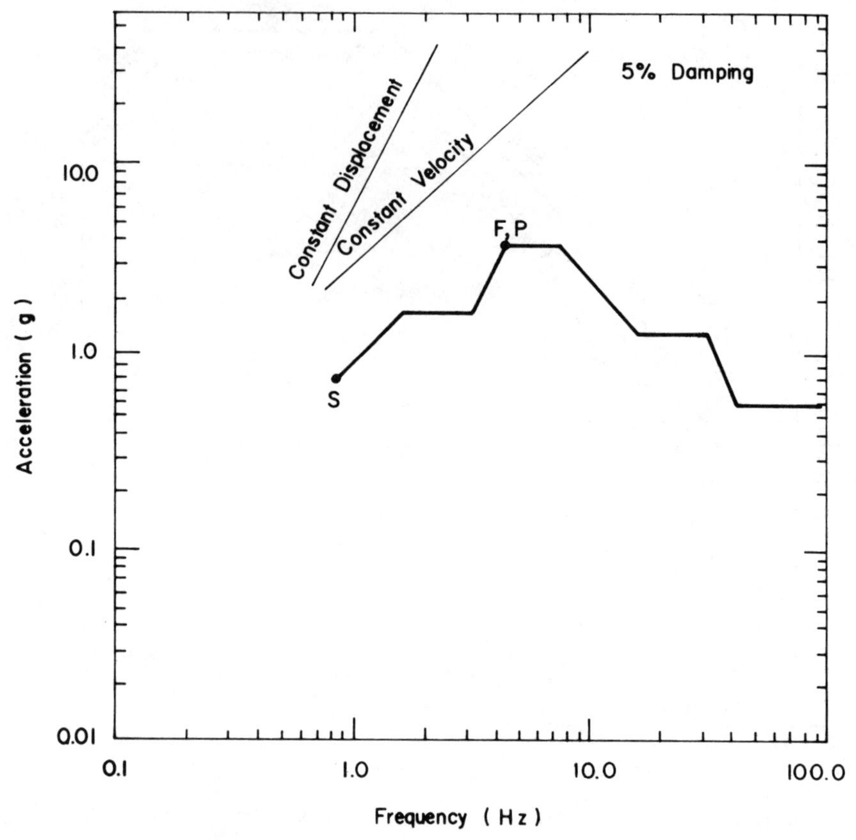

Figure 9-3. Acceleration RRS Example

Stroke Rating. The amplitude of the shaker-head displacement is given by

$$S = \frac{v}{\omega} \tag{9.8}$$

or

$$S = \frac{a}{\omega^2} \tag{9.9}$$

The stroke rating is given by the highest constant-displacement line touching the RRS. Equation (9.8) should be used if the acceleration RRS is available and equation (9.9) if the velocity RRS is available.

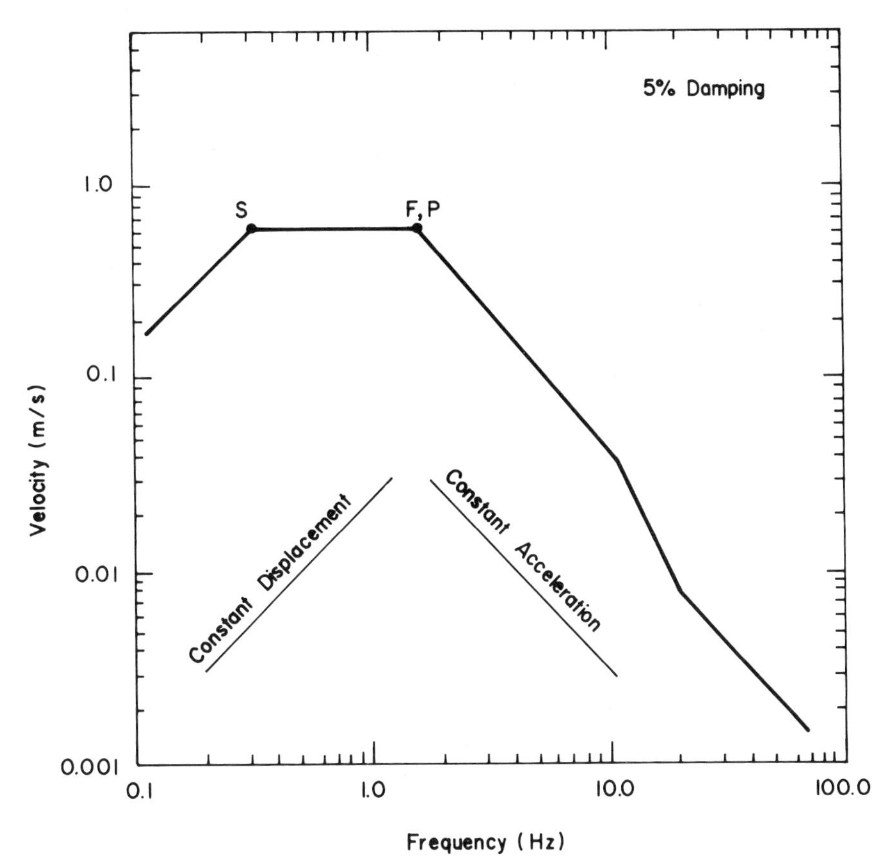

Figure 9-4. Velocity RRS Example

It should be clear from the foregoing results that the point of output-power rating on the RRS curve always corresponds to the point of force rating. The converse, however, is not generally true. Furthermore, the point of stroke rating is usually different from the points of force and power ratings.

The RRS curves for a particular test are given for a number of damping values. The proper RRS curve to be used in this method is the one that has a damping value equal to or less than the estimated damping of the test object.

Numerical Examples

Use of the Acceleration RRS. A device of 100 kg mass is to be tested using the acceleration RRS curve shown in figure 9-3. The estimated damping of the device is 5 percent. It is necessary to determine the shaker specifications for the test. The device is assumed fairly rigid.

From the development presented in the previous section, it is clear that the point F, P in figure 9-3 corresponds to the force and output power ratings and the point S corresponds to the stroke rating. The coordinates of these critical points are $F, P = (4.2 \text{ Hz}, 4.0 \text{ g})$ and $S = (0.8 \text{ Hz}, 0.75 \text{ g})$. Using equation (9.1), the force rating is determined to be

$$F_r = 100 \times (4.0 + 1) \times 9.81 N = 4905.0 N$$

Using equation (9.4), the output rating is obtained:

$$P_r = 100 \times \frac{4}{4.2 \times 2\pi}(4.0 + 1) \times (9.81)^2 \text{ watts} = 7.3 \text{ kw}$$

Using equation (9.9), the stroke rating is obtained:

$$S_r = \frac{0.75 \times 9.81}{(0.8 \times 2\pi)^2} m = 0.29 m$$

Use of the Velocity RRS. An equipment of 10kg mass is estimated to have a damping ratio of 6 percent. For seismic qualification testing, the specified velocity RRS curve at 5 percent damping is shown in figure 9-4. The equipment is assumed to be reasonably rigid. The specifications of the dynamic shaker that is suitable for the test are determined as follows. From the analysis given in the previous section, it follows that the point F, P in figure 9-4 corresponds to the force and power ratings and the point S corresponds to the stroke rating. The coordinates of the critical points are $F, P = (1.5 \text{ Hz}, 0.6 \text{ m/s})$.

From equation (9.2),

$$F_r = 10(1.5 \times 2\pi \times 0.6 + 9.81) N = 154.6 N$$

From equation (9.5),

$$P_r = 10 \times 0.6(1.5 \times 2\pi \times 0.6 + 9.81) \text{ watts} = 92.8 \text{ watts}$$

From equation (9.8),

$$S_r = \frac{0.6}{(0.3 \times 2\pi)} m = 0.32 m$$

Appendix 9A: Sample Seismic Qualification Report

SEISMIC QUALIFICATION ANALYSIS AND
TESTING OF A 10″ STAINLESS-STEEL
SOLENOID VALVE SYSTEM

Qualification performed at:
Alpha Test Laboratories, Inc.
Birmingham, CA

Author: ______, Qualification Engineer

Approved by: ____________
Manager I

____
Manager II

December 1982

Page 1:

Abstract

This report presents the results of a seismic qualification analysis and single-frequency sine-beat test performed on a 10″ stainless-steel solenoid valve system.

A finite-element model of the valve system was developed. The natural frequencies and mode shapes determined by use of this model were compared with the test results from a resonance search performed on the prototype valve that was supplied. Dynamic response and critical stresses were determined analytically, using the dynamic model. A single-frequency OBE test was conducted by using two sine sweeps in each of three orthogonal axes of the valve. This test was followed by an SSE test in each of the three axes, using sine-beat excitations. Analytical responses were compared with test results.

Test and analysis established the functional operability and structural integrity of the valve system during and after subjecting it to OBE excitations followed by an SSE excitation. It was concluded that the valve-actuator system is seismically qualified for use in the Beta Nuclear Power Plant for safety-related functions.

Page 2:

Table of Contents

Section	*Title*	*Page*

Page 3:

List of Illustrations

Figure No.	*Title*	*Page*

Page 4:

List of Tables

Table No.	*Title*	*Page*

Page 5:

Section 1: Introduction

Seismic qualification of safety-related equipment may be done by analysis, by testing, or by combined analysis and testing. The method of combined analysis and testing is employed in the present case to qualify a 10″ stainless-steel solenoid valve system, model no. SV-165, manufactured by Alpha Valve Co. for use at the Beta Nuclear Power Plant.

Analysis and testing of the valve system were performed according to the general guidelines provided in the following documents:

1. American National Standard for Qualification of Safety-Related Valve Actuators, IEEE 382, ANSI 41-6, Draft 3, Rev. 6, Nov. 1978.
2. x x x
3. x x x

Section 2: Equipment Description

A schematic of the solenoid valve unit used in the qualification program is shown in figure 1 [figure 9A-1]. The model number is SV-165. The valve is a plunger-disc type of unit operated by an electric solenoid actuator. This is one of a series of emergency flow-control valves used in the reactor-coolant system in the Beta Nuclear Power Plant. The pressure rating is 100 psi and the flow rating is 20 gal/min.

Section 3: Exploratory Resonance Search

Using the test setup shown schematically in figure 2 [figure 9A-2], a sine-sweep resonance-search test was conducted over 1-50 Hz and backwards over 50-1 Hz. The sweep rate was 1 octave/min, and the amplitude of input excitation (at the control accelerometer) was maintained at 0.2 g. The system

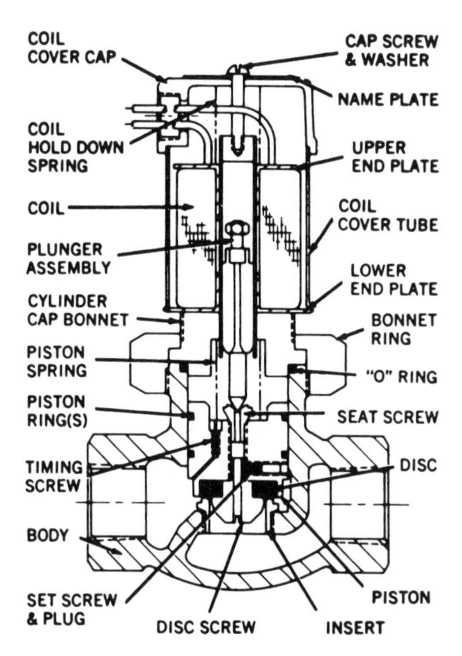

Source: J.L. Lyons and C.L. Askland, Jr., *Lyons Encyclopedia of Valves* (New York: Van Nostrand Reinhold, 1975). Copyright © 1975 by Van Nostrand Reinhold Co. Reprinted by permission of the publisher.

Figure 9A-1. Schematic of a Solenoid Valve

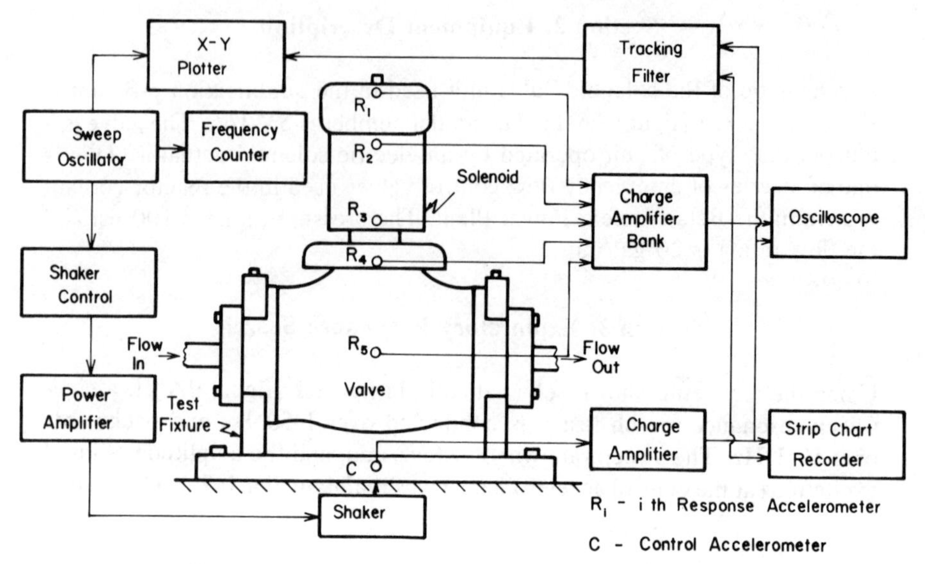

Figure 9A-2. Schematic of a Dynamic Test Setup

was pressurized, but operability was not tested. The test-apparatus log sheet, giving such details as make, model, serial number, specifications, and the date of calibration is included in appendix 2.

The response time histories were observed on the oscilloscope and were plotted using a strip-chart recorder. These records are included in appendix 4. Frequency-response functions for several crucial locations of the valve unit were obtained by Fourier analysis and plotted on the *X-Y* plotter. They are shown in figure 3 [figure 9A-3]. Resonant frequencies were determined from the frequency-response plots. The mode shapes were obtained by determining the magnitude and phase shift of each response-accelerometer output, in comparison to the control-accelerometer output, as obtained from the oscilloscope records shown in figure 4 [figure 9A-4].

The test was performed in the horizontal direction only, because the resonant frequencies in the vertical direction would be much higher. Only one resonance below 35 Hz (18Hz) was observed in the lateral mode that was tested.

Section 4: Structural Analysis

4.1: *Finite-Element Model*

A finite-element model was developed for dynamic analysis of the solenoid valve unit. The valve body, plunger, and solenoid coil were represented by

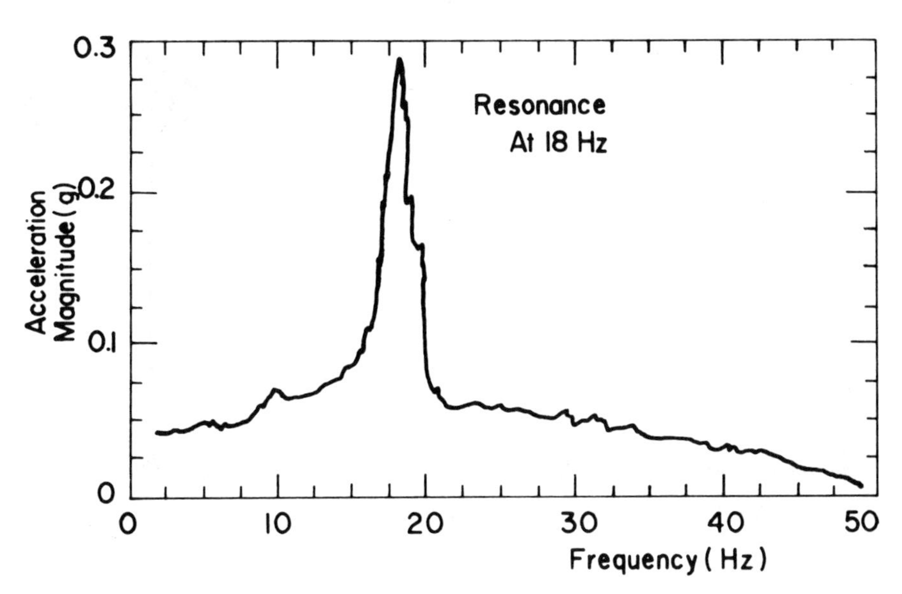

Figure 9A-3. Fourier Analysis Results

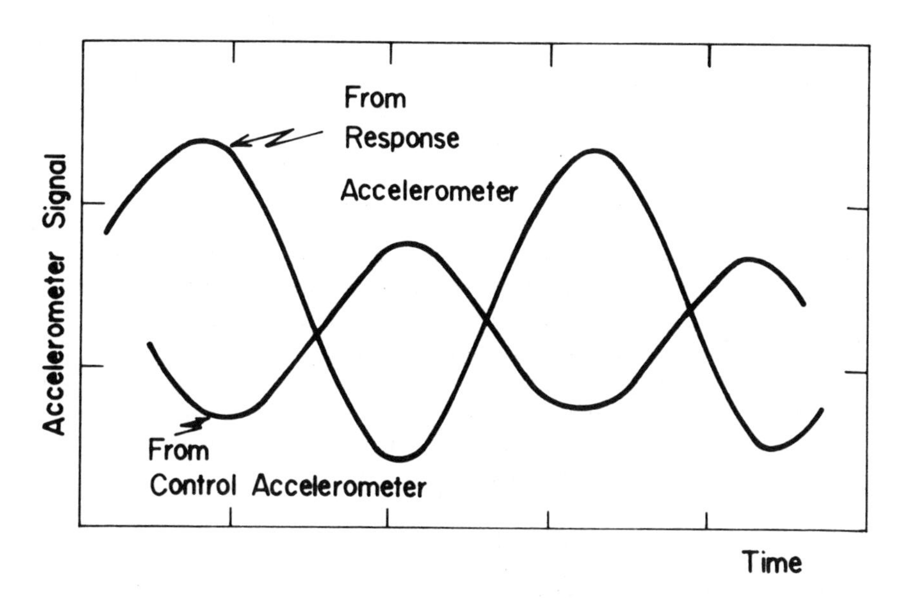

Figure 9A-4. Oscilloscope Records

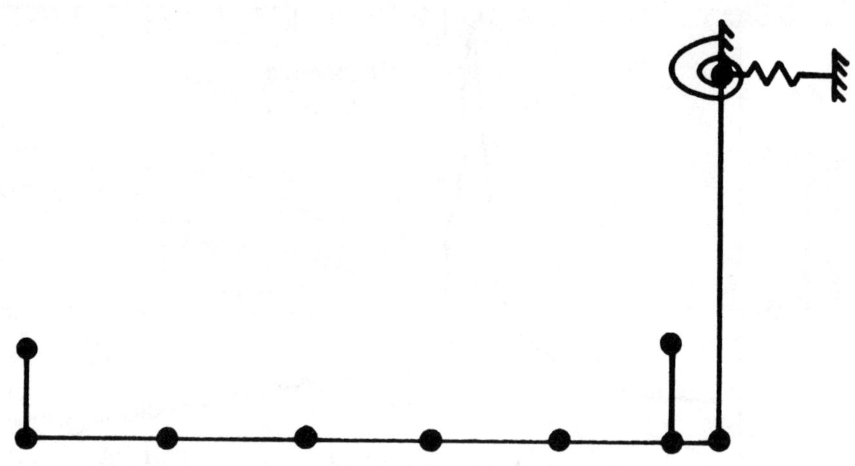

Figure 9A-5. A Finite-Element Model for the Solenoid Valve Unit

using flexible-beam elements. The disc and the end plates were represented by using bending-plate elements. Flexibility in the supporting pipeline was represented by using direct and torsional spring elements. Various auxiliary nonstructural components attached to the solenoid valve unit were accounted for by means of concentrated-mass elements at appropriate nodes of the finite-element model. The model is shown schematically in figure 5 [Figure 9A-5]. Element and node descriptions are tabulated in tables [give table nos.]. Geometric and material parameters (cross-sectional area, moments of inertia, tension, bending and torsional stiffness values, elastic moduli, plate thickness, and so on) are given in tables [give table nos.].

4.2: *Modal Analysis*

A normal mode analysis was performed on the finite-element model of the solenoid valve unit, using the dynamic structural-analysis program [give program name and reference]. The analytical natural frequencies below 50 Hz are listed and compared with the test values in table [give table no.]. The analytical mode shapes are compared with the test results in figure 6 [Figure 9A-6]. The error in the analytical value of the fundamental natural frequency is less than 10 percent. Agreement in the analytical and test mode shapes is also satisfactory. Hence, the analytical model is verified.

4.3: *Seismic Qualification Analysis*

A static stress analysis was performed using seismic inertia loads for OBE and SSE conditions separately, nozzle loads at flow inlets and outlets, dead loads, and other normal operating loads. The axial force, shear force, torsional moment, and bending moment values corresponding to these loads at various loading locations are given in table [give table no]. The yield stress of the material was taken as 36,000 psi, and the allowable stress was taken as 60 percent of this value. The combined maximum stresses are significantly below the allowable stress values.

For time-history dynamic analysis, the input motions at the pipe-support locations were developed using sine-beat components to satisfy the levels specified by the required input motion (RIM) curve shown in figure [give figure no]. Three orthogonal input excitations were applied at each support location. The dynamic-response analysis was performed using the computer program [give program name and reference]. Some representative response

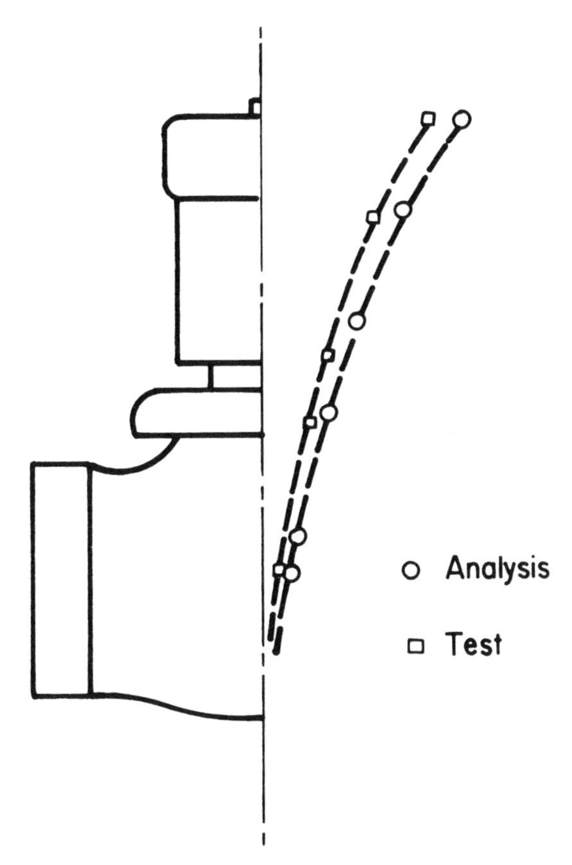

Figure 9A-6. Comparison of Analytical and Test Mode Shapes of the Solenoid Valve Unit

time histories at crucial locations of the valve system and the corresponding response spectra are given in figure [give figure no.] These will be used for comparison with the test results.

Section 5: Qualification Testing

Single-frequency seismic qualification tests were performed using the test setpup illustrated in figure [give figure no.]. Test-instrument check-out log sheets are included in appendix 2. The valve was pressurized, and operability was checked by performing several open-close valve-operation cycles and observing the pressure drop across the inlet and the outlet before conducting the shaker tests.

The OBE testing was performed as follows: Two sine sweeps were conducted at each of three orthogonal axes of the valve system, at 70 percent of the SSE required input motion (RIM) amplitude levels [give figure no.] at a sweep rate of 1 octave/min. One sweep was conducted with the valve in the fully open position, the other sweep with the valve in the fully closed position. Each sweep covered the frequency cycle 1 to 35 to 1 Hz. The operability of the valve system was checked during at least one OBE test cycle.

The SSE test was carried out, after the OBE test, along each axis. The SSE test consisted of applying a series of sine beats over the frequency interval 1-35 Hz at one-third-octave frequency values (that is, 1, $2^{1/3}$, $2^{2/3}$, 2, . . . Hz). The number of oscillations per beat was 10. The dwell times used for each frequency are given in table [give table no.]. Dwell times were taken to be at least 15 sec. In addition, a sine-beat dwell was conducted at each resonance of the valve system in the 1-35 Hz interval. The amplitude levels of the sine beats at each dwell frequency satisfied the SSE RIM curve shown in figure [give figure no.]. Two open-close valve cycles were performed at each dwell frequency, and the operability (pressure drop across inlet and outlet) was monitored. In addition, structural integrity was checked during and after each test.

The test-input time histories response time histories, and response spectra at crucial locations of the valve system are shown in figures [give figure nos.].

Section 6: Conclusion

Analytical and single-frequency-test seismic qualification results reveal that structural integrity and functional operability were maintained during and after the application of specified OBE inputs followed by an SSE input. The solenoid valve unit is therefore seismic qualified for use in safety-related operations at the Beta Nuclear Power Plant.

References

1. x x x
2. x x x

Appendixes

Appendix 1: Approved Qualification Procedure
Appendix 2: Instrument Data and Log Sheets
Appendix 3: Sine-Beat Acceleration Test Input Data
Appendix 4: Strip-Chart Records of Accelerometer Signals

Appendix 9B: Sample Seismic Qualification Report Review

Cover Letter:

[Reviewer Letterhead]

Your Ref.: ______ Sept. 10, 1982

Our Ref.: ______

File: ______

Project Engineer: ______

[Purchaser Address]

Dear Sir:

BETA NUCLEAR POWER PLANT
Review of Seismic Category I Analysis of 10″ Stainless-Steel Gate Valve and Gear Actuator, Report No. ______, Jan 1, 1982

The subject seismic qualification report was reviewed by us to evaluate it with respect to the guidelines provided by the [give purchaser specifications reference]. Provided that the comments given in the attachment are satisfactorily resolved, the qualification analysis procedure would be found adequate.

Please contact us in case of any questions regarding the comments made in the attachment.

Very Truly Yours,
[Reviewing Company]

____

Manager
cc: x x x
x x x

Page 2 (Attachment):

1. Analysis of Section 1-1, shown in figure 1 [figure 9B-1], as summarized on page [give page no.] of the report, appears to be in error. First, simply supported ends do not represent realistic valve-pipeline interface conditions. Moments and slope constraints exist at these locations. An assumption of equal reactions (R) at the two joints is also unrealistic. In any event, the joint loads, as obtained from the analysis, do not satisfy the dynamic equilibrium (Newton's second law). Furthermore, M_{max} should contain a $w_5 y_5$ term, and T_{max} should not contain the $W_1 X_4$ term. Consequently, $M_{max} \neq T_{max}$. Also, there must be axial end forces to resist seismic loads. A more realistic analysis is given below.

Initially, the maximum seismic inertia loads in the x and y directions are replaced by equivalent loads W in the pipe-run and body-run directions at the point of intersection of the axes, along with moments T_{max} and M_{max}, as shown in figure 1 [Figure 9B-1]. The equivalent loads are given by

$$W = (W_1 + W_2 + W_3 + W_4 + W_5) g_{max} + O_p$$

$$T_{max} = (W_1 y_1 + W_2 y_2 + W_3 y_3 + W_4 y_4 + W_5 y_5) g_{max} + O_p y_1$$

$$M_{max} = (W_1 y_1 + W_2 y_2 + W_3 y_3 + W_4 y_4 + W_5 y_5 + W_1 x_4) g_{max} + O_p y_1$$

Pipe-run end loads are assumed to be equal. For dynamic equilibrium, they

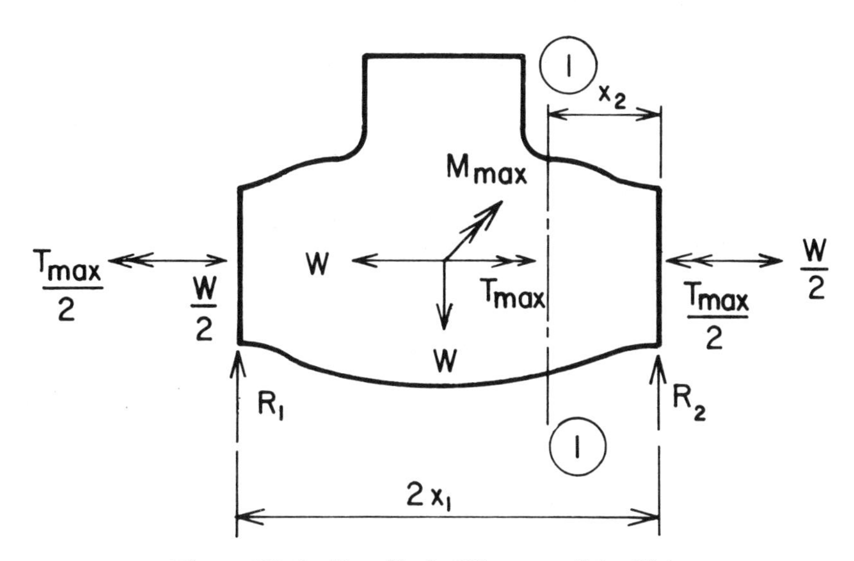

Figure 9B-1. Free-Body Diagram of the Valve

should be equal to $W/2$. Also, the end moments are neglected. Then, for dynamic equilibrium,

$$R_1 = \frac{W}{2} - \frac{M_{max}}{2x_1}$$

$$R_2 = \frac{W}{2} + \frac{M_{max}}{2x_1}$$

At Section 1-1, thrust $= W/2$ (identical to the value in the report), shear force $= R_2$, Bending moment $= R_2 x_2$, Torque $= T_{max}/2$ (assume to be equally shared at the ends).

2. On page [give page no.], the moment expression should contain the $-W_5(y_8 + y_5)$ term in order to account for seismic loading caused by the disc.

3. Give proper reference for the design specifications used in the analysis (for example, for values in table [give table no.]). Also, literature references should be given (for example, for stem factor and stem thread on page [give page no.]).

4. Expressions for thrust, shear force, torque, and bending moment, given on pages [give page nos.] are obtained directly from dynamic equilibrium. There is no need to assume a cantilever-beam configuration, as suggested on page [give page no.].

5. Justify lumping one-quarter (yoke and stem mass) at the centroid of the gear operator in obtaining the moment equation for Section X-X on page [give page no.]. Also, seismic load caused by disc mass should be included in this moment expression, unless the stem is also sectioned by X-X (which is not the case, as seen from the sketches given in the report).

6. The values given in table [give table no.] are not consistent with the sketch shown on page [give page no.]. For instance, according to the sketch, number of bolts = 6, a = b = 3.1″, $I_3 = 1.5I_1$, and $I_2 = 2I_1$.

7. Bolts subjected to both axial tension and shear should be evaluated based on the shear-tension interaction equation:

$$\left\{\frac{f_t}{F_t}\right\}^2 + \left\{\frac{f_v}{F_v}\right\}^2 \leq 1$$

in which f_t and f_v are actual tension and shear stresses, and F_t and F_v are the allowable tension and shear-stress values.

8. The shear-stress equation on page [give page no.] should read

$$= \frac{\text{shear}}{2\pi r_0 T} + \frac{\text{torque}}{2\pi r_0^2 T}$$

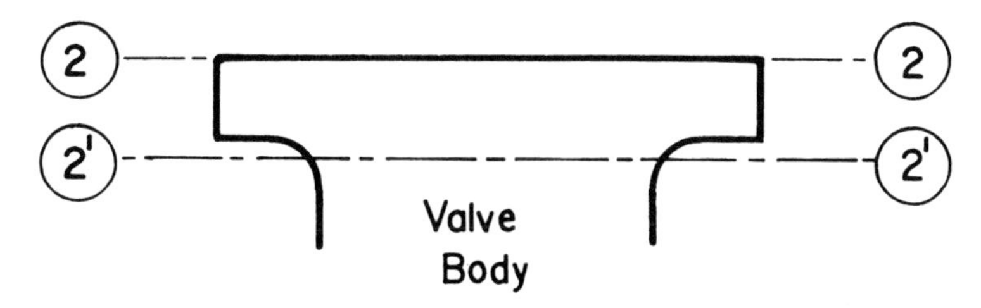

Figure 9B-2. Valve Body-Section Nomenclature

to account for the true area of shear

9. According to table [give table no.], the valve-body cross-sectional properties are identical for Sections 2-2 and 3-3. This is not consistent with figure [give figure no.]

10. In the first paragraph of page [give page no.], it is stated that the force results previously obtained did not include the effects caused by pipe reaction. This does not follow from the analysis on page [give page no.]. What does reaction *R* represent in that case? Also, how would the dynamic equilibrium be maintained in that case? (See comment #1.)

11. From the development on pages [give page nos.], it appears that Section 2-2 is not a section through the bolted joint (flange face) but is a section just below the flange (Section 2′-2′), as shown in figure 2 [figure 9B-2].

12. The finite-element model given on page [give page no.] should be verified by testing. This may be done by a resonance search to determine the fundamental resonant frequency and the corresponding mode shape. If the test results agree with the analytical results from the finite-element model (within 10 percent error), then the model is acceptable. One concern is the assumption of fully restrained translation and rotation boundary conditions. When realistic flexibility properties of the supporting pipeline are incorporated the estimated natural frequencies could become significantly lower.

13. The finite-element program [give program name] and other computer programs [give program names] should be verified on the same computer by solving baseline problems for which the solutions are already known.

14. The conclusion that the valve design is ensured of operability during and after the prescribed seismic disturbances, as noted on page [give page no.], must be justified. It should be shown that the components in the valve that were not analyzed are not important for guaranteeing its functional operability. In particular, components such as the bevel-gear mechanism and considerations of gate-valve disc seating should be analyzed. It should be shown that structural integrity of the analyzed components alone will ensure functional operability. Otherwise, operability testing would be required.

15. Justify the assumption that the two end moments are equal in the analysis of yoke bending, given on page [give page no.].

10 Optimization of Seismic Qualification Tests

The IEEE-recommended practice for seismic qualifications favors the use of three uncorrelated excitation inputs applied simultaneously in the vertical and two orthogonal horizontal directions of the test object, to provide a realistic characterization of earthquake motions. In view of the difficulty of realizing such triaxial testing, the IEEE standard also recommends a biaxial alternative, with simultaneous inputs in vertical and principal horizontal axes. In general, two or four tests are performed, depending on whether biaxial inputs are uncorrelated or correlated. A more popular test, however, used extensively by seismic qualification test laboratories, is the symmetrical rectilinear test (see chapter 7). In this case, a single excitation is first applied along the line of symmetry with respect to the equipment principal axes. This is followed by three more tests, with the equipment rotated through 90°, 180°, and 270°, respectively, about the vertical axis, starting from the initial orientation. The test-input intensity along the drive axis is larger than the intensity of the three individual components by a factor of $\sqrt{3}$.

The sequence of four tests in different test-object orientations that is employed in symmetrical rectilinear testing can produce significant fatigue (overtesting), at least in some directions. In this chapter, we shall develop a rectilinear test procedure whereby a single-degree-of-freedom test performed in a certain optimal direction with least excitation intensity can subject the critical components of the equipment to the levels of acceleration that will maximize the risk of component failure.[1] This procedure has the advantage of replacing the sequence of four tests by a single test, so that overtesting of the test object is avoided. To apply this technique, frequency-response functions at crucial locations of the test object should be available by either prior testing or analysis.

Excitation-Input Representation

The classic response-spectrum concept has been widely used to represent seismic motions. The response spectrum is based on the peak response of a linear, hypothetical, single-degree-of-freedom, damped system subjected to a representative seismic time history (see chapter 5). Clearly, the response spectrum lacks direct physical interpretation in the case of multi-degree-of-freedom and distributed-parameter systems for which the peak response cannot be directly determined from the response spectrum. If the modal-participation factors are known, however, a conservative upper bound for the peak response can be determined. Since only the peak values of the time response are considered, the correlation with the actual time-history response is limited, and the cyclic (fatigue) behavior requires separate treatment. Furthermore, the classic response-spectrum concept is based on deterministic system theory, rather than on a more appropriate random characterization of seismic motions and the system.

Seismic disturbances essentially are random processes, and a statistical description would be most appropriate in representing them. In view of this, the theoretical development in this chapter is based on a stochastic foundation. Once the test procedure is developed, the response spectra may be used, if so desired, to represent the actuator-input signals during the actual seismic qualification test. The specific statistical representation used in this development is the power spectral density (psd) of the excitation input. For the psd to exist, the signal must be covariance-stationary (see chapter 5). This property does not hold strictly for earthquake motions, because they possess the time-decaying characteristic, with a definite beginning and an ending. Their correlation functions depend on the time origin as well as on the time interval, but it is possible to construct a stationary process that retains the important characteristics, such as the intensity and the frequency content of a given seismic motion. To accomplish this, a significant segment of the actual motion over which the major energy content occurs is chosen. The statistical characteristics of the random process are not expected to vary significantly over this time interval. Consequently, the selected record segment may be thought of as a finite-length cut from a stationary sample function. Next, simulated stationary sample functions that have approximately the same statistical properties as the chosen principal sample segment are repeated continuously before and after the principal segment. The result is a sample function from a stationary process. The stationary processes normally encountered in engineering practice are at least weakly ergodic. In numerical terms, the spectral error introduced by this process of input synthesis is negligible, provided that the record length used in digital processing is not much larger than the synthesized record.

For a realistic seismic qualification, the equipment should be tested for the seismic floor motions at its actual location in the building. Because of the structural dynamics of the building, the floor motions will have different characteristics from the ground motions. In addition, because of coupled motions of the building, the floor response in the principal directions correlates to some extent even if the ground motions in these directions are uncorrelated.

By analyzing the strong-motion part of a transient disturbance record over which the major energy content occurs as if it were a covariance-stationary record, it is possible to extract the required information to some degree of accuracy. The psd curves from many such records that are representative of the vibrations of interest are then plotted on the same graph paper and, using engineering judgment, a single psd curve that could conservatively represent the collection of time-history records is obtained. This is termed the required psd. The procedure is demonstrated in figure 10-1. Then, for a given vibration test, the input time history that is used in the vibratory-motion simulator is generated by synthesizing a signal whose psd (test psd) envelops the required psd over the significant frequency range.

Test-Severity Measure

The optimization problem is formulated in terms of a test-severity measure (TS), which reflects the objective of the test. This measure is a function of the parameters that are variable during the process of optimization. The primary purpose of a seismic qualification test is to ensure that component failure does not occur during normal operation in a strong seismic environment. Consequently, it is desirable to maximize the risk of component failure during testing, so that, in service, the chances of failure are less. In the present formulation, the test-severity measure employed is

$$TS = \sum_{i=1}^{N} p_i \Phi_i(\omega) \tag{10.1}$$

in which N is the number of critical components in the equipment, $\Phi_i(\omega)$ is the acceleration psd at the ith component along its direction of maximum sensitivity, and p_i is the weighting parameter for the ith location. It is assumed that the components fail independently, and there is no hardware redundancy among the components selected in the test-severity measure. If the direction of maximum sensitivity is not known for a given component, three orthogonal directions should be used for that location. Consequently, each such component is represented by three terms in the summation of equation (10.1).

Among a set of test-input directions, the one that maximizes the TS is considered the best configuration, because it represents the worst test severity for a given level of excitation input. Consequently, the problem of input-direction design for seismic qualification testing corresponds to the optimization problem of maximizing the test-severity measure.

Optimal Solution

The cartesian frame $OXYZ$ is fixed in the equipment, as shown in figure 10-2. The origin O is the base point at which the rectilinear input acceleration

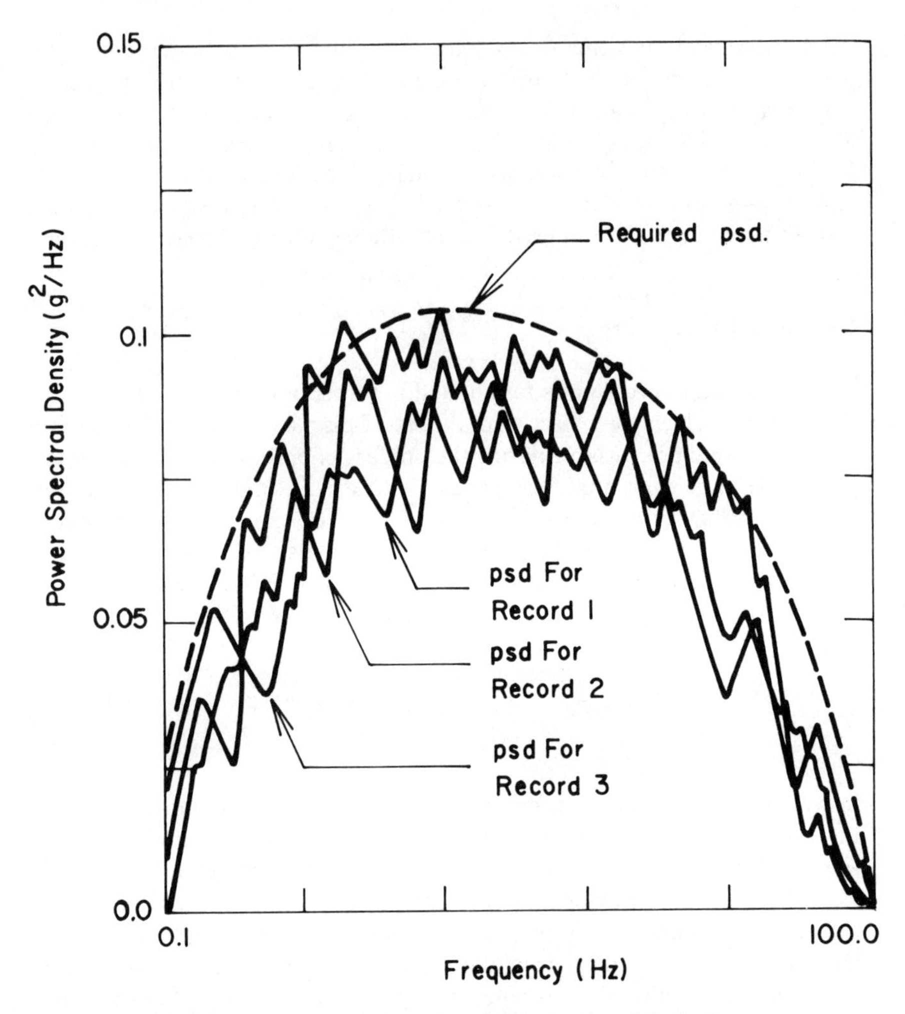

Figure 10-1. Required psd Generation Method

$u(t)$ is applied. The direction consines α_X, α_Y, and α_Z of the excitation-input vector are given by $\alpha_X = \cos\theta_X$, $\alpha_y = \cos\theta_y$, and $\alpha_Z = \cos\theta_Z$. The Fourier transform of acceleration time history $y_i(t)$ at the ith component is given by

$$\mathscr{F}[y_i(t)] = [\alpha_X H_{Xi}(\omega) + \alpha_Y H_{Yi}(\omega) + \alpha_Z H_{Zi}(\omega)]\mathscr{F}[u(t)] \qquad (10.2)$$

in which H_{Xi}, H_{yi}, and H_{Zi} are the frequency-response functions between the output $y_i(t)$ and the input components in the X, Y, and Z directions, respectively, and $\mathscr{F}[\ ]$ denotes the Fourier integral transform operator. This

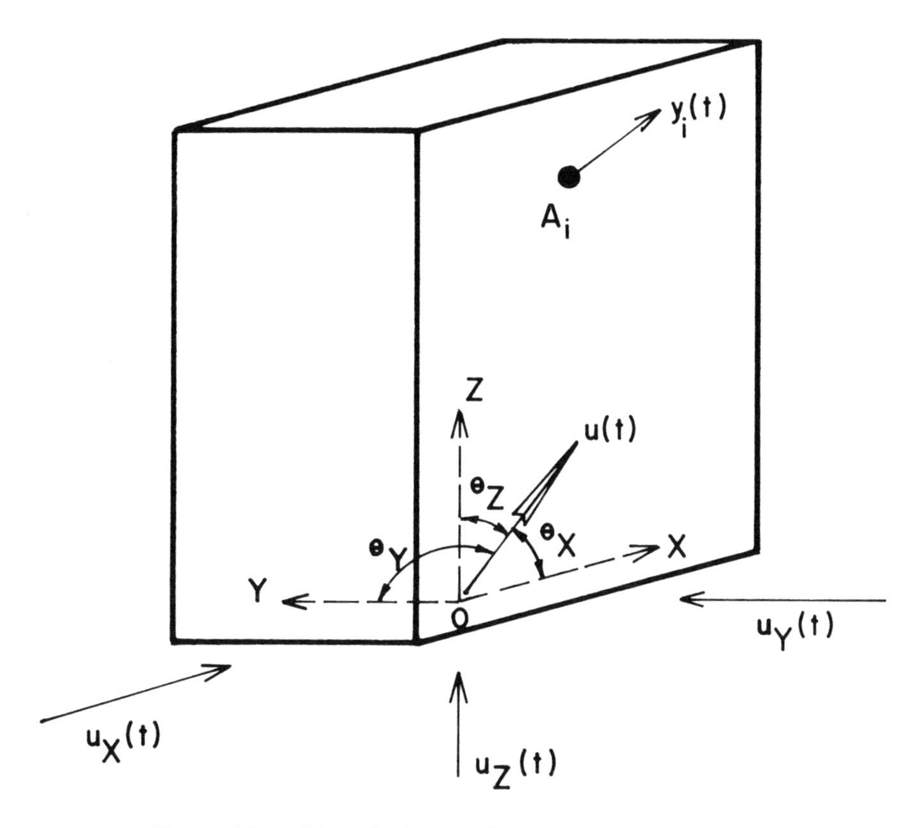

Figure 10-2. Test Excitation-Response Nomenclature

input-output configuration is shown by the block diagram in figure 10-3. With the usual assumptions of linear, time-invariant system behavior, the psd of $y_i(t)$ is given by

$$\Phi_i(\omega) = | \alpha_X H_{Xi}(\omega) + \alpha_Y H_{Yi}(\omega) + \alpha_Z H_{Zi}(\omega) |^2 \Phi_{uu}(\omega) \quad (10.3)$$

in which $\Phi_{uu}(\omega)$ is the psd of the input acceleration $u(t)$.

The real and imaginary parts of the frequency-response function are defined by

$$H_{XI}(\omega) = R_{Xi}(\omega) - jI_{Xi}(\omega)$$

$$H_{Yi}(\omega) = R_{Yi}(\omega) - jI_{Yi}(\omega)$$

$$H_{Zi}(\omega) = R_{Zi}(\omega) - jI_{Zi}(\omega). \quad (10.4)$$

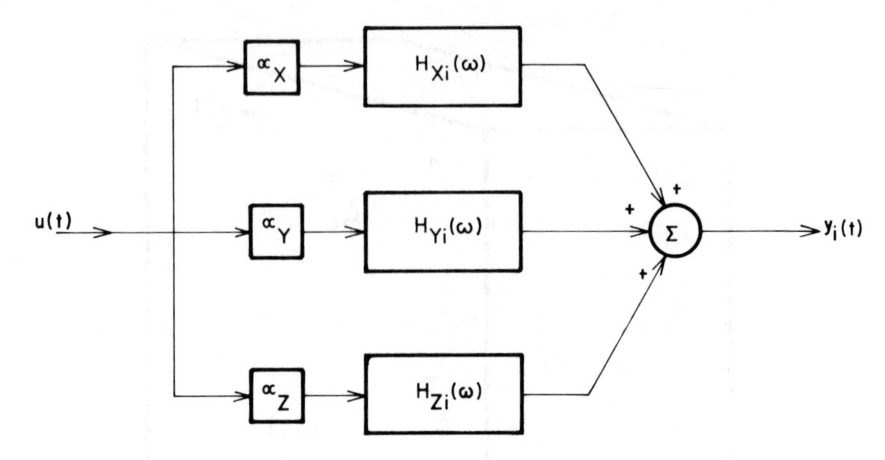

Figure 10-3. Combination of Responses Caused by Orthogonal Excitation Components in Rectilinear Testing

in which $j = \sqrt{-1}$. By substituting equation (10.3) in equation (10.1) and using equation (10.4), we obtain

$$J = \sum_{i=1}^{N} p_i[\alpha_X R_{Xi}(\omega) + \alpha_Y R_{Yi}(\omega) + \alpha_Z R_{ZI}(\omega)]^2$$

$$+ \sum_{i=1}^{N} p_i[\alpha_X I_{Xi}(\omega) + \alpha_Y I_{Yi}(\omega) + \alpha_Z I_{Zi}(\omega)]^2 \qquad (10.5)$$

subject to

$$\alpha_X{}^2 + \alpha_Y{}^2 + \alpha_Z{}^2 = 1 \qquad (10.6)$$

Note that the common term $\Phi_{uu}(\omega)$ has been omitted from the test-severity measure in equation (10.5).

Solution Using Matrix-Eigenvalue Formulation

The optimization consists of selecting the direction cosine vector $\boldsymbol{\alpha} = [\alpha_X, \alpha_{Yb}, \alpha_Z]^T$ so as to maximize the test-severity measure in equation (10.5), subject to the condition in equation (10.6). This constrained optimization is accomplished by employing the Lagrange multiplier method. The augmented function to be maximized is

$$\widetilde{J} = \sum_{i=1}^{N} p_i[\alpha_X R_{Xi}(\omega) + \alpha_Y R_{Yi}(\omega) + \alpha_Z R_{Zi}(\omega)]^2$$

$$+\sum_{i=1}^{N} p_i[\alpha_X I_{Xi}(\omega) + \alpha_Y I_{Yi}(\omega) + \alpha_Z I_{Zi}(\omega)]^2$$

$$+ \lambda(1 - \alpha_X{}^2 - \alpha_Y{}^2 - \alpha_Z{}^2) \qquad (10.7)$$

in which λ is the Lagrange multiplier. At this stage, it is shown that the optimization problem can be reduced to a matrix-eigenvalue problem. The eigenvalues of a certain real symmetrical matrix are the stationary values of $\widetilde{J}$. The normalized eigenvector corresponding to the maximum eigenvalue gives the optimal direction cosines for the test input. To show this, the necessary conditions for optimization

$$\frac{\partial \widetilde{J}}{\partial \alpha_X} = \frac{\partial \widetilde{J}}{\partial \alpha_Y} = \frac{\partial \widetilde{J}}{\partial \alpha_Z} = 0$$

are determined. The resulting three equations may be written as the matrix equation

$$[\mathbf{P} - \lambda \mathbf{I}]\boldsymbol{\alpha} = \mathbf{0} \qquad (10.8)$$

in which

$$\mathbf{P} = \sum_{i=1}^{N} p_i \begin{bmatrix} (R_{Xi}{}^2 + I_{Xi}{}^2) & (R_{Xi}R_{Yi} + I_{Xi}I_{Yi}) & (R_{Zi}R_{Xi} + I_{Zi}I_{Xi}) \\ (R_{Xi}R_{Yi} + I_{Xi}I_{Yi}) & (R_{Yi}{}^2 + I_{Yi}{}^2) & (R_{Yi}R_{Zi} + I_{Yi}I_{Zi}) \\ (R_{Zi}R_{Xi} + I_{Zi}I_{Xi}) & (R_{Yi}R_{Zi} + I_{Yi}I_{Zi}) & (R_{Zi}{}^2 + I_{Zi}{}^2) \end{bmatrix} \qquad (10.9)$$

and $\mathbf{I}$ is the identity matrix. The matrix $\mathbf{P}$ is termed the test matrix. For nontrivial $\boldsymbol{\alpha}$, the determinant of the left-hand-side matrix in equation (10.8) must vanish. The corresponding solutions for λ are, in fact, the eigenvalues of $\mathbf{P}$. For each distinct solution for λ, the direction cosine vector $\boldsymbol{\alpha}$ is determined up to a single unknown parameter. The unknown parameter is obtained using the normalizing relation equation (10.6). These solutions $\boldsymbol{\alpha}$ are the normalized eigenvectors of $\mathbf{P}$.

Significance of Eigenvalues

To verify that the eigenvalues of **P** are the values of J that correspond to the stationary points of $\tilde{J}$, the three equations denoted by equation (10.8) are multiplied by α_X, α_Y, and α_Z, respectively, and the resulting equations are added together. This gives

$$\sum_{i=1}^{N} p_i(\alpha_X R_{Xi} + \alpha_Y R_{Yi} + \alpha_Z R_{Zi})\alpha_X R_{Xi}$$

$$+ \sum_{i=1}^{N} p_i(\alpha_X I_{Xi} + \alpha_Y I_{Yi} + \alpha_Z I_{Zi})\alpha_X I_{Xi}$$

$$+ \sum_{i=1}^{N} p_i(\alpha_X R_{Xi} + \alpha_Y R_{Yi} + \alpha_Z R_{Zi})\alpha_Y R_{Yi}$$

$$+ \sum_{i=1}^{N} p_i(\alpha_X I_{Xi} + \alpha_Y R_{Yi} + \alpha_Z R_{Zi})\alpha_Y I_{Yi}$$

$$+ \sum_{i=1}^{N} p_i(\alpha_X R_{Xi} + \alpha_Y R_{Yi} + \alpha_Z R_{Zi})\alpha_Z R_{Zi}$$

$$+ \sum_{i=1}^{N} p_i(\alpha_X I_{Xi} + \alpha_Y I_{Yi} + \alpha_Z I_{Zi})\alpha_Z I_{Zi}$$

$$- \lambda(\alpha_X^2 + \alpha_Y^2 + \alpha_Z^2) = 0$$

In vew of equation (10.6), the desired result is obtained following straightforward manipulation of the last expression:

$$\lambda = \left[\sum_i p_i(\alpha_X R_{Xi} + \alpha_Y R_{Yi} + \alpha_Z R_{Zi})^2 + \sum_i p_i(\alpha_X I_{Xi} + \alpha_Y I_{Xi} + \alpha_Z I_{Zi})^2 \right]_{st} \tag{10.10}$$

The subscript $[\ ]_{st}$ denotes the stationary values, because the equations (10.8) correspond to the stationary values of $\tilde{J}$. It follows that the maximum eigenvalue of the test matrix **P** corresponds to the constrained global maximum J_{max} of J, subject to equation (10.6). The associated normalized eigenvector gives the optimal direction cosines $\boldsymbol{\alpha}_{opt}$.

Existence of a Global Maximum

In the foregoing development it has been assumed that a well-defined global maximum exists for J, subject to equation (10.6). It is established in this section that the assumption holds for the present problem.

The parameter vector $\boldsymbol{\alpha}$ is defined in the finite domain $\mathscr{D}$ described by equation (10.6), which is the unit sphere centered at the origin of the cartesian frame $(OXYZ)$. The objective function J is a continuous function of $\boldsymbol{\alpha}$ and has continuous derivatives, as is evident from equation (10.5). Furthermore, in the domain $\mathscr{D}$, the following inequality is satisfied:

$$J = \sum_{i=1}^{N} p_i \mid \alpha_X H_{Xi} + \alpha_Y H_{Yi} + \alpha_Z H_{Zi} \mid^2$$

$$\leq \sum_{i=1}^{N} p_i (\mid \alpha_X \mid \mid H_{Xi} \mid + \mid \alpha_Y \mid H_{yi} \mid + \mid \alpha_Z \mid \mid H_{Zi} \mid)^2$$

$$< \sum_{i=1}^{N} p_i (\mid H_{Xi} \mid + \mid H_{Yi} \mid + \mid H_{Zi} \mid)^2 \qquad (10.11)$$

Consequently, a conservative upper bound exists for J. This guarantees a well-defined global maximum. Note that the formulation itself assumes that the component-transfer functions $H_{Xi}(\omega)$, $H_{Yi}(\omega)$, and $H_{Zi}(\omega)$ have finite magnitudes. This is necessarily satisfied in the practical range of frequencies. As a design requirement for the equipment, the peak values of the transfer functions must not be large. This is usually guaranteed by adequate system damping.

Minimum Excitation Intensity

To determine the minimum allowable intensity of the excitation input, the proposed test procedure is compared with the three-degree-of-freedom qualification test with uncorrelated inputs. Suppose that the uncorrelated excitation inputs $u_X(t)$, $u_Y(t)$, and $u_Z(t)$ are applied simultaneously in X, Y, and Z directions at the equipment-support location (see figure 10-2). The Fourier transform of the time response $\tilde{y}_i(t)$ at the ith component is given by

$$\mathscr{F}[\tilde{y}_i(t)] = H_{Xi}(\omega)\mathscr{F}[u_X(t)] + H_{Yi}(\omega)\mathscr{F}[u_Y(t)] + H_{Zi}(\omega)\mathscr{F}[u_Z(t)] \tag{10.12}$$

The psd of $\tilde{y}_i(t)$ may be expressed as

$$\widetilde{\Phi}_i(\omega) = |H_{Xi}(\omega)|^2 \Phi_{u_X u_X}(\omega) + |H_{Yi}(\omega)|^2 \Phi_{u_Y u_Y}(\omega) + |H_{Zi}(\omega)|^2 \Phi_{u_Z u_Z}(\omega) \tag{10.13}$$

The criterion for assuring that, in the proposed rectilinear test, the disturbance at the component expected to fail is not less than that in the three-degree-of-freedom test with uncorrelated input is

$$\sum_{i=1}^{N} p_i \Phi_i(\omega) \geq \sum_i p_i \widetilde{\Phi}_i(\omega) \tag{10.14}$$

The inputs $u(t)$, $u_X(t)$, $u_Y(t)$, and $u_Z(t)$ must have identical probability distributions, because they are necessarily generated by the same random process. Their intensities need not be identical, however. Consequently,

$$\Phi_{uu}(\omega) = k^2 \Phi(\omega) \tag{10.15}$$

$$\Phi_{u_X u_X}(\omega) = a_X^2 \Phi(\omega) \tag{10.16}$$

$$\Phi_{u_Y u_Y}(\omega) = a_Y^2 \Phi(\omega) \tag{10.17}$$

$$\Phi_{u_Z u_Z}(\omega) = a_Z^2 \Phi(\omega) \tag{10.18}$$

in which k, a_X, a_Y, and a_Z are the corresponding scaling factors for the input intensities. The minimum intensity $k_{\min}$ is determined by substituting equations (10.3) and (10.13) into equation (10.14), in conjunction with equations (10.15) to (10.18). Finally, in view of equation (10.10),

$$k_{\min} = \left[\Sigma p_i\{(R_{Xi}^2 + I_{Xi}^2)a_X^2 + (R_{Yi}^2 + I_{Yi}^2)a_Y^2 + (R_{Zi}^2 + I_{Zi}^2)a_Z^2\}/\lambda_{\max} \right]^{1/2} \tag{10.19}$$

in which $\lambda_{\max}$ is the maximum eigenvalue of the test matrix $\mathbf{P}$. For most situations, it is sufficient to use $a_X = a_Y = a_Z = 1$. If the critical frequency band is less than 3.5 Hz, however, it is recommended that a_Z be given a value smaller than unity but greater than 0.67.

Seismic Qualification Procedure

Determination of the Critical Frequency

The transfer functions $H_{Xi}(\omega)$, $H_{Yi}(\omega)$, and $H_{Zi}(\omega)$ are generally dependent on the frequency of excitation ω. Consequently, the value of the test-severity measure J depends on the frequency point at which it is computed. For a given transfer function, the critical magnitude is not necessarily its peak. If the peak falls outside the frequency band of interest, for example, the highest magnitude in the critical-frequency band is more significant than the peak value of the transfer function.

The optimal test consists of first, computing the optimal test-input direction $\boldsymbol{\alpha}_{\text{opt}}$ for each test frequency ω and then, adjusting the shaker-actuator orientation accordingly. This requires a capability of automatic adjustment of the actuator orientation continuously, depending on the test frequency. This may not be feasible with conventional shaker tables. Furthermore, in multifrequency and wide-band random excitations, it is not possible to select a single-frequency component of interest at a given instant.

A more convenient procedure would be to select one frequency that is most suitable for computing the optimal test-input direction. The frequency at which the optimal test parameters are computed must reflect the influence of the severity of the frequency-response functions corresponding to the individual response components in various orientations. Also, the critical frequencies of the components must be weighted according to their likelihood of failure during the test. A satisfactory way of accomplishing this is as follows.

A critical frequency is assigned to each transfer-function component. This is typically the resonant frequency if it falls within the frequency band of interest. Otherwise, a choice must be made using engineering judgment, past experience, and available data. Let $\bar{\omega}_{Xi}$, $\bar{\omega}_{Yi}$, and $\bar{\omega}_{Zi}$ be the critical frequencies of the ith component associated with the frequency-response functions H_{Xi}, H_{Yi}, and H_{Zi}. The corresponding maximum values of the test-severity measure in equation (10.5)—that is, the maximum eigenvalues of the test matrix $\mathbf{P}$—are λ_{Xi}, λ_{Yi}, and λ_{Zi}, respectively. Then, the critical test frequency $\bar{\omega}$ is determined using

$$\bar{\omega} = \frac{1}{\sum_{n=1}^{N} p_n} \sum_{i=1}^{N} p_i \left[\frac{\lambda_{Xi}\bar{\omega}_{Xi} + \lambda_{Yi}\bar{\omega}_{Yi} + \lambda_{Zi}\bar{\omega}_{Zi}}{\lambda_{Xi} + \lambda_{Yi} + \lambda_{Zi}} \right] \tag{10.20}$$

This frequency favors the component that is most likely to fail. It is also biased toward the excitation-input orientation that produces the largest response in the frequency range of interest. The optimal direction cosines and the minimum-input-intensity scaling factor for performing the qualification test are computed using $\bar{\omega}$.

Test Plan

The major steps of the rectilinear seismic qualification test procedure developed in this chapter are as follows.

Step 1 Assign weighting parameters to each component or component orientation.

Step 2 Locate accelerometers at these component locations or orientations.

Step 3 Using a frequency-response test, determine the frequency-response functions $H_{Xi}(\omega)$, $H_{Yi}(\omega)$ and $H_{Zi}(\omega)$ at each critical component location or orientation.

Step 4 By examining the frequency-response functions, choose the critical frequencies $\bar{\omega}_{Xi}$, $\bar{\omega}_{Yi}$, and $\bar{\omega}_{Zi}$ and note the corresponding real and imaginary parts of the transfer functions. Form the test matrix **P** [equation (10.9)] for each chosen frequency, and compute the corresponding maximum eigenvalues λ_{Xi}, λ_{Yi}, and λ_{Zi}.

Step 5 Compute the test frequency $\bar{\omega}$ using equation (10.20). Note the corresponding values of the transfer functions. Form the test matrix **P**, and compute its maximum eigenvalue $\lambda_{\max}$ and the associated normalized eigenvector $\mathbf{a}_{\text{opt}}$.

Step 6 Using equation (10.19), compute the minimum intensity scaling factor $k_{\min}$.

Step 7 Synthesize the excitation input using the procedure for a standard three-degree-of-freedom test. Scale it by $k_{\min}$. Conduct the rectilinear test using this input in the $\boldsymbol{\alpha}_{\text{opt}}$ direction.

Illustrative Examples

Example 1

Consider a piece of equipment that possesses fully symmetrical dynamic characteristics with respect to the cartesian frame ($OXYZ$). For this case,

$$H_{Xi} = H_{Yi} = H_{Zi} = H_i$$

or

$$R_{Xi} = R_{Yi} = R_{Zi} = R_i \quad \text{and} \quad I_{Xi} = I_{Yi} = I_{Zi} = I_i$$

The test matrix for this case becomes

$$\mathbf{P} = \sum_{i=1}^{N} p_i |H_i|^2 \begin{bmatrix} 1 & 1 & 1 \\ 1 & 1 & 1 \\ 1 & 1 & 1 \end{bmatrix}$$

The eigenvalues of **P** are

$$\lambda_1 = \lambda_2 = 0, \qquad \lambda_3 = 3 \sum_{i=1}^{N} p_i |H_i|^2$$

Consequently,

$$J_{\max} = \lambda_{\max} = 3 \sum_{i=1}^{N} p_i |H_i|^2$$

The normalized eigenvector corresponding to $\lambda_{\max}$ is determined as

$$\boldsymbol{\alpha}_{\text{opt}} = \left[\frac{1}{\sqrt{3}}, \frac{1}{\sqrt{3}}, \frac{1}{\sqrt{3}} \right]^T$$

This gives the optimal direction of testing. The minimum-excitation-intensity scaling factor is obtained using equation (10.19):

$$k_{\min} = \left[\sum_{i=1}^{N} p_i |H_i|^2 (a_X^2 + a_Y^2 + a_Z^2)/\lambda_{\max} \right]^{1/2}$$

or

$$k_{\min} = [(a_X^2 + a_Y^2 + a_Z^2)/3]^{1/2}$$

If equal intensities are specified for the standard three-degree-of-freedom test with uncorrelated inputs, then $a_X = a_Y = a_Z = 1$. Then, $k_{\min} = 1$. This tells us that the optimal rectilinear test should be performed at the same intensity as the uncorrelated-input three-degree-of-freedom test. This intensity is lower by a factor of $\sqrt{3}$ than that used in the conventional symmetrical rectilinear test.

Example 2

Consider a piece of equipment that is sensitive to excitations in one direction (for example, the X direction) only. The corresponding test matrix is

$$\mathbf{P} = \sum_{i=1}^{N} p_i |H_{Xi}|^2 \begin{bmatrix} 1 & 0 & 0 \\ 0 & 0 & 0 \\ 0 & 0 & 0 \end{bmatrix}$$

The maximum eigenvalue of $\mathbf{P}$ is

$$\lambda_{\max} = \sum_{i=1}^{N} p_i |H_{Xi}|^2$$

The corresponding normalized eigenvector is

$$\boldsymbol{\alpha}_{\text{opt}} = [1,0,0]^T$$

From equation (10.19), with equal intensities being stipulated for the standard three-degree-of-freedom test with uncorrelated inputs, we obtain

$$k_{\min} = a_X = 1$$

Thus, the optimal rectilinear test input should be applied in the direction of sensitivity (the X direction for this example). Furthermore, the test intensity need not be greater than the intensity of one input component in the three-degree-of-freedom test with uncorrelated inputs. As in the previous example, an intensity reduction by a factor of $\sqrt{3}$ over the conventional symmetrical rectilinear test is obtained.

Example 3

Consider a simple test object represented by a single component that has three direct frequency-response functions $H_{xx}(\omega)$, $H_{yy}(\omega)$, and $H_{zz}(\omega)$ only. All the cross-frequency-response functions are assumed to vanish. Note that the response locations 1, 2, and 3 are replaced by the directions x, y, and z, respectively. The test-severity measure, as given by equation (10.1), is

$$\text{TS} = p_x \Phi x(\omega) + p_y \Phi_y(\omega) + p_z \Phi_z(\omega) \tag{10.21}$$

For the present model, the test matrix [equation (10.9)] becomes

$$\mathbf{P} = \begin{bmatrix} p_x |H_{xx}|^2 & 0 & 0 \\ 0 & p_y |H_{yy}|^2 & 0 \\ 0 & 0 & p_z |H_{zz}|^2 \end{bmatrix} \tag{10.22}$$

This has eigenvalues

$$\begin{aligned} \lambda_1 &= p_x |H_{xx}|^2 \\ \lambda_2 &= p_y |H_{yy}|^2 \\ \lambda_3 &= p_z |H_{zz}|^2 \end{aligned} \tag{10.23}$$

The corresponding eigenvectors are

$$\begin{aligned} \boldsymbol{\alpha}_1 &= [1, 0, 0]^T \\ \boldsymbol{\alpha}_2 &= [0, 1, 0]^T \\ \boldsymbol{\alpha}_3 &= [0, 0, 1]^T \end{aligned} \tag{10.24}$$

From equation (10.19), the minimum-input intensity is obtained as

$$k_{\min} = \left[\frac{p_x |H_{xx}|^2 a_x^2 + p_y |H_{yy}|^2 a_y^2 + p_z |H_{zz}|^2 a_z^2}{\lambda_{\max}} \right]^{1/2} \tag{10.25}$$

The input intensities for three-degree-of-freedom testing are taken to be

$$a_x = a_y = a_z = 1$$

Equal sensitivity weightings in the three directions are assumed:

$$p_x = p_y = p_z = 1/3$$

The test psd is taken as Gaussian white whose intensity is given by

$$\Phi(\omega) = 0.01 \text{ g}^2/\text{Hz}$$

For this numerical example, the transfer functions $H_{xx}(\omega)$, $H_{yy}(\omega)$, and $H_{zz}(\omega)$ are shown in figure 10-4.

For comparison purposes, the response measure is determined for three distinct cases:

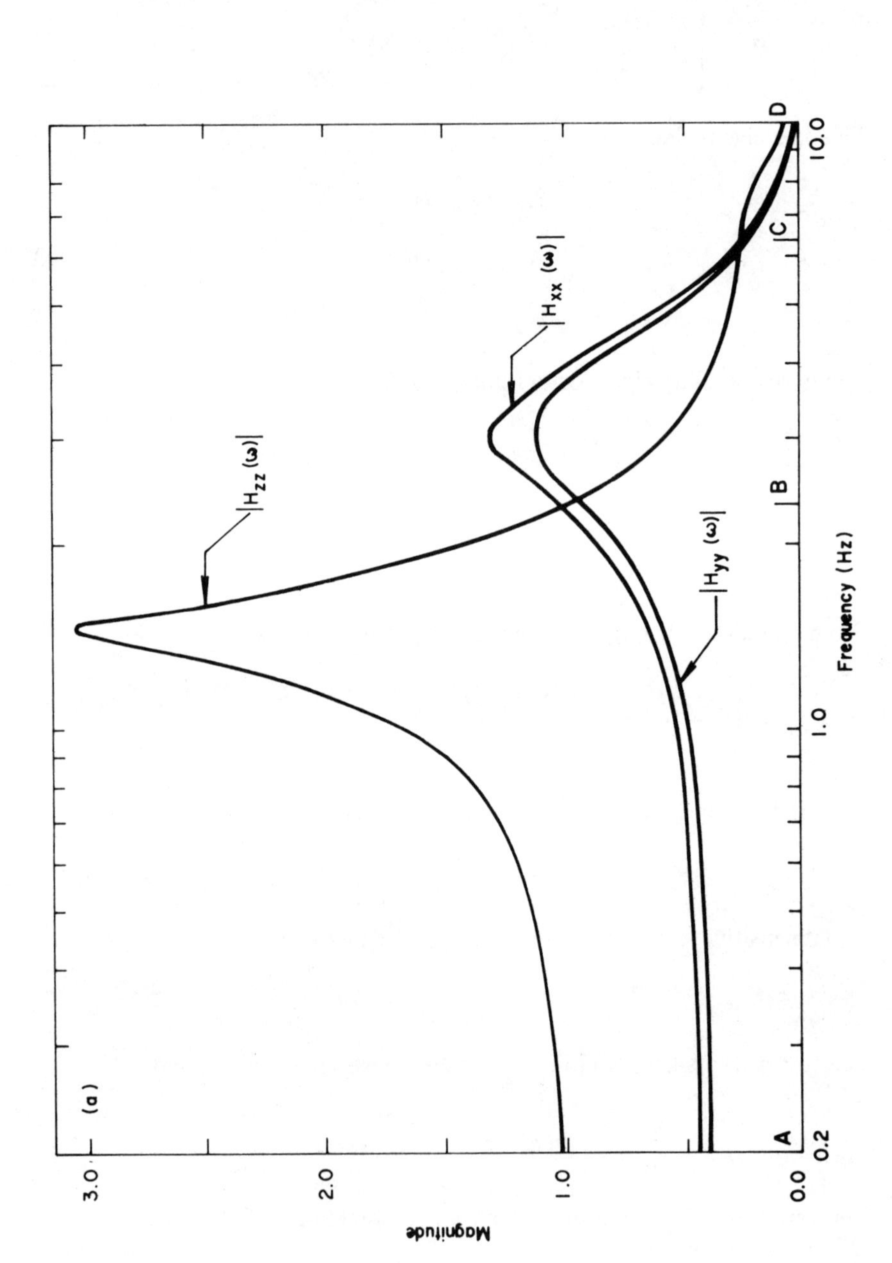
(a)
Magnitude
3.0
2.0
1.0
0.0
0.2
1.0
10.0
Frequency (Hz)
A
B
C
D
$|H_{zz}(\omega)|$
$|H_{xx}(\omega)|$
$|H_{yy}(\omega)|$

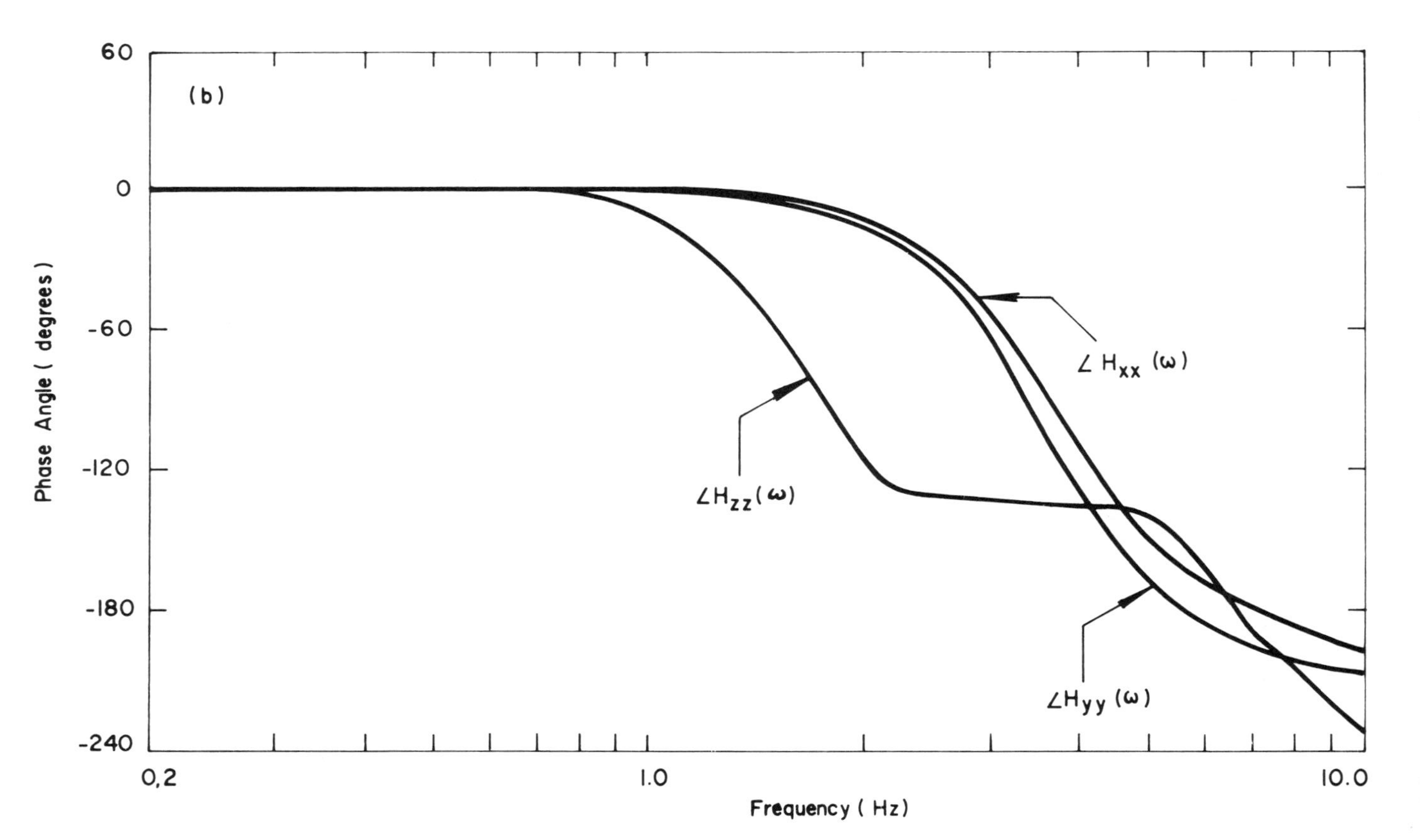

Figure 10-4. Frequency-Response Functions for the Optimal Test Example

Case 1 Using the procedure developed in this chapter (single-degree-of-freedom test applied along $\boldsymbol{\alpha}_{\text{opt}}$ at each frequency)

Case 2 Using the classical three-degree-of-freedom test procedure with three uncorrelated inputs

Case 3 Using the single-degree-of-freedom test, but with the input applied along the unit vector $\boldsymbol{\alpha} = [1/\sqrt{3},\ 1/\sqrt{3},\ 1/\sqrt{3}]^T$ instead of along $\boldsymbol{\alpha}_{\text{opt}}$ (symmetrical recilinear test).

Note from figure 10-4(a) that, for the frequency intervals *A-B* and *C-D*, H_{zz} dominates. For these frequencies,

$$\lambda_{\max} = p_z |H_{zz}|^2$$

and

$$\boldsymbol{\alpha}_{\text{opt}} = [0,\ 0,\ 1]^T$$

For the frequency interval *B-C*, H_{xx} dominates. In that case,

$$\lambda_{\max} = P_x |H_{xx}|^2$$

and

$$\alpha_{\text{opt}} = [1,\ 0,\ 0]^T$$

The test-severity measure was computed for three cases described, using equation (10.1). These computations are plotted in figure 10-5. Note that the test severity is identical for the optimal recilinear test and the classic three-degree-of-freedom test with uncorrelated inputs (cases 1 and 2, respectively). The test severity is significantly less, however, for the symmetrical rectilinear test (case 3).

General Discussion

The foregoing test procedure is based on an assumption of asymptotically stable, linear, time-invariant-parameter system behavior. In the presence of strong nonlinearities, the analytical development is valid for a narrow band of parameter variations about an operating point. In such cases, the results are usually reliable, provided that the test-input intensity is of the same order of magnitude as the anticipated earthquakes at the equipment location. At lower seismic intensities, the damaging effects of the seismic disturbances are negligible, although the accuracy of the analytical results is lost under these conditions.

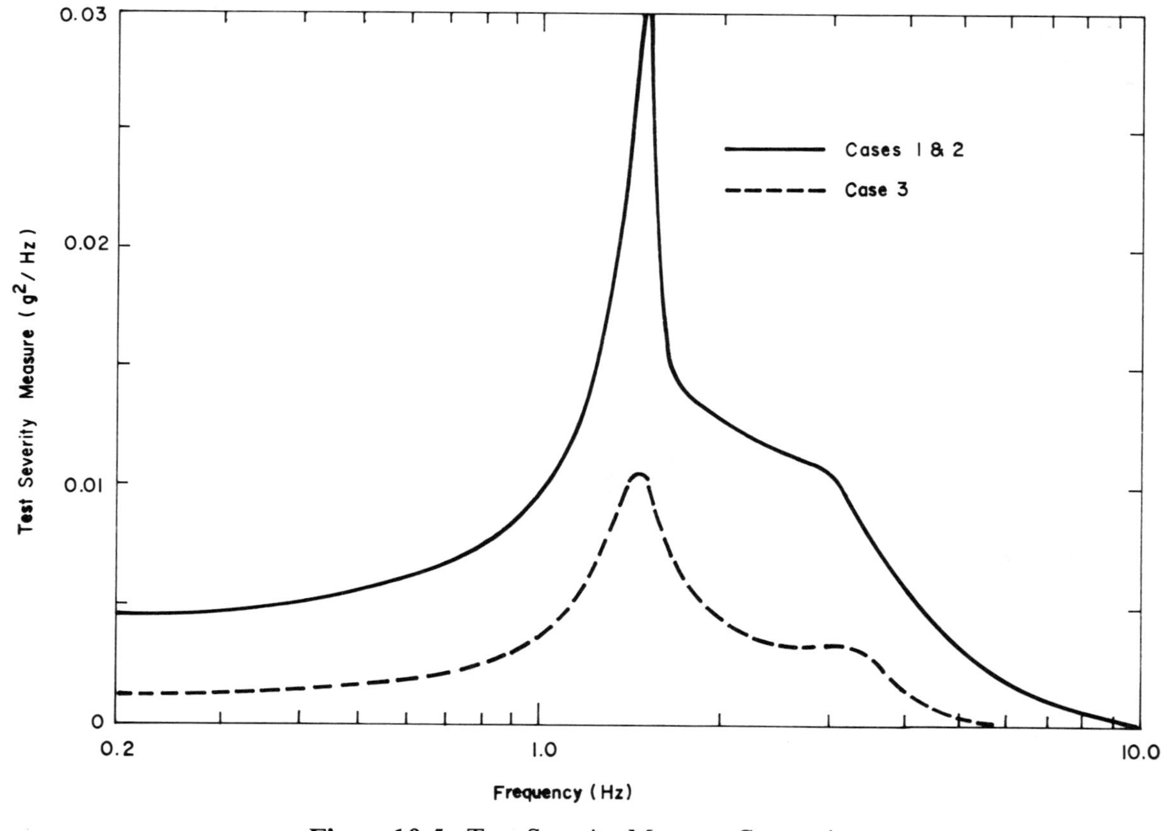

Figure 10-5. Test-Severity Measure Comparison

Fatigue effects caused by testing must be given due consideration. Increasing the test duration increases the fatigue of the equipment because of prolonged stressing of various components. This is the case when the test is repeated one or more times at the same intensity as that prescribed for a single test. The conventional symmetric rectilinear test, for example, employs four separate tests at an input intensity that is higher by a factor of $\sqrt{3}$ than that recommended for the three-degree-of-freedom test with uncorrelated inputs. Because of the particular test-input configuration, overtesting results in the vertical direction. The degree of overtesting can be minimized if only three tests are performed in orthogonal directions. In any event, care must be exercised to avoid overtesting or overfatiguing in sequential tests. In this respect, the single rectilinear test developed in this chapter has an obvious advantage over sequential tests.

It can be shown that the result obtained in example 3 in the previous section is generally valid if the direction of excitation is changed continuously with frequency during the optimal rectilinear test. The equality will be applied in equation (10.14) in this case. To verify this, we note that, for the optimal rectilinear test,

$$(TS)_{opt} = \lambda_{max}\Phi_{uu}(\omega) = \lambda_{max}k^2_{min}\Phi(\omega) \tag{10.26}$$

which follows by applying equations (10.10) and (10.15) in equation (10.1).

For the three-degree-of-freedom test with uncorrelated inputs,

$$(TS)_{tdf} = \sum_{I=1}^{N} p_i\{|H_{Xi}|^2 a_X^2 + |H_{Yi}|^2 a_Y^2 + |H_{Zi}|^2 a_Z^2\}\Phi(\omega) \tag{10.27}$$

which follows by using equation (10.13) in conjunction with equations (10.16) through (10.18) in equation (10.1). In view of equation (10.19), it is seen that

$$(TS)_{opt} = (TS)_{tdf} \tag{10.28}$$

As expected, we note that, if the capability of continuously adjusting the actuator along the optimal direction is available, then the optimal rectilinear test is equivalent to the classical three-degree-of-freedom test with uncorrelated inputs.

Note

1. C.W. de Silva, F. Loceff, and K. M. Vashi, "Consideration of an Optimal Procedure for Testing the Operability of Equipment Under Seismic Disturbances," *Shock and Vibration Bulletin* 50 (Part 4): 149–158, 1980. Reprinted with permission.

Bibliography

System Modeling and Automatic Control

Blackman, P. F. *Introduction to State Variable Analysis.* London: Macmillan, 1977.

Brebbia, C. A., and Connor, J. J. *Fundamentals of Finite Element Techniques.* New York: Wiley, 1974.

Bryson, A. E., and Ho, Y. C. *Applied Optimal Control.* Blaisdell, Waltham, Mass.: 1969.

Cannon, R. H., Jr. *Dynamics of Physical Systems.* New York: McGraw-Hill, 1967.

Casti, J. L. *Dynamical Systems and their Applications.* New York: Academic Press, 1977.

Close, C. M., and Frederick, D. K. *Modeling and Analysis of Dynamic Systems.* Boston: Houghton Mifflin, 1978.

Clough, R. W., and Penzien, J. *Dynamics of Structures.* New York: McGraw-Hill, 1975.

Crandall, S. H., Karnopp, D. C.; Kurtz, Jr., E. F.; and Pridmore-Brown, D. C. *Dynamics of Mechanical and Electromechanical Systems.* New York: McGraw-Hill, 1968.

D'Azzo, J. J., and Houpis, C. H. *Linear Control System Analysis and Design.* New York: McGraw-Hill, 1975.

Desai, C. S., and Abel, J. F. *Introduction to the Finite Element Method.* New York: Van Nostrand Reinhold, 1972.

de Silva, C. W. "A Technique to Model the Simply Supported Timoshenko Beam in the Design of Mechanical Vibrating Systems." *International Journal of Mechanical Sciences* 17: 389–393, 1975.

______. "Dynamic Beam Model with Internal Damping, Rotatory Inertia and Shear Deformation." *AIAA Journal* 14(No. 5): 676–680, 1976.

______. "Optimal Estimation of the Response of Internally Damped Beams to Random Loads in the Presence of Measurement Noise." *Journal of Sound and Vibration* 47(No. 4): 485–493, 1976.

______. "An Algorithm for the Optimal Design of Passive Vibration Controllers for Flexible Systems." *Journal of Sound and Vibration* 74(No. 4): 495–502, 1981.

Dorf, R. C. *Modern Control Systems,* 3rd ed. Reading, Mass.: Addison-Wesley, 1980.

Franklin, G. F., and Powell, J. D. *Digital Control of Dynamic Systems.* Reading, Mass.: Addison-Wesley, 1980.

Gajda, W. J., and Biles, W. E. *Engineering Modeling and Computation.* Boston: Houghton Mifflin, 1979.

Harrison, T. J. *Minicomputers in Industrial Control.* Pittsburgh: Instrument Society of America, 1978.

Karnopp, D. C., and Rosenberg, R. C. *System Dynamics: A Unified Approach.* New York: Wiley, 1975.

Kelley, J. M., and Sackman, J. L. "Conservatism in Summation Rules for Closely Spaced Modes." *Earthquake Engineering and Structural Dynamics* 8: 63–74, 1980.

Luenberger, D. G. *Introduction to Dynamic Systems.* New York: Wiley, 1979.

Ogata, K. *System Dynamics.* Englewood Cliffs, N.J.: Prentice-Hall, 1978.

Shearer, J. L.; Murphy, A. T.; and Richardson, H. H. *Introduction to System Dynamics.* Reading, Mass.: Addison-Wesley, 1967.

Van de Vegte, J., and de Silva, C. W. "Design of Passive Vibration Controls for Internally Damped Beams by Modal Control Techniques." *Journal of Sound and Vibration* 45(No. 3): 417–425, 1976.

Zienkiewicz, O. C. *The Finite Element Method in Engineering Science.* London: McGraw-Hill, 1971.

Signal Processing and Fourier Analysis

Beauchamp, K. G. *Signal Processing.* London: George Allen & Unwin, 1973.

Beauchamp, K. G., and Yuen, C. K. *Digital Methods for Signal Analysis.* London: George Allen & Unwin, 1979.

Bendat, J. S., and Piersol, A. G. *Random Data: Analysis and Measurement Procedures.* New York: Wiley-Interscience, 1971.

Blackman, R. B., and Tukey, J. W. *The Measurement of Power Spectra.* New York: Dover, 1959.

Brigham, E. O. *The Fast Fourier Transform.* Englewood Cliffs, N.J.: Prentice-Hall, 1974.

Champeney, D. C. *Fourier Transforms and Their Physical Applications.* London: Academic, 1973.

Cooley, J. W., and Tukey, J. W. "An Algorithm for the Machine Computation of Complex Fourier Series," *Mathematics of Computation* 19 (April): 297–301, 1965.

Cooley, J. W.; Lewis, P. A. W.; and Welch, P. D. "Application of the Fast Fourier Transform to Computation of Fourier Integrals, Fourier Series, and Convolution Integrals." *IEEE Transactions on Audio and Electronics* AU-15(June): 79–84, 1967.

Goldberg, R. R. *Fourier Transforms.* Cambridge, England: Cambridge University Press, 1961.

Peled, S., and Liu, B. *Digital Signal Processing.* New York: Wiley, 1976.

Randall, R. B. *Application of B&K Equipment to Frequency Analysis,* 2nd ed. Naerum, Denmark: Brüel & Kjaer, 1977.

Stearns, S. D. *Digital Signal Analysis.* Rochelle Park, N.J.: Hayden, 1975.

Thong, T., and Liu, B. "Fixed-Point Fast Fourier Transform Error Analysis." *IEEE Transactions on Acoustics, Speech, and Signal Processing* ASSP-24 (December): 563–573, 1976.

Welch, P. D. "A Fixed-Point Fast Fourier Transform Error Analysis." *IEEE Transactions on Audio and Electronics* AU-17(June): 151–157, 1969.

Damping and Vibration Control

Baker, W. E. "Inherent Vibration Damping of Helicopter Blades." *International Journal of Mechanical Sciences* 13: 157–170, 1971.

Beranek, L. L. *Noise and Vibration Control.* New York: McGraw-Hill, 1971.

Blevins, R. D. *Flow-Induced Vibration.* New York: Van Nostrand Reinhold, 1977.

Caughey, T. K., and O'Kelly, M. E. J. "Classical Normal Modes in Damped Linear Dynamic Systems." *Journal of Applied Mechanics* (Transactions of the ASME) 32 (September): 583–588, 1965.

Crocker, M. J. *Noise and Vibration Control Engineering.* West Lafayette, Ind.: Purdue University Press, 1972.

Freudenthal, A. M. *The Inelastic Behavior of Engineering Materials and Structures.* New York: Wiley, 1950.

Irwin, J. D., and Graf, E. R. *Industrial Noise and Vibration Control.* Englewood Cliffs, N.J.: Prentice-Hall, 1979.

Johnson, R. D. "The Damping Characteristics of Certain Steels, Cast Irons, and Other Metals." *Journal of Sound and Vibration* 23(No. 2): 199–216, 1972.

Lazan, B. J. *Damping of Materials and Members in Structural Mechanics.* London: Pergamon, 1968.

Paidoussis, M. P., and Des Trois Maisons, P. E. "Free Vibration of a Heavy, Damped, Vertical Cantilever." *Journal of Applied Mechanics* (Transactions of the ASME) 93(June): 524–526, 1971.

Petrusewicz, S. A., and Longmore, D. K. *Noise and Vibration Control for Industrialists.* New York: American Elsevier, 1974.

Ruzicka, J. E., and Derby, T. F. *Influence of Damping in Vibration*

Isolation. Washington, D.C.: The Shock and Vibration Information Center, 1971.

Sevin, E., and Pilkey, W. D. *Optimum Shock and Vibration Isolation.* Washington, D.C.: The Shock and Vibration Information Center, 1971.

Snowdon, J. D. *Vibration and Shock in Damped Mechanical Systems.* New York: Wiley, 1968.

Vibration and Random Vibration

Bishop, R. E. D., and Johnson, D. C. *The Mechanics of Vibration.* Cambridge, England: Cambridge University Press, 1960.

Crandall, S. H., and Mark, W. D. *Random Vibration in Mechanical Systems.* New York: Academic Press, 1963.

Den Hartog, J. P. *Mechanical Vibrations.* New York: McGraw-Hill, 1956.

Dimarogonas, A. D. *Vibration Engineering.* St. Paul: West, 1976.

Meirovitch, L. *Elements of Vibration Analysis.* New York: McGraw-Hill, 1975.

Papoulis, A. *Probability, Random Variables, and Stochastic Processes.* New York: McGraw-Hill, 1965.

Parzen, E. *Stochastic Processes.* San Francisco: Holden-Day, 1962.

Robson, J. D. *Random Vibration.* Edinburgh: Edinburgh University Press, 1964.

Thompson, W. T. *Theory of Vibration with Applications,* 2nd ed. Englewood Cliffs, N.J.: Prentice-Hall, 1981.

Timoshenko, S., and Young, D. H. *Vibration Problems in Engineering.* New York: Van Nostrand Reinhold, 1955.

Volterra, E., and Zachmanoglou, E. C. *Dynamics of Vibrations.* Columbus, Ohio: Merrill, 1965.

Seismic Environment Representation and Response Analysis

Ang, A. H. S. "Probability Concepts in Earthquake Engineering." *Applied Mechanics in Earthquake Engineering,* AMD-Vol. 8. New York: American Society of Mechanical Engineers, 1974.

Bolotin, V. V. "Statistical Theory of the Aseismic Design of Structures." *Proceedings of the Second Conference on Earthquake Engineering,* Tokyo and Kyoto, pp. 1365–1374, 1960.

Hart, G. C., and Vasudevan, R. "Earthquake Design of Buildings: Damping." *Journal of the Structural Division, ASCE* ST 1(January): 11–30, 1975.

Idriss, I. M. "Characteristics of Earthquake Ground Motions." *Earthquake Engineering and Soil Dynamics* (Proceedings of the ASCE Geotechnical Division Special Conference) 3(June): 1151–1265, 1978.

Joyner, W. B.; Chen, A. T. F.; and Doherty, P. C. "Nonlinear Calculations of Ground Response in Earthquakes." *Wind and Seismic Effects.* Washington, D.C.: National Bureau of Standards, Publication No. 444, pp. III-1-12, May 1974.

Kana, D. D. "Seismic Response of Flexible Cylinderical Liquid Storage Tanks." *Nuclear Engineering and Design* 52: 185–199, 1979.

Krinitzky, E. L., and Chang, F. K. "Design Earthquakes." *Wind and Seismic Effects.* Washington, D.C.: National Bureau of Standards, Publication No. 477, pp. VI-179-191, May 1977.

Lin, C. W. "OBE Design Effects on Nuclear Power Plant Pressure Vessels." ASME paper 78-PVP-14, 1978.

McDonald, C. K. "Seismic Analysis of Vertical Pumps Enclosed in Liquid Filled Containers." ASME Paper 75-PVP-56, 1975.

Newmark, N. M. "Earthquake Response Analysis of Reactor Structures." *Nuclear Engineering and Design* 20: 303–322, 1972.

Newmark, N. M., and Rosenblueth, E. *Fundamentals of Earthquake Engineering.* Englewood Cliffs, N.J.: Prentice-Hall, 1971.

Ohashi, M., Iwasaki, T.; Susumu, W.; and Takida, K. "Statistical Analysis of Strong-Motion Acceleration Records." *Wind and Seismic Effects.* Washington, D.C.: National Bureau of Standards, Publication No. 523, pp. IV-48-77, September 1978.

Ohtani, K., and Kinoshita, S. "On a Method for Synthesizing the Artificial Earthquake Waves by Using the Prediction Error Filter." *Wind and Seismic Effects.* Washington, D.C.: National Bureau of Standards, Publication No. 523, pp. IV-28-47, May 1977.

Ruiz, P., and Penzien, J. *Probabilistic Study of the Behavior of Structures During Earthquakes.* Report No. EERC 69-3, College of Engineering, University of California, Berkeley, March 1969.

Stevenson, J. D., and Lapay, W. S. "Amplification Factors to Be Used in Simplified Seismic Dynamic Analysis of Piping Systems." ASME Paper 74-PVP, 1974.

URS/John A. Blume & Associates, *Effects Prediction Guidelines for Structures Subjected to Ground Motion.* Report JAB-99-115, San Francisco, July 1975.

Vashi, K. M. "Computation of Seismic Response from Higher Frequency Modes." ASME Paper 80-C2/PVP-50, 1980.

Yamazaki, Y., and Koizumi, Y. "Study on Earthquake Response of Structures by Considering Non-deterministic Variables." *Wind and Seismic Effects.* Washington, D.C.: National Bureau of Standards, Publication No. 470, pp. VI-1-21, May 1975.

Young, D. "Response of Structural Systems to Ground Shock." *Shock and Structural Response.* New York: American Society of Mechanical Engineers, pp. 52–68, 1960.

Dynamic Testing and Seismic Qualification

Bessey, R. L., and Kana, D. D. "Some Research Needs for Improved Seismic Qualification Tests of Electrical and Mechanical Equipment." *Nuclear Engineering and Design* 50: 71–82, 1978.

Blume, J.A. "A Machine for Setting Structures and Ground into Forced Vibration." *Bulletin of the Seismological Society of America* 25, October 1935.

Bouwkamp, J.G., and Rea, D. "Dynamic Testing and the Formulation of Mathematical Models." *Earthquake Engineering.* University of California, Berkeley, pp. 151–165, 1965.

Broch, J. T. *Mechanical Vibration and Shock Measurements.* Naerum, Denmark: Brüel & Kjaer, 1980.

Craig, R.R., Jr., and Su, Y.W.T. "On Multiple-Shaker Resonance Testing." *AIAA Journal* (July):924–931, 1974.

Curtis, A.J.; Tinling, N.G.; and Abstein, H.T., Jr. *Selection and Performance of Vibration Tests.* Washington, D.C.: The Shock and Vibration Information Center, 1971.

de Silva, C.W. "Seismic Qualification of Electrical Equipment Using a Uniaxial Test." *Earthquake Engineering and Structural Dynamics* 8: 337–348, 1980.

______. "Matrix Eigenvalue Problem of Multiple Shaker Testing." *Journal of the Engineering Mechanics Division, ASCE,* April 1982.

de Silva, C.W.; Loceff, F.; and Vashi, K. M. "Consideration of an Optimal Procedure for Testing the Operability of Equipment Under Seismic Disturbances." *Shock and Vibration Bulletin* 50(Part 5): 149–158, 1980.

Dodds, C.J. "Environmental Testing Under Random Loading." *Journal of Testing and Evaluation* (ASTM) 7(No. 4): 232–237, 1979.

Fackler, W.C. *Equivalence Techniques for Vibration Testing.* Washington, D.C.: The Shock and Vibration Information Center, 1972.

Fisher, E.G. "Seismic Qualification of Systems, Structures, Equipment and Components." *Nuclear Engineering and Design* 46: 151–167, 1978.

Halvorsen, W.G., and Brown, D.L. "Impulse Technique for Structural Frequency Response Testing." *Sound and Vibration* 11(No. 11): 8–21, 1977.

Hoerner, J.B., and Jennings, P.C. "Modal Interference in Vibration Tests."

Journal of the Engineering Mechanics Division, ASCE 8: 827–839, 1969.

Hudson, D.E. "Dynamic Tests of Full-Scale Structures." *Earthquake Engineering.* University of California, Berkeley, pp. 127–149, 1965.

Hyati, S.A., and Auslander, D. "Control of a Single-Actuator Shaking Table." ASME Paper 77-WA/Aut-11, 1977.

Inaba, S. "Dynamic Tests of Structures for Oil Tanks and Nuclear Power Stations." *Wind and Seismic Effects.* Washington, D.C.: National Bureau of Standards, Publication No. 470, pp. VII-1-6, May 1975.

______. "Dynamic Tests of Structures Using a Large Scale Shake Table." *Wind and Seismic Effects.* Washington, D.C.: National Bureau of Standards, Publication No. 444, pp. V-25-34, May 1974.

Inaba, S., and Kinoshita, S. "Dynamic Test of a Circuit Breaker for Transformer Station." *Wind and Seismic Effects.* Washington, D.C.: National Bureau of Standards, Publication No. 477, pp. VI–50–60, May 1977.

Kana, D.D. *A Determination of Seismic Facility Capabilities.* San Antonio: Southwest Research Institute, 1976.

Kana, D.D., and Le Blanc, R.W. *An Evaluation of Seismic Qualification Tests for Nuclear Power Plant Equipment.* San Antonio: Southwest Research Institute, 1979.

Kimmel, J. "Accelerated Life Testing of Paper Dielectric Capacitors." *Proceedings of the Fourth National Symposium on Reliability and Quality Control* IRE, pp. 120–134, January 1958.

Magrab, E.B., and Shinaishin, O.A. *Vibration Testing—Instrumentation and Data Analysis*, AMD-Vol. 12. New York: American Society of Mechanical Engineers, 1975.

Miner, M.A. "Cumulative Damage in Fatigue." *Journal of Applied Mechanics* 12: A159, 1945.

National Academy of Engineering, *Earthquake Environment Simulation,* Report PB-240 404, Washington, D.C., 1974.

Ratz, A.G. and Barlett, F.R. "Vibration Simulation Using Electrodynamic Exciters," *Vibration Testing - Instrumentation and Data Analysis,* AMD - Vol 12, New York: American Society of Mechanical Engineers, 1975.

Roberts, C.W. "Seismic Testing Capabilities of a Typical Commercial Laboratory." *Earthquake Environment Simulation,* Report No. PB-240 404. Washington, D.C.: National Academy of Engineering, pp. 213–217, 1974.

Tustin, W. "Quantized Goals for the Design of Vibration and Shock Test Fixtures." *Proceedings of the 1973 Annual Reliability and Maintainability Symposium.* Philadelphia, January 1973.

Instrumentation

Collacott, R.A. *Vibration Monitoring and Diagnosis.* New York: Wiley, 1979.

Gibson, J.E., and Tuteur, F. B. *Control System Components.* New York: McGraw-Hill, 1958.

Harris, C.M., and Crede, C.E. *Shock and Vibration Handbook* 2nd ed. New York: McGraw-Hill, 1976.

Neubert, H.K.P. *Instrument Transducers.* Oxford: Clarendon Press, 1975.

Norton, H.N. *Handbook of Transducers for Electronic Measuring Systems.* Englewood Cliffs, N.J.: Prentice-Hall, 1969.

Oliver, F.J. *Practical Instrumentation Transducers.* London: Pitman, 1972.

Index

About the Author

Clarence W. de Silva is an assistant professor of mechanical engineering at Carnegie-Mellon University. He received the B.Sc. in mechanical engineering from the University of Ceylon, the M.A.Sc. in mechanical engineering from the University of Toronto, the M.S. in aerospace engineering and applied mechanics from the University of Cincinnati, and the Ph.D in mechanical engineering (system dynamics and control) from the Massachusetts Institute of Technology.

Before joining the faculty at Carnegie-Mellon University, Dr. de Silva taught at the University of Ceylon, Miami University, and Southern Ohio College. He also has been a consultant on seismic qualification of equipment to Westinghouse Electric Corporation. He is the author of *Modern Mathematics* and a number of technical papers on dynamic modeling and analysis, failure detection, automated guideway transit, dynamic testing, and seismic qualification.